Chemicals in the Oil Industry:
Developments and Applications

Chemicals in the Oil Industry: Developments and Applications

Edited by
P.H. Ogden
Akzo Chemie UK Ltd

The Proceedings of a Symposium Organised by the North West Region of the Industrial Division of the Royal Society of Chemistry with the Royal Netherlands Chemical Society as part of the Annual Chemical Congress held 8–11 April 1991 at Imperial College, London

Special Publication No. 97

ISBN 0-85186-466-X

A catalogue record of this book is available from the British Library

Published by The Royal Society of Chemistry,
Thomas Graham House, Science Park, Cambridge CB4 4WF

Printed by Redwood Press Ltd, Melksham, Wiltshire

Preface

The Northwest Region of the Society's Industrial Division held its first Chemicals in the Oil Industry Symposium at Manchester University in 1983. At that time the Oil Industry was considered by many major chemical companies to be an attractive market offering opportunities for diversification in co-operation, or competition with existing oilfield service companies. During the intervening eight years, collapsed oil prices together with the effects of tightening environmental regulations have caused several of these companies to reconsider their position and latterly there has been a degree of rationalisation. Consequently, although much of the fourth symposium addressed technical and environmental issues, attention was also given to a review of recent market development and how this is expected to continue in the foreseeable future. The scope of the conference was extended to include some aspects of refinery operations where environmental pressures are also leading to significant changes.

The meeting was jointly organised by The Royal Society of Chemistry and the Royal Netherlands Chemical Society. It consisted of six distinct sessions, each with its own convener and I wish to express gratitude to my co-conveners, T.A.B.M. Balsman (Shell Chemicals), M. Fielder (B.P. Exploration), D. Graham (B.P. Research), G. Pusch (Technische Universitat Clausthal) and P.M. van der Velden (Servo Delden).

The RSC is also grateful to the management of Akzo Chemicals, BP, Servo, and Shell Chemicals for their generous sponsorship of the symposium.

Paul H. Ogden
Akzo Chemicals Ltd.
Littleborough
Lancashire

Contents

Future Market Aspects of Oilfield Specialty Chemicals

Christopher W. Houston

COLIN A. HOUSTON AND ASSOCIATES, INC, 689 MAMARONECK AVENUE, PO BOX 416, MAMARONECK, NEW YORK 10543, USA

The oilfield specialty chemical market has undergone many changes over the past 10 years. Looking back to 1980-1981, the oilfield chemical market was robust as oilfield activity was peaking at 4,500 rigs in the U.S. and 5,800 rigs worldwide. The following Figure 1 shows the effect oil prices have on the rig count.

At that time, a fever-pitched industry sought new specialty chemicals for improved drilling, completion and workover, cementing, stimulation, production, and particularly enhanced oil recovery (EOR) techniques. After all, crude oil was reaching $40/barrel and the outlook showed $100/barrel by the end of the decade.

Complicating matters, however, was the fact that while higher oil and gas prices were spurring oilfield activity they were also pushing the world into a recession. And, as commodity chemical usage was declining rapidly, many chemical companies, seeking relief from the cyclicality of commodity products, ventured into the oilfield specialty chemical segment.

Figure 2 shows oilfield chemical usage as it stood in 1980 by segment.

As oilfield chemical usage grew, demand for improved chemicals was also growing. Industry sought new products from chemical and oil company R&D centers, but as this work was being accomplished, the first signs of cracks in the oilfield bubble were being seen.

The subsequent decline in oil price and concurrently oilfield activity caught most people by surprise. The oilfield service and chemical industry, in particular, had built new plants, warehouses, and labs to handle the burgeoning market. Oilfield personnel were at an all-time high at most service companies such as Dresser, NL Baroid, Halliburton, Dowell, and Baker.

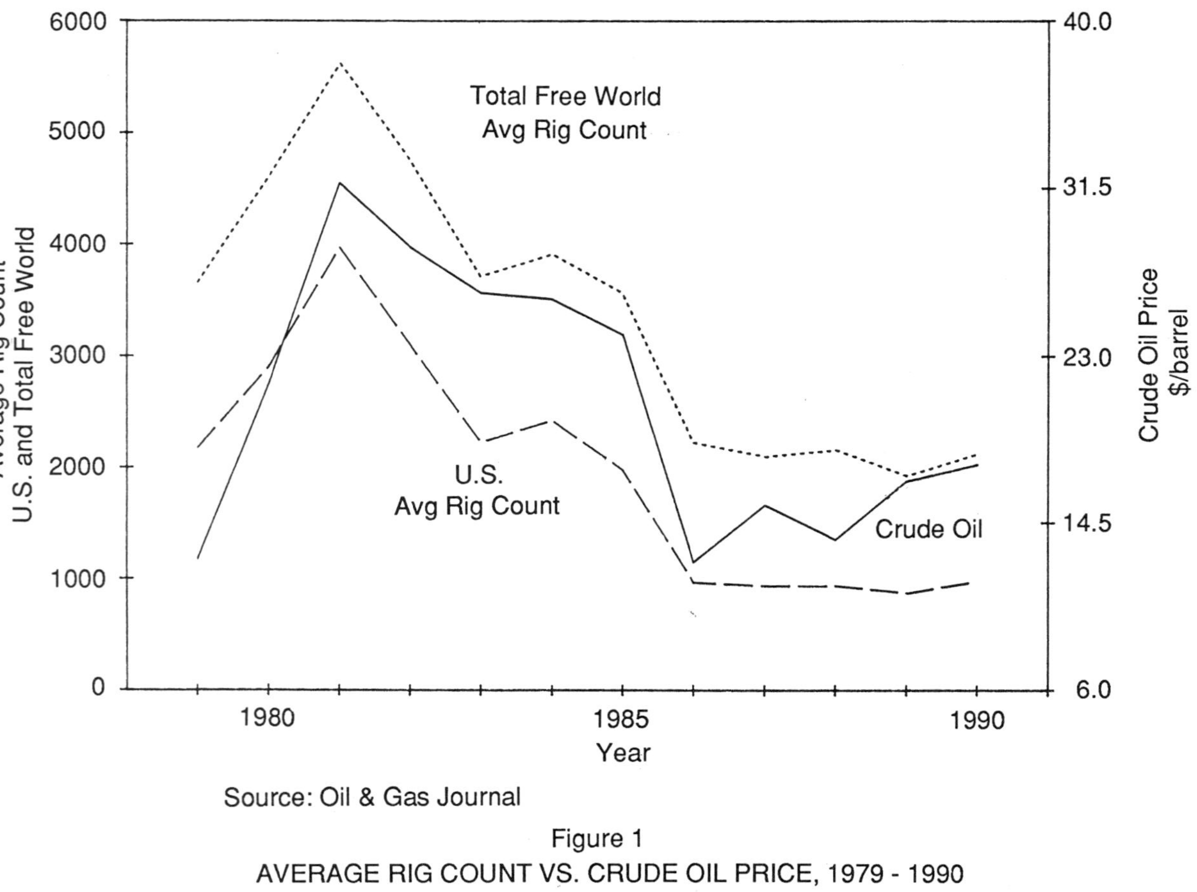

Source: Oil & Gas Journal

Figure 1
AVERAGE RIG COUNT VS. CRUDE OIL PRICE, 1979 - 1990

Figure 2

WORLDWIDE OILFIELD CHEMICAL MARKET, 1980
(million dollars)

	Service Value
Drilling	2400
Completion and Workover	300
Cementing	1800
Stimulation	1500
Production	600
EOR	150
	6750

Source: Colin A. Houston & Associates, Inc.

Figure 3 shows total oilfield service company employment from 1982 to 1990.

Needless to say, the resulting decline in oilfield activity was a devastating shock to those companies that had invested heavily in the growth of oilfield chemicals. Oilfield service companies were forced to sell equipment, close stock points, and lay off personnel to compensate for a rapidly decreasing market. The scars left from this experience are still evident in most companies which participated in the surge of the late 1970s into the early 1980s.

Of course, the sobering experience of 1983-1985 was but a mere shadow of the forthcoming devastation of the 1986-1987 oil price collapse. While $25-30/barrel oil had dampened enthusiasm for new entrants, most surviving oilfield service companies were still profitable during the period of 1983 to 1985.

The 1986-1987 oil price collapse was certainly a negative factor in determining oilfield chemical selection. Most operators had thought sufficient cut-backs had been made in 1983 to 1985 to compensate for $25 to $30/barrel oil. When $10/barrel oil arrived, staff reductions were again undertaken but additional focus was placed on slashing oilfield chemical costs. (See Figure 4.) Whenever and wherever reductions in operating costs were possible, operators made them. The easiest way to drive costs lower was to competitive bid, where price was the key selection criteria for choosing any oilfield chemical. Bidding became so popular in all phases of oilfield chemical usage that it quickly took its toll on

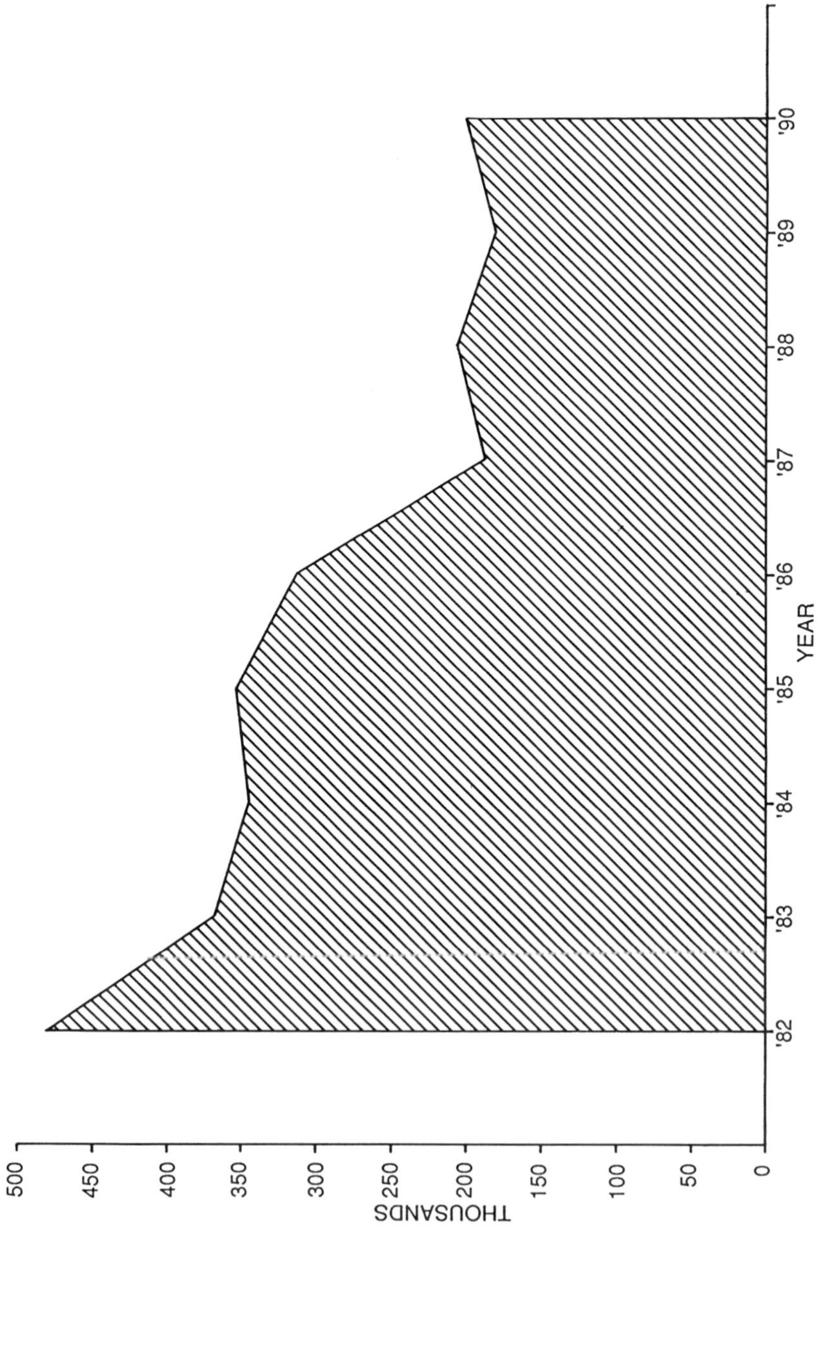

Figure 3
U.S. OIL & GAS FIELD SERVICES - TOTAL EMPLOYEES, 1982-1990

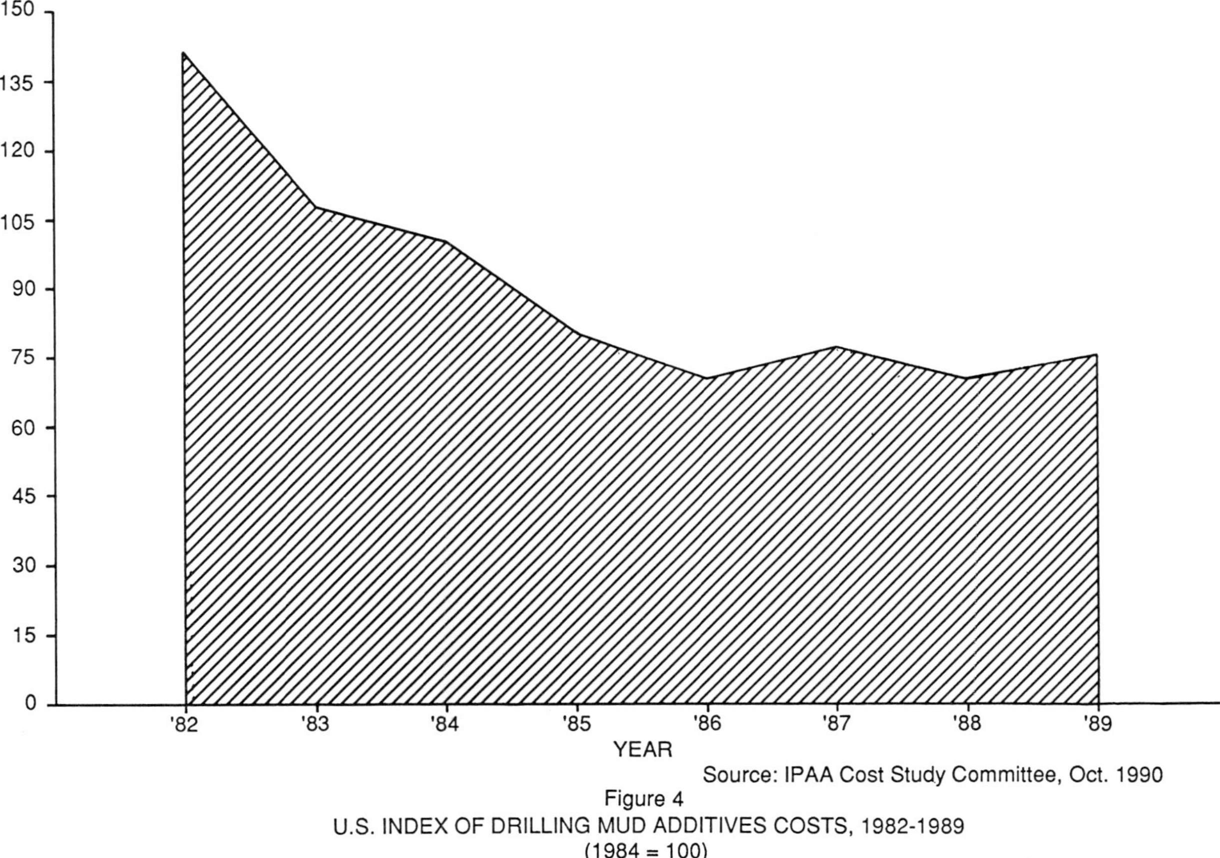

Figure 4
U.S. INDEX OF DRILLING MUD ADDITIVES COSTS, 1982-1989
(1984 = 100)

the oilfield chemical service companies as consolidation was necessary in order to compensate for lower oilfield activity and substantially reduced prices. Figure 4 traces oilfield drilling fluid costs from 1982 to 1989. As can be seen in the figure, drilling fluid costs dropped 47 percent over the period in the U.S. Other oilfield chemical services had similar price erosion.

Thus by 1988 the oilfield chemical market took on quite a different tone as bankruptcies, divestitures, LBOs and mergers created a new set of players in the oilfield chemical service marketplace.

From 1988 to 1990 a new tone of cautious optimism has been present as oil prices have mostly fluctuated between $15 and $20 per barrel. Rig activity has gradually increased during the period from a post-World War II low of 650 rigs in 1986/1987 in the U.S. to 1050 rigs in March of 1991. The Iraqi invasion of Kuwait and subsequent run-up in oil prices has had a negligible impact on the oilfield chemical market. Most operators see oil prices hovering in the $15 to $25/barrel range over the next ten years and are planning accordingly.

That is a brief history of the oilfield chemical market over the past 10 years. Currently, (see Figure 5) the worldwide market for oilfield chemicals and services is $5.4 billion. The market is split 55:45 in favor of the international segment. Of course, the international market will continue to grow during the 1990s while the U.S. market will remain relatively flat.

Each oilfield chemical segment presents unique marketing challenges for the specialty chemical manufacturer. Depending on the segment, specialty chemicals may be sold directly to the service company, to the operator, or to a blender. This matrix of end users is one of the chief marketing problems for the specialty chemical company.

As mentioned previously, the current worldwide market for oilfield chemicals with service is $5.4 billion. Drilling fluids represent 26 percent of the total market, while cementing and stimulation take up another 57 percent. The bulk of the products used in these two segments are commodity chemicals such as barite, bentonite, and hydrochloric acid. However, the need for specialty chemicals is increasing as oilfield work moves into more hostile and environmentally sensitive areas. Figure 6 summarizes the current worldwide oilfield chemical market with respect to sales by segment and the amount of specialty chemicals being used.

While at first a large dynamic specialty chemical market appears, the complexities of the usage of these products is vast. (See Figures 7A and 7B.)

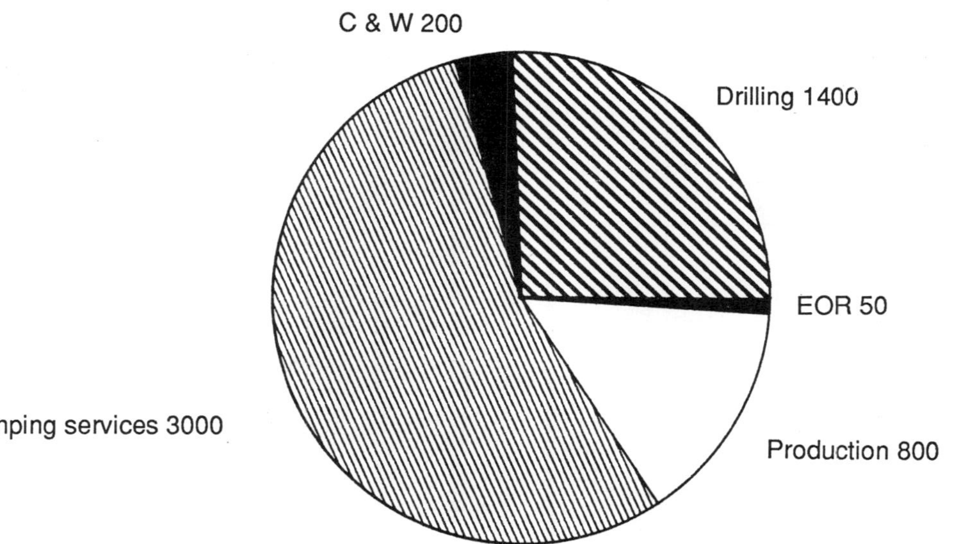

Total Market = $5400

Figure 5
WORLD OILFIELD CHEMICAL MARKET - 1990
(million dollars)

Figure 6

WORLDWIDE OILFIELD CHEMICAL MARKET - 1990

	Total Service Market (billion $)	% Specialties (of total service market)	Manufacturers Value of Spec. Chemicals (million $)
Drilling	$ 1.4	20%	$275
C&W	$ 0.2	5%	$10
Cementing/ Stimulation	$ 3.0	10%	$300
Production	$ 0.8	45%	$360
EOR (chemical)	$ <0.1	90%	$45
Total	$ 5.4	18%	$990

Figure 7A Drilling

- Approximately 50,000 wells drilled worldwide in 1990 (peak was 100,000 in 1981).

- 500+ end users worldwide with markets ranging from $100,000 to $500 million.

Figure 7B Drilling

- Specialty chemicals are generally not sold to the end user/oil company. They must be first sold to a drilling fluid service company (e.g. M-I, Baroid, Milpark, IDF) which will blend the product into its own formulation.

- New specialty chemical products must be evaluated by both the oil company and the drilling fluid service company prior to field usage.

- Improved viscosifiers, shale control additives, oil-based alternatives are being sought.

In drilling, oilfield service companies such as M-I, Baroid, Milpark and IDF are not chemical manufacturers. While they do process barite and bentonite, and have ventures in lignosulfonates, in most other cases they depend heavily on specialty chemical suppliers to provide polymers, corrosion inhibitors, biocides and defoamers.

Prior to 1980, chrome lignosulfonate, gelled bentonite, and oil-based drilling muds were used in 85+ percent of all the wells drilled. (See Figure 8.) Specialty chemical usage was limited. However, as drilling ventured into more hostile environments (deep, offshore, horizontal, inclinate), better performing products were required.

Spearheading many of the changes were environmental concerns regarding disposal of chrome lignosulfonates and oil muds (especially offshore). Additionally, fine clay formation damage potential has resulted in a trend towards polymer muds.

Bringing new products into the drilling fluid segment requires a great deal of patience. Two approval procedures are required: one by the DFSC and another by the oil company operator. Each has its own priorities for evaluating new products which are very complicated. The procedures to assess a new product can be expedited if an immediate need is determined.

Specialty drilling chemical oilfield marketing requires a multi-pronged effort whereby numerous contacts are necessary to determine a product's true value. This web of contacts will produce varying opinions on a product's potential which range from extremely low to very high. (See Figure 9.)

Assessing the real potential for a new drilling fluid additive is especially difficult because of a complex evaluation process. Specialty chemical companies typically utilize a push/pull strategy of marketing for drilling fluid additives.

The push/pull strategy requires contacting both the DFSC and the oil companies to test the product against existing products. The specialty chemical company needs both the DFSC to help formulate the product and the operator to provide another evaluation of the product's worthiness. It takes a substantial amount of patience to introduce a new drilling fluid additive because of the complexity of the evaluation process.

Will this change in the future? Possibly, as drilling conditions become more hostile or environmentally sensitive, new drilling fluid specialty chemicals will be required. (See Figure 10.)

Over the past three years, for example, a number of new specialty drilling fluid additives have been introduced. A wide range of needs has developed for new specialty products which are environmentally safe, offer high temperature, pressure and divalent cation resistance, and are, of course, cost-effective. The products introduced have substantially improved the drilling practice via lower overall costs to drill difficult wells. The table below shows a brief list of new products for the drilling industry.

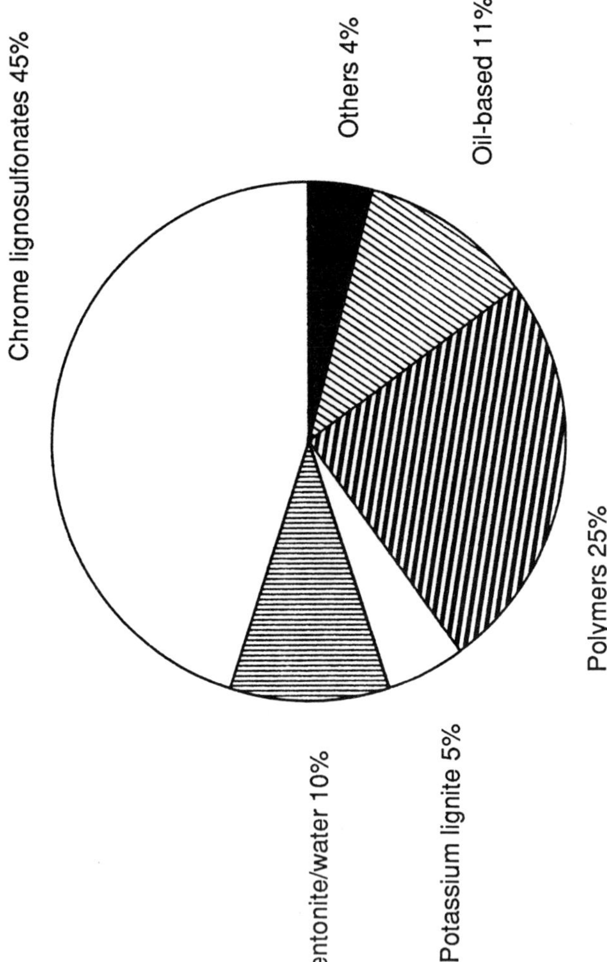

Figure 8
U.S. DRILLING FLUID SYSTEMS - 1990
(Percent of Total Drilling Fluid Market)

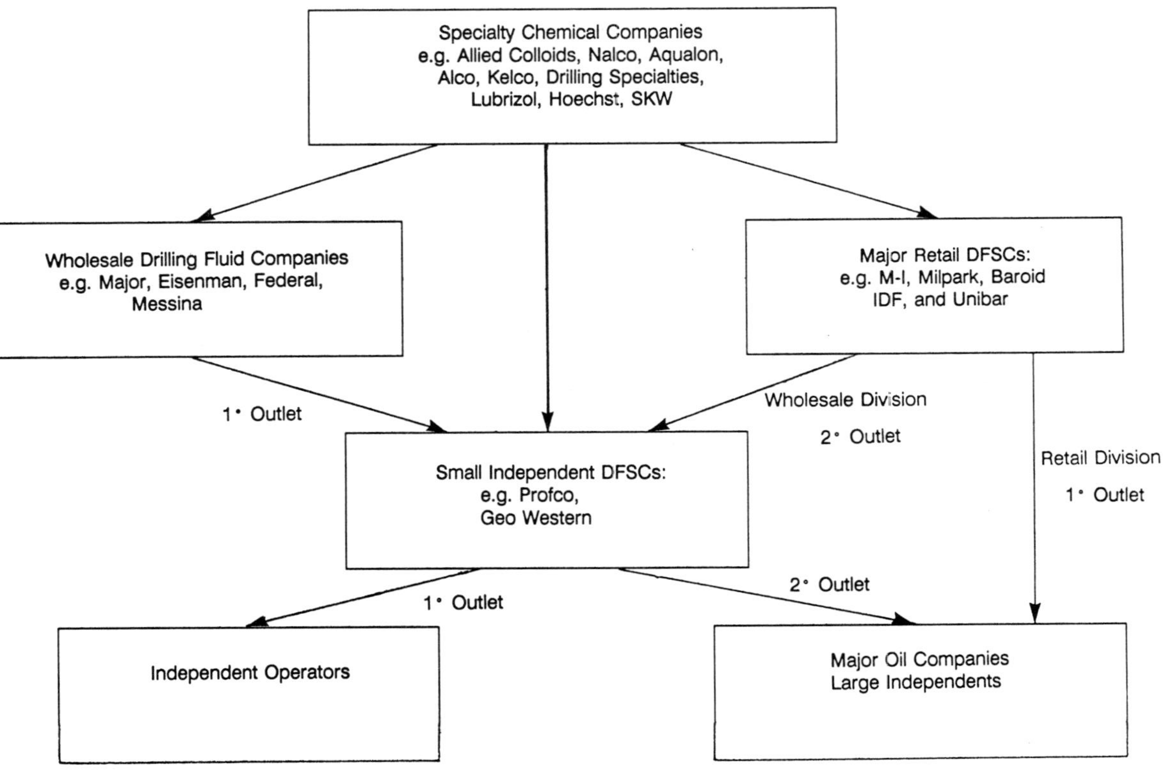

Figure 9
U.S. DRILLING FLUID MARKETING PATHS - 1990

Figure 10

NEW DRILLING FLUID PRODUCTS

Product	Application	Specialty Chemical Supplier/Service Company
Polydrill	Fluid loss control	SKW
MMH	Cationic viscosifier	Dow
Petrofree	Oil-based mud alternative	Henkel/Baroid
1402	Oil-based mud alternative	Milpark
HF 100	Glycol-based mud system	Hydrafluids
Ancoquat	Cationic viscosifier	Anchor
Novadril	Oil-based mud alternative	M-I
CAT - I	Cationic-based mud system	Baroid
M CAT	Cationic-based mud system	M-I

Similar marketing methods for drilling fluid additives are utilized in the completion and workover, cementing, and stimulation segments. (See Figures 11A and 11B.) The oilfield chemical service companies are not chemical manufacturers and hence rely heavily on their chemical suppliers to assist in their R&D efforts.

Figure 11A Completion and Workover (C&W)

- Small but dynamic market which is receiving more attention.

- Similar customer base to drilling fluids but market size for total C&W market is generally 15 percent of drilling market.

- Specialty chemicals are sold to C&W service companies such as Baroid, OSCA, BW Mud and Tetra. Limited specialty chemical market to date.

Figure 11B Completion and Workover (C&W)

- New specialty chemical product evaluation is similar to drilling fluids.

- Non-calcium-based fluids 12.6 to 15.1 PPG are being sought.

<u>Figure 12A</u> Cementing/Stimulation

- Large market via number of wells drilled (cementing) and number of acidizing/fracturing jobs performed.

- End user base similar to drilling and completion and workover.

<u>Figure 12B</u> Cementing/Stimulation

- Specialty chemicals sold to service companies such as Halliburton, Dowell-Schlumberger, BJ Services, and Western.

- New specialty chemical product evaluation directed by service companies and to a less extent oil companies.

Companies such as Halliburton, Dowell-Schlumberger, BJ Services and most specialized completion and workover service companies typically approach specialty chemical companies with their new product requirements. (See <u>Figures 12A and 12B</u>.)

The major difference in these segments is that specialty chemical functions and testing methods for these applications are extremely complex and often more difficult to evaluate. In general, most new products for the cementing and stimulation markets are brought forward by either the service companies or oil company R&D centers, which have a thorough understanding of both procedures. Specialty chemical companies rarely actively promote stand-alone products in these oilfield segments. This is an important distinction because without a service company champion, it is extremely difficult for a specialty chemical company to introduce new specialty chemical products into this segment.

Specialty chemical needs in completion and workover, cementing, and stimulation are certainly present. Close working relationships are considered essential between the service company and the specialty chemical supplier. Such exclusives are very prevalent. Service companies virtually demand this arrangement due to heavier front-end costs to test and more importantly to introduce a new product. These barriers to the introduction of new products into the cementing and stimulation oilfield segments are substantial, and unless one has a good existing working relationship with a service company, it can be a very tedious process.

The future needs of these oilfield segments will be driven by the service companies and the oil company R&D centers. These companies will in most cases be the primary initiators of specialty chemical R&D for these oilfield segments.

Figure 13 SPECIALTY PRODUCTION CHEMICALS
Emulsion breakers
Defoamers
Corrosion inhibitors
Gas sweetening aids
Biocides
Scale inhibitors
Pour-point depressants
Flow improvers

Production

Production chemicals in the oilfield area represent the largest application for specialty chemicals. Production chemicals are generally tailored for each individual oilfield. Figure 13 lists these specialty production chemicals.

As shown in a previous table, the worldwide market for production chemicals is $800 million. Of the $800 million, approximately 45 percent (at the manufacturers level) is considered specialty chemicals using $360 million as the chemical value of the products without service. Of this $360 million, about 75 percent of the products are manufactured by the individual production chemical service company. Thus, only about $90 million of specialty chemicals are actually purchased annually from non-oilfield specialty service companies.

Production chemicals are also typically manufactured by specialty chemical service companies in drum quantities. These service companies generally ship concentrated (75 percent plus active) products in bulk containers to a site near the oilfields where the product can be diluted to field strength (30 to 40 percent active) prior to application. (See Figure 14A.)

Production chemical costs are typically calculated on a cents/barrel of oil produced basis. For example, in 1985, production chemical treating costs in the U.S. were about 15¢ per barrel of oil produced. As oil prices declined, treating costs were brought under greater scrutiny and programs were altered, reduced, and even discontinued. Additionally, the initiation of bidding practices lowered costs even further such that by 1990 treating costs have been reduced to 12¢/barrel of oil produced. Treating

<u>Figure 14A</u> Production

- 600,000+ oil wells worldwide and no two are the same. Very fragmented market due to sheer numbers.

- End user base is 500+ oil company and national oil company operators.

- Specialty chemical service companies typically manufacture their own products for this segment.

<u>Figure 14B</u> Production

- New specialty chemical product evaluation is conducted by both the specialty chemical service companies and end user operator.

- Improved downhole corrosion and scale inhibitors, biocides, H_2S scavengers are being sought.

costs are highest in the U.S. because water production in association with the crude oil is higher here than any other area in the world.

International treating costs range from under 1¢/barrel of produced in parts of the Middle East to 6¢/barrel in Canada. North Sea treating cost are generally about 4¢/barrel of oil produced. As can be seen by these numbers, there is significant room for growth in the international production chemical/treating market. However, it is unclear whether higher associated costs to produce oil in areas such as the North Sea will result in earlier abandonment of fields (than the U.S.).

Oilfield production chemical companies typically carry out their own R&D for new products. The companies are generally sophisticated specialty chemical manufacturers which can conduct most of their own R&D though basic chemical synthesis is less active.

<u>EOR</u>

Specialty chemical usage in EOR (See <u>Figure 15</u>) is still very limited as low oil prices have dampened enthusiasm for chemical EOR. High front-end costs attached with most chemical EOR processes place substantial burden on the operator since chemicals are typically injected for several years before increased oil production is seen. To date, few chemical EOR projects have been both technical and economic successes. However, substantial promise is foreseen for steam assisted oil production which utilizes surfactants to improve reservoir sweep efficiency.

Figure 15 EOR

- Similar market and end user numbers as production.

- Specialty chemicals sold direct to operator in most cases.

EOR chemicals are typically supplied directly from the manufacturer to the operator. A modest market exists for service companies in EOR but most operators prefer to design and implement their own EOR projects.

New specialty chemicals for the oilfield are being sought continually. As the number of wells which are drilled and produced in more hostile conditions grow and environmental awareness and compliance factors receive greater scrutiny, specialty chemical demand will increase rapidly. Quality and safety programs will also have a major impact on chemical selection. Over the past 5 years, oilfield chemical selection has been driven by price. In the 1990s, performance, environmental compliance, and safety will be increasingly important in the purchasing decision.

Chemicals in Oil Production: A Producers Perceived Needs

D. Callaghan

SHELL INTERNATIONALE PETROLEUM MAATSCHAPPIJ, THE HAGUE, THE NETHERLANDS

1. INTRODUCTION

Following the identification of a potential hydrocarbon bearing structure by seismic survey, the subsequent development of an oil or gas field is progressed in a number of discrete stages. An exploration well is drilled to confirm the presence of hydrocarbon, followed by a number of appraisal wells to establish the extent of the find. Following the assessment that a find is sufficiently large for commercial exploitation, production wells are drilled and process facilities are installed to separate and condition the produced fluids to commercial specification products and transport them to a consumer, or to a storage terminal for subsequent shipment. Finally, when a field is depleted, it must be abandoned in a technically safe and environmentally acceptable manner.

The current state of technology does not allow production of only those hydrocarbon components from the reservoir for which there is a market. In particular, large quantities of water are often produced together with the hydrocarbon and must be separated from the oil and disposed of. The associated gas phase may contain components (e.g. H_2S, CO_2) which must be removed in order to meet the product sales specification. When non associated gas is produced, water present in the vapour phase condenses as the temperature drops through the process facilities.

Three different recovery methods are used to produce oil. In "Primary" production, oil is lifted to the surface by natural drive (provided by expanding gas or water influx) or artificially with pumping or gas sparging (gas lift). "Secondary" recovery implies enhancement of the natural drive mechanism by gas and/or water injection. From suitable reservoirs, "Tertiary", or Enhanced Oil

Recovery (EOR) techniques allow additional quantities of oil to be produced, by modifying the physical properties of the oil or of the driving fluid in the reservoir. The most common EOR techniques are steam flooding, polymer or surfactant/polymer flooding and CO_2 flooding. The production cost of EOR oil is generally significantly higher than that of oil from primary recovery and demand for tertiary recovery is therefore related to movements in crude oil price.

Exploration and production activities consume large quantities of both commodity and speciality chemicals. The estimated worldwide annual expenditure on drilling and production chemicals for Shell companies is approching 100 million pounds sterling, split approximately 3:1 between drilling and production. Previous reviews[1,2,3] have surveyed the application and types of chemicals used in the upstream oil sector. Little has changed in the meantime, with the notable exception of growing concern within the Industry for the impact which these chemicals may have on the environment.

A brief review only is given below of the application of chemicals in the upstream oil business, for a more detailed review reference should be made to the previous papers on this subject.[1,2,3.]

2. CHEMICALS USED IN THE UP-STREAM OIL SECTOR

2.1 Drilling, completion and maintenance of wells

Drilling. Drilling a well requires the use of a drilling fluid or "mud". The fluid serves a number of functions, including:

- transportation of cuttings to surface
- control/balancing of downhole pressure
- lubrication for the drill string and cooling of the bit
- protection of sensitive formations
- formation of an impermeable barrier to restrict fluid loss to permeable zones

Two principal types of drilling muds are used:

i) Water-based muds, which may typically contain in addition to water:

- weighting agents to provide the required density. Barytes, iron oxides and calcium carbonate are most commonly used.
- viscosifiers, for which bentonite clay and polymers such as Xanthan gum and Guar gum are frequently used.
- flocculants and clay stabilisers, for which polymers, such as

hydrolysed polyacrylamides, cellulose ethers and co-polymers of vinyl acetate and maleic anhydride are used.

- dispersants eg ferrochrome lignosulphonates, chrome free lignosulphonates and lignites, sodium polyacrylates and phosphates are used to disperse solid particles.
- fluid loss additives eg carboxymethyl cellulose, polyacrylates and pre-gelatinised starch are used to reduce fluid loss to permeable and porous formations.
- other chemicals are added to control pH and correct ionic balance. Small quantities of biocides, corrosion inhibitors, surfactants and defoamers are sometimes used.

ii) Oil-based muds, of which there are two types; true oil muds containing less than 5% of an aqueous phase and Invert-Oil-Emulsion Mud (IOEM), where a saline aqueous phase of 10-40% is dispersed (emulsified) in a continuous oil phase. The early use of diesel oil has been superseded by low aromatic mineral oils. Vegetable and other non-mineral oils are now being tested as environmentally friendly replacements for the mineral oils previously used. Similar weighting agents and commodity chemicals to those used in water based muds are used, with additives such as emulsifiers, wetting agents and detergents to produce a stable oil based formulation.

Drilling mud used on both onshore and offshore locations is often collected for subsequent reconditioning and re-use, or disposal in an environmentally acceptable manner. However the cuttings recovered from the well are inevitably contaminated with mud. Traditionally, cuttings from offshore drilling have been discharged to the sea, which has given rise to concern over the environmental impact, particularly when mineral oil-based mud has been used. In some areas, eg the North Sea, strict limits have been set for the amount of oil which can be discharged with the cuttings. There is continuous pressure to reduce or even eliminate the use of oil based muds, or at least stop the disposal of oil contaminated cuttings to the sea. The declaration from the North Sea Ministers Conference of March 1990, confirms this trend by indicating a time schedule for reduction, which may ultimately lead to a ban on the discharge of oil contaminated cuttings to the North Sea.

Cementing. In the course of drilling a well, a number of cementations are carried out using a special oil well cement. This is normally based on specially ground Portland cement with additives to control fluidity and setting characteristics. Cement constitutes the second largest market for chemicals in the upstream sector after drilling fluids.

Completion Fluids. Completion of a well as an oil or gas producer requires the use of a clean fluid which will not damage the permeability of the reservoir rock. Typically, filtered brines are used, sometimes with polymers such as hydroxy-ethylcellulose or polysaccharides added to impart the desired viscosity, eg for gravel carrier in gravel pack operations.

Stimulation. The productivity of a well may be unsatisfactory due to impairment caused by particle invasion during drilling/completion, deposition of scale/corrosion products, or by the properties and condition of the reservoir rock itself. Stimulation may be carried out in such cases, using hydrochloric acid alone, or in combination with hydrofluoric acid. Corrosion inhibitors and surfactants are included in acid formulations. To improve productivity from low permeability rock, the latter can be fractured by pumping in a highly viscous fluid containing a proppant which keeps open the fracture.

2.2 Production Chemicals

Demulsifiers. Oil is normally produced together with dissolved or free gas and with water. Traditionally, the gas is separated first at one or more pressure levels, followed by the separation of water at, or close to, atmospheric pressure. The water cut may vary from less than 0.5% to over 90% and normally increases during the producing life of a field. When a large proportion of water is present, some will be in the form of "free" water which separates easily from the oil. Some water will be present as an emulsion in the oil, which is stabilised by naturally occurring surface active compounds in the crude, by added chemicals such as corrosion inhibitor and by finely divided solids which adsorb at the oil-water interface. To dehydrate the crude effectively, the following sequential steps are necessary:

- destabilisation of the emulsion
- coalescence of the destabilised water droplets
- separation of the water from the oil phase.

Demulsifiers are blends of surface active agents, normally supplied in a solvent and used to destabilise water-in-oil (w/o) emulsions. Anionic, cationic and non-ionic demulsifiers can be used depending on the specific nature of the emulsion. The optimum formulation is normally selected by laboratory evaluation, followed by field trials.

De-oilers or reverse emulsion breakers. Water which is separated from the oil often contains finely dispersed oil droplets, which have been carried over as a reverse (o/w) emulsion. It is normally necessary to reduce the residual amount of oil prior to disposal to the sea or re-injection into a waterflood or disposal well. Re-injection of water containing dispersed oil together with suspended solids, often leads to plugging of the formation and consequent loss of injectivity. Process water discharged to the sea from offshore platforms is often subject to a maximum oil content by legislation and in the absence of this, a responsible Operator will adhere to a similar self imposed standard. The countries bordering the North Sea have adopted the limit proposed by The Paris Commission, which is currently a maximum daily average of 40 mg/l of oil.

The first stage of de-oiling is normally gravity separation, using typically API separators onshore and Plate separators offshore. To reduce the surface tension, enabling the oil droplets to coalesce, polyamine or polyamine quaternary compounds may be added. Coagulants and flocculants may also be added to improve the separation of oil to low levels and to remove suspended solids from water before injection into the reservoir. Chemicals used for this duty are generically similar to demulsifiers. Lower levels of oil are often achieved with a flotation unit, which may require the addition of a polyelectrolyte to maintain a stable froth which can be skimmed off. Other de-oiling techniques, such as coalescers and filters, may also use chemicals to improve the separation.

Corrosion Inhibitors. Production facilities are subject to corrosive attack by carbon dioxide, hydrogen sulphide and other aggressive constituents which may be present in the reservoir fluids. Corrosion inhibitors are added to the fluids, either continuously or batchwise, in order to maintain the rate of corrosion at an acceptable level. Depending on whether the steel is water or oil wet, oil soluble/water dispersable or water soluble/oil dispersable compounds are used to form a thin protective layer on the metal surface, or by a physio-chemical reaction/neutralisation. Much research has been conducted into oil field corrosion and whilst the basic mechanisms are well understood, surprises still occur due to the lack of appropriate wellstream composition data when facilities are designed, due to changes in fluid composition, or changes in flow regimes during the producing lifetime of a field. Depletion of the "easy" oil and gas reserves will be partly compensated in future by increasing production of more corrosive oil and particularly gas.

The surface active nature of corrosion inhibitors present in a well stream, together with large pressure drops in the process train, often leads to the production of stable water in oil emulsions, from which it is difficult to separate the water. Subsequently, it is equally difficult to remove the residual oil from the water to a level acceptable for subsequent disposal of the water. Depending on the type of inhibitor used, varying amounts will remain in the water phase. Many of the formulations used in the past are considered to be environmentally "unfriendly", leading to a desire on the part of Operators to reduce and eventually eliminate the use of these compounds in the future. A strong incentive exists for the Service Industry to develop environmentally friendly inhibitors, which will partition only to a negligible extent into the water phase, as an economic alternative to the use of more exotic, corrosion resistant materials for the construction of facilities.

Scale Inhibitors. Scale deposition inside flowlines and equipment occurs most commonly when the formation water becomes super saturated with the carbonate or sulphate of Barium, Calcium or Strontium. This may occur due to changes in temperature and pressure of the fluids in well tubulars, flowlines and process equipment, or when non-compatible waters are mixed. The latter phenomenon can occur when sea water is injected into a reservoir where Barium or Strontium ions are present in the formation water, causing either injectivity impairment or scaling of tubulars and equipment when sea water breakthrough occurs in producing wells. Typical chemical types used to inhibit scale formation are organic phosphates, phosphonates and polyacrylates. Computer simulation programs are available to predict the scaling tendency of various waters and their mixtures. As with most oil field chemicals, the optimum choice of inhibitor is often made only after screening tests.

The volume of water projected to be handled by the oil industry (both produced and injected) will increase in the future, with the ageing of producing fields. Scale inhibitors, most of which are subsequently disposed with water, must be demonstrated not to have a negative impact on the environment into which they are discharged, i.e. they must be of low toxicity to marine life and must not be bioaccumulative.

Wax Inhibitors and Pourpoint Depressants. Many crude oils exhibit crystallisation of aliphatic long chain hydrocarbons (wax) as the temperature decreases from the reservoir to the stock tank. Reduction in flow and eventually complete plugging of pipelines and facilities can occur. As an alternative to removal of wax using solvents or mechanical means, growth of paraffin crystals can be restricted, reducing their tendency to deposit, by addition of wax dispersants or pourpoint depressants.

Gas Drying and Hydrate Suppression. To reduce pipeline corrosion in the presence of carbon dioxide and/or to suppress hydrate formation at low temperatures, the water dew point of the gas is controlled, frequently using glycol in a contacting tower. Glycol is normally recovered by boiling off the water. Aromatic components present in the gas (principally benzene, toluene and xylene) have an affinity to dissolve in glycol and are boiled off with the water during separation. This may lead to an atmospheric emission problem, or in the case where the vapour is condensed, to a water stream for disposal, containing a high level of dissolved aromatics. Methanol is also used for hydrate suppression by direct injection into pipelines. Methanol is not normally recovered and tends to be disposed with the water.

Secondary Oil Recovery. Significant quantities of chemicals are used in the treatment of water which is injected to improve oil recovery. Depending on the source of the water, filtration aids may be required, which must be disposed of and oxygen scavenger, corrosion inhibitor, scale inhibitor and biocides may be added to the water. When water breakthrough occurs, these chemicals, or their degradation products, may be present in the water which must be disposed.

Enhanced Oil Recovery. The enhanced (tertiary) oil recovery techniques of steam injection, chemical flooding and CO_2 flooding require the use of chemicals. In steam injection, chemicals are used for regeneration of the ion exchange resins used for softening the boiler feed water, as chelating agents, for pH control, scale suppressants and oxygen scavengers. In chemical floods using surfactants and/or polymers, in addition to the latter two chemicals, oxygen scavengers, radical scavengers and biocides are added to protect the stability of the polymers in the reservoir. Oil prices in recent years have not been conducive to the development of large scale EOR projects, although this situation will inevitably change when crude prices rise to the level at which such projects can be economically implemented.

Oil produced using EOR techniques, particularly with chemical methods, is more difficult to dehydrate than primary recovery oil. The traditional thinking of the Industry will imply using more chemicals to effect adequate dehydration and water clean up. Disposal of water from EOR projects into the environment may not be acceptable in future and alternative disposal options will have to be considered for the development of such projects.

3. THE ENVIRONMENTAL IMPACT OF OIL FIELD CHEMICALS

The use of chemicals in the drilling and completion of wells is shown in a simplified manner in figure 1. Traditionally, the bulk of these chemicals have been lost to the environment during drilling operations, or have been disposed to the environment when they have served their function. Drill rigs are now being equipped with contained drainage systems to prevent mud losses to the environment. Currently there is a move away from those chemical formulations which are considered to be most damaging to the environment.

In the Dutch sector of the North Sea, cuttings contaminated with oil based mud are no longer discharged in environmentally sensitive areas and a maximum residual oil content of 100 g/kg is now allowed elsewhere.Following the North Sea Ministers Conference in 1990, it has been agreed not to use oil based mud for single well exploration sites from 1993 and to study the feasibility of, and a date for implementation of a total phase-out for the discharge of mineral oil-

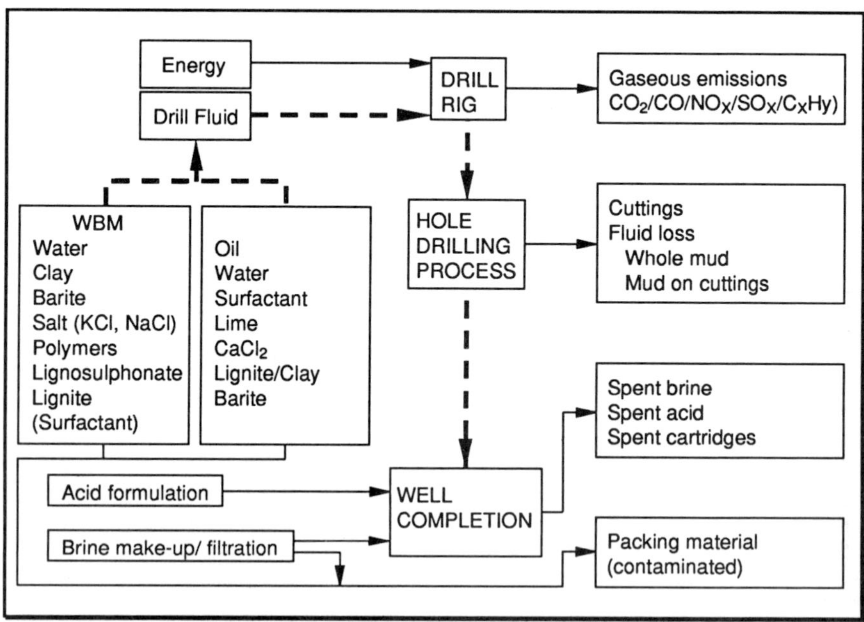

Figure 1 - Schematic showing principal chemicals used and emissions from well drilling and completion

based mud and contaminated cuttings. Technology is being developed to remove and recover oil from contaminated cuttings, to a level which allows the cuttings to be disposed as industrial waste. Some operators are pro-actively changing to the use of water-based muds to drill wells, where oil based mud would have been the preferred choice on technical grounds.

A typical oil or gas production process is shown schematically in figure 2. With the current state of technology, the desired hydrocarbon cannot be selectively produced from the reservoir. Water, containing naturally occurring chemical compounds may be co-produced in increasing quantities during the producing life of a field. To process and transport oil and gas economically, various chemicals mentioned in the preceding sections may be added to the produced fluids and will partition between the oil and water phases. Chemicals may also accumulate in emulsion and sludge residues, which require periodic disposal. Unless it can be proven otherwise, it is often assumed by authorities that all chemicals added during the process are disposed to the environment.

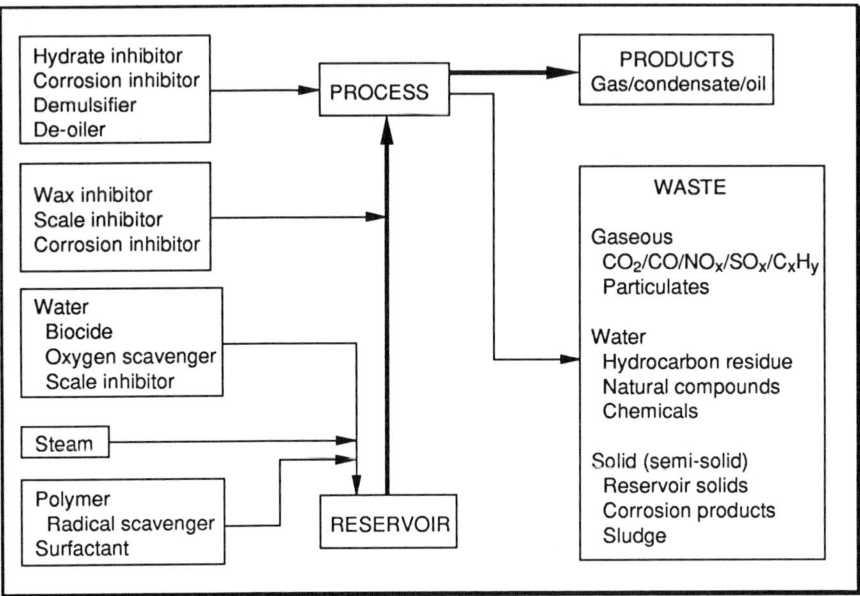

Figure 2 - Schematic showing principal chemicals used and waste emissions from an oil and gas production process

Most of the water produced from on-shore oil fields operated by Shell companies is disposed by injection, either back into the producing formation or into a shallower, non potable aquifer. The high cost associated with re-injection offshore has precluded this option in the past on economic grounds, particularly as each disposal well would occupy a slot which can be used for a producing well. Generally, water produced offshore is disposed into the sea after reducing the residual free (dispersed) oil content to an acceptable level. The maximum oil content in discharged water allowed by legislation varies in different countries, but typical average maximum values for dispersed oil are in the 40-50 mg/l range. There is currently no other generally applied restriction on the composition of water which may be discharged from offshore platforms, although in the North Sea for example, discharges from production platforms are covered by the Paris Convention for the Prevention of Marine Pollution from Land-based Sources.

It must be recognised that chemical constituents and organic residues other than the dispersed oil which is currently monitored, may be present in varying amounts in separated production water. Some Operators are currently engaged in extensive monitoring programmes to identify those components which are produced naturally from the reservoir. The responsibility for implementing appropriate technology to limit the negative environmental impact of naturally occurring materials rests with the Oil Industry. The co-operation of the Chemical Suppliers is essential in tackling the problem produced by the chemicals which are intentionally added during the production process.

The technical performance of chemicals has been the principal criterion for selection in the past; the impact of subsequent discharge to the environment has been of secondary concern. Little relevant ecotoxicological data has been available in the past. The chemical supply industry has been reluctant to provide information on the formulation of products, in the belief that disclosure may damage their commercial interests. It should also be recognised that proprietary chemicals are often complex mixtures, or the reaction products, of several commercially available components, containing additional impurities.

There has been no generally and internationally accepted requirement, or universally applicable procedure for toxicity testing of chemicals and drilling muds. The four countries with offshore operations in the North Sea have required toxicity testing of oil based muds during the 1980's, although different parameters have been specified in the various countries. The reluctance of the Oilfield

Chemical Supply Companies to become involved in such a multiplicity of testing is understandable. Data which has been supplied is not in a consistent form, leading to difficulty in interpretation. A common protocol, acceptable in different countries and requiring only one set of tests, is now being developed.

Safe Handling of Chemicals (SHOC)

Operators in the upstream oil sector recognise their responsibility towards the health and safety at work of their employees and the protection of the environment. Operators need reliable information on the chemicals used and Suppliers have a responsibility to provide such information. On behalf of the Industry, the E&P Forum*) is setting up a data base, or Minimum Data Set (MDS).[4] The Chemical Suppliers were represented in the E&P Forum Chemicals Data Base Working Group, which was responsible for setting up the data base guidelines. The proposal has been acknowledged by the Paris Commission Group on Oil Pollution and by the equivalent regional organisations in the Arabian Gulf and the Mediterranean. (At the time of preparation of this paper, some UK suppliers appear to be reluctant to provide all the data requested for the SHOC MDS). It is intended that data obtained through the MDS will be stored in the E&P Forum data bank. Members of the E&P Forum may have access to the MDS for the chemicals they use or intend to use and the data may be passed on to the appropriate regulatory authorities. Chemical Suppliers will not have access to data on competitors products. The data base should avoid the wasteful duplication of effort by Chemical Suppliers in providing the same information to several Operators. Within the Shell companies, the data for a number of SHOC or SHOC MDS questionnaires has been validated and the result for each chemical translated onto a single page A4 SHOC card. The information is presented in the form of standard phrases, each covering a range of results, which can be easily understood and used in the field. The SHOC cards are being issued to all Shell E&P companies worldwide.

Information provided to the data base should be updated when product formulations are changed, or recertified annually. A completed MDS questionnaire will contain the following information:

- Suppliers information.
- Product characterisation, composition, regulatory requirements, physical properties.

---

*) E&P Forum - The Oil Industries International Exploration & Production Forum

- Safety data.
- Toxicological data.
- Environmental data.
- Accompanying documents and certification statements.

(Each submission must be accompanied by either a completed SHOC questionnaire or a Material Safety Data Sheet.)

The E&P sector of the Oil Industry has taken the initiative in setting up this data base and is now looking to the Chemical Suppliers for their continued full support. The environmental data requirements of the MDS do not place an additional strain on the ecotoxicological testing capacity of existing laboratories, as they are based on existing data, accepted by Contracting Parties to the Paris Convention, the US authorities, or data produced in accordance with EEC regulations. It should be recognised that the time will come when Operators will not purchase chemicals without the requisite data and those suppliers who can demonstrate that their products are safe and environmentally friendly will be in a competitive position. Shell companies have adopted the SHOC MDS for collecting HSE data on chemicals used in, or intended for use in its operations. The data base will be a key element in a harmonised procedure for the approval, evaluation and testing of offshore chemicals and drilling muds; guidelines for which were adopted by the Paris Commission in June 1990, for a two year trial period.

There is a strong incentive for the Oil Industry and the Chemical Suppliers to work together in pro-actively ensuring that the procedures and controls regarding the use of chemicals are in order, before such controls are imposed by legislation.

4. LIKELY FUTURE TRENDS IN THE USE OF OIL FIELD CHEMICALS

Large quantities of chemicals are handled by the up-stream sector of the Oil Industry and some are discharged intentionally or accidentally into the environment. It is a business policy for Shell companies to conduct their activities in such a way as to take foremost account of the health and safety of their employees and of other persons and to give proper regard to the conservation of the environment. In pursuit of the latter policy there is an aim to eliminate,over time, all emissions to the environment which are known to have a negative impact on the environment. In so far as this relates to the discharge of chemicals which are used in the drilling and production processes, the objective can be achieved in various ways:

i) Removal of residual chemicals from waste streams using "end of pipe technology". This route is not favoured as a long term solution, since the offending components, where they cannot be recycled, still have to be disposed of, but in a more concentrated form.

ii) Substitution of harmful chemicals with alternatives which have been demonstrated to be environmentally friendly.

iii) Solution of the problem at source, using processes which do not rely on chemicals. This implies a change from traditionally used physio-chemical processes to physical processes.

iv) Injection of all waste streams back into the producing formation. In some countries legislation only allows the injection of those substances which were produced from the reservoir and the addition of chemicals therefore prohibits reinjection in these cases.

In practice, the long term goal may be achieved by a combination of the possible solutions listed, depending on the particular circumstances. The Chemical Suppliers face a challenge, to develop products which will satisfy the requirement of alternative ii) above and be economically more attractive than alternatives iii) or possibly iv), which may be developed by the Operators.

In the past, Operators have bought chemicals based on cost per unit quantity and it has been in the commercial interest of Chemical Suppliers to sell the maximum amount possible. The future may show a change in contracting strategy, with the emphasis on buying performance rather than chemicals. As examples:

- Dehydration. Rather than awarding contracts for the supply of demulsifier and de-oiling chemicals, the future trend may be to contract for an agreed residual water content in the oil and an agreed water quality for disposal. This should provide the incentive to Service Companies to minimise the use of chemicals, which together with the other environmental safeguards mentioned above, should minimise the impact of our operations on the environment.
- Drilling fluids. Purchase a completely engineered drilling fluid system in future, which will allow the well to be drilled optimally, rather than buying separate mud chemicals and a mud engineering service.

- Well stimulation. Following selection of stimulation candidates by the Operator, the design and execution of the job are obvious areas for incentive contracts.
- Corrosion inhibition. This is an area which may be more difficult to resolve contractually, since the consequence of a major corrosion problem may be enormous and the effect of poor inhibitor performance may not be immediately obvious.
- Packaging. Disposal of contaminated packaging materials (particularly drums) used to supply chemicals is becoming increasingly difficult. Movement away from the traditional single journey, throw-away packages to reusable bulk containers can be anticipated and should be developed to the mutual benefit of oil field operators and the chemical supply industry.

5. CONCLUSIONS

Growing concern over the environmental impact of all types of emissions, will lead to a reappraisal of the way in which chemicals are used during the exploration and production of oil and gas. Large quantities of chemicals are used in drilling and production operations and unless it can be proven otherwise, it may be assumed that they will be eventually discharged into the environment. Unwanted, naturally occurring compounds are also separated from the reservoir fluids during processing and must be disposed of. The latter problem is being dealt with by the Oil Industry and the problems associated with the additional chemicals which are used, must be addressed jointly by the Industry and the Chemical Suppliers.

Health, safety and environmental data must be made transparent through a universally recognised system, for all chemicals used in the exploration and exploitation of oil and gas reserves. The Industry, through the E&P Forum, has taken the lead in setting up such a system. Based on the data collected, chemicals will be chosen which have minimum impact on the environment in future.

If Operators in the upstream oil sector cannot satisfy themselves on the health, safety and environmental aspects of chemicals which are used in the traditionally employed physio-chemical processes, the incentive will exist to accelerate the implementation of alternative processes and engineering solutions, which will not rely on the use of chemicals.

REFERENCES

1 R. C. Parker
Chemicals in the Oil Industry
Royal Soc. Chem. Industrial Div. Symposium Proc. March 1983, p.206.

2. R. B. Hough
Chemicals in the Oil Industry - Where Next?
Royal Soc. Chem. Industrial Div. Symposium Proc. March 1985, p 1.

3. D. Antheunis
Changing Chemical Needs in the Oil Industry
Royal Soc. Chem. Industrial Div. Symposium Proc. March 1989.

4. *Safe Handling of Chemicals Minimum Data Set (MDS) Questionnaire with Associated Guidelines.*
E&P Forum Report No.6.17/163 Oct.1990. London.

Chemical Enhanced Oil Recovery Methods in the East European Countries and in the USSR

J. Tóth, J. Török,

RESEARCH LABORATORY FOR MINING CHEMISTRY, HUNGARIAN ACADEMY OF SCIENCES, MISKOLC-EGYETEMVÁROS, H-3515, HUNGARY

and Y.G. Mamedov

ALL-UNION SCIENTIFIC-RESEARCH OIL AND GAS INST, 10 DMITROVSKY PROEZD, MOSCOW, 125422, USSR

1. INTRODUCTION

Over the last decades all conventional EOR techniques under various geological and enviromental conditions have been tested and applied on different scales. Figure 1 presents statistical data[1] on world EOR oil production and on the number of active projects in 1989.

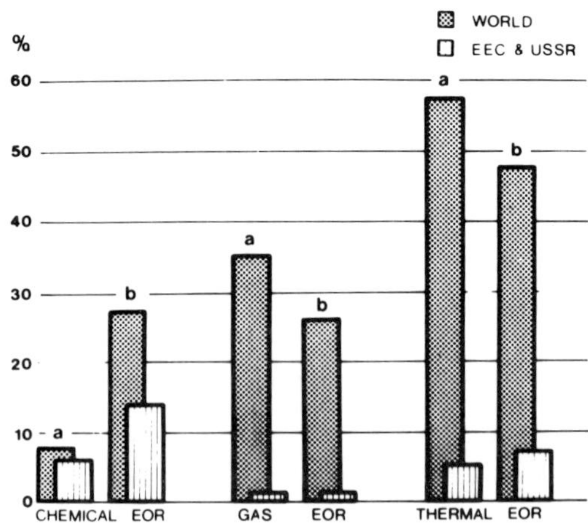

Figure 1 World EOR production (a) from active projects (b) in 1989 (Total EOR production: 93,313 t/d from 742 projects)

As for the chemical EOR production in the East European countries and in the USSR shown in Figure 2, most of the results are coming from the projects in the Soviet Union. However, remarkable results have also been accumulated in Albania, Bulgaria, Czechoslovakia, Hungary, Poland, Romania and Yugoslavia, as well.

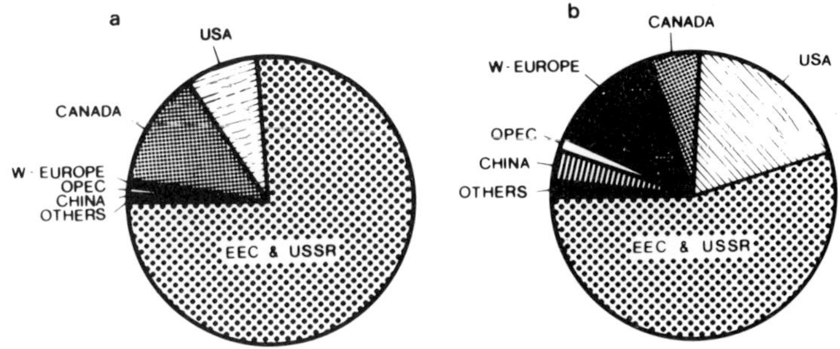

Figure 2 World chemical EOR production (a) from active projects (b) in 1989 (Total chemical EOR production: 7136 t/d from 196 projects)

The practice of Soviet oil production proves the advantages of non-conventional water-flooding over the conventional one in increasing volumetric sweep efficiency. Thermal and chemical EOR methods also contribute to the additional oil recovery in the Soviet fields as is shown in Figure 3. It is worth mentioning that according to one forecast the proportion of chemical and thermal methods will increase there by the end of this century.

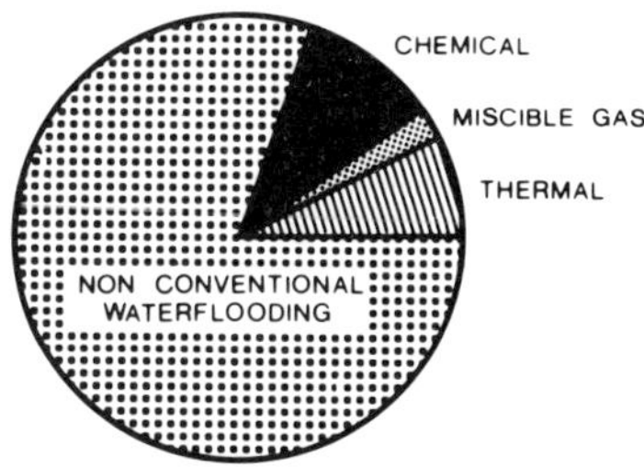

Figure 3 Contribution of main IOR methods to the oil production in the Soviet Union in 1989. (Total EOR production: 146,000 t/d)

In the East-European countries it is waterflooding which is generally applied, however, thermal projects in Czechoslovakia, Hungary and Romania, as well as carbonated natural gas injection in Hungary, are also successful EOR processes resulting in a remarkable additional oil production. Chemical flooding is of less importance in these countries, however, attractive results have been reported concerning the use of chemicals for near-well-bore treatment in Hungary[2] and in the USSR as well[1].

As for the EOR methods used on a large scale in the East European countries and in the USSR attention will be focused here on those in which additional oil recovery is the result of chemicals applied. Therefore, no remarks will be made in connection with water flooding, conventional thermal methods, gas flooding and microbiological EOR techniques. Their combinations, however, where chemicals are also involved, will also be discussed.

EOR methods in general, and chemical methods especially are complicated, costly, risky and usually sensitive ones. The return of capital invested is slow and the economy of these processes is strongly affected by the world oil price. This is why their economic efficiency cannot be easily estimated and forecasted. Therefore, the main goals of this paper are: to summarize chemical methods applied in the East European countries and in the Soviet Union with special respect to their characteristics and their effect on recovery efficiency without any remarks on economic questions.

2. CONVENTIONAL CHEMICAL EOR PROCESSES

Among the conventional EOR processes studied on laboratory scale and applied in field practice in the East European countries and in the USSR the following chemicals are widely used: polymers, surfactants and alkalines. Results are summarized below.

Polymers

Enhanced oil production due to the application of polymers in 16 projects was near 930 thousand tons in the East European countries and the USSR in 1989.

In practice only partially hydrolized polyacrylamides (PAA) have been used for flooding, mobility control, and for profile modification . The average concentration of polymers in displacement processes was 0.05-0.1%wt. in 20-50 PV slugs. The PAA molecular mass ranged from 6 to 20 million depending on the reservoir conditions. Polymers were used in sandstone formations up to about 80°C reservoir temperatures where porosity and permeability were in the 20-35% and 0.1-1.8 μm^2 range, respectively. The viscosity and density of oil to be displaced were 0.9-75 mPa.s and 820-930 kg/m^3 under reservoir conditions[3]. The average TDS in the formation waters was: 10,000-15,000 ppm and much higher in the Volga-Ural regions.

Polymer solutions were generally prepared using formation waters the composition of which did not affect the behaviour of polymers used. Preslugs to displace formation brine or to control cation exchange and thus inhibiting the appearance of undesired cations were applied only in rare cases.

The feasibility studies of biopolymers have been completed in the Soviet Union and pilot tests are going on concerning their application under conditions unfavourable for polyacrylamides. No information is available from other East European countries about the future application of biopolymers for oil displacement.

Based on successful field scale projects in Romania and mostly in the Soviet Union it was found that injection of 1 ton PAA (in dry base) resulted in 160-250 tons of additional oil produced. It is worth mentioning that in earlier years about 7% of field applications produced less than the economic limit of 75 t oil/t PAA . No success has been reported in one Hungarian pilot test where reservoir temperature was nearly 100°C[4] . Here the salinity of the formation water was about 10 ppm only.

Surfactants

Enhanced oil production, as a result of the application of surfactants in 27 projects, was near 1,200,000 tons since the early 1960s in the East European countries and in the USSR.

Anionic, cationic and non-ionic surfactants have been tested and used in field projects.

Besides commercially available petroleum sulfonates domestic products were the active materials for mobilizing residual oil. Narrow range high molecular mass petroleum sulfonates with conventional co-surfactants were used for the preparation of microemulsions supplied by a Tatarian chemical plant in the Soviet Union. Broad molecular mass petroleum sulfonates were produced in Hungary for testing the displacement process by a micellar slug[5], cationic and non-ionic surfactants applied in certain projects in the Soviet Union were partly by-products of their chemical industry. Early single well tests in Hungary with sodium sulfosuccinate were unsuccessful. A polyethyleneglycol ester and a polyglycol ester of fatty acids as well as mixtures of alkyl-aryl sulfonates with hydrocarbon solvent additive resulted in additional oil recovery.

It is worth mentioning that all field projects were performed under highly waterflooded reservoir conditions, characterized by about 90-95% water cut.

As for the design, highly concentrated small size microemulsions and relatively large size micellar solutions have been used with various types of surfactants. Usually preslugs were injected to realize optimum salinity in the reservoir and also to control cation exchange. In many cases surfactant slugs were driven by graded polyacrylamide slugs chased by water. Preslugs and chasing slugs generally also contained tracers.

It is worth mentioning that in the case of expensive non-ionic surfactants polyglycerol as a special additive with different proportion has been used in the Soviet Union. In agreement with laboratory tests field application proved that in this way adsorption of the non-ionic surfactants is drastically decreased . This decrease resulted in 20,000 t additional oil recovered in the Uzen oil field in 1990 where the method was applied first. Here 98 t oil/t surfactant recovery has been reported[6].

Surfactant flooding has been tested not only in sandstone formations but also in carbonate reservoirs. Most of these projects were realized in the Soviet Union, in sandstone formations[3] , however, successful Albanian projects in fractured, dual porosity limestone formation have also been reported[7].

The temperature of reservoirs involved was 20-90 $^{\circ}$C in the Soviet Union, while it was around 100 $^{\circ}$C in the Hungarian project. Porosity and permeability of the reservoir rocks were in the 20-37% and 0.005-1.4 μm^2 range, respectively. The viscosity and density of oil to be displaced were 1-31 mPa.s and 780-900 kg/m^3 under reservoir conditions, where the extreme (186-240 mPa.s and 889-933 kg/m^3) data refer to the Albanian field test.

Based on the successful field scale projects in the Soviet Union it was found that 1 ton active material injected resulted in 60-120 tons of additional oil production. It is worth mentioning, however, that most of the projects gave gloomy results. In every case it was proved that surfactants mobilize the target oil but the additional oil production was far below that which was predicted. That is why surfactant technology is considered as one which needs further improvement. Additional studies are needed to clear the mechanism and to understand unfavourable factors affecting these processes, prior to starting new projects. It is predicted in the Soviet Union that as a result of ongoing fundamental research many of the problems will be solved in this decade. At this time the number of surfactant projects could increase.

Comparing the surfactant processes with other chemical EOR methods, surfactant enhanced oil recovery is the most expensive one, hence they are more sensitive to the world oil price. It is obvious that less expensive and more efficient surfactants are needed for this method. In this respect it is worth referring to laboratory studies which show promise using a combination of anionic and cationic surfactants.

Alkalines

Enhanced oil production as a result of the application of alkalines in 8 projects was 180,000 tons in the East European countries and the USSR in 1989.

Alkaline flooding was tested under different reservoir conditions in fractured, karstic limestone-dolomite and also in sandstone formations. Porosity and permeability of sandstone reservoir rocks were in the 22-35 % and 0.07-1.2 μm^2 range, respectively. These petrophysical characteristics cannot be properly defined for karstic formations. The viscosity and density of oil to be displaced were generally 1-50 mPa.s and 800-910 kg/m^3 under reservoir conditions.

Ammonium hydroxide solution was used in the Nagylengyel oil field, Hungary, in 1975 with moderate success[8]. Here the karstic reservoir was partially oil wet. The analysis of the produced fluids proved that asphaltic compounds were displaced from the rock-water interface by the alkaline solution, but it hardly affected the heavy, viscous oil trapped in the karst domes. Post-flood analysis highlighted the unfavourable selection of the reservoir for the test.

Lately a successful project in Albania has been reported[7] where the acidity of the heavy oil was 1.4-1.7 mg KOH/g oil and where the limestone reservoir rock was oil-wet, similar to the conditions in the Nagylengyel oil field, Hungary. It was concluded that surfactant was more efficiently used than the alkaline solution in mobilizing target oil probably due to the greater pore volume injected.

Concerning the application of alkaline solutions in the Soviet Union for improving oil recovery different techniques have been used. In the Trekzerskoe field, 39% PV alkaline solution was injected till the breakthrough with periodic water injection for mobility control. It resulted in 3.1% increase in the oil recovery and also an increase in the injectivity[9] .

It was found that 1 t injected alkaline material resulted in 5-16 t additional oil production. Alkaline projects are characterized by less than 10% recovery efficiency.

Although laboratory studies indicate that the combination of alkalines with surfactants in a polymer driven EOR process shows great promise, no such field tests in East European countries or the USSR have yet been reported.

3. NON-CONVENTIONAL EOR PROCESSES

Oil recovery methods where special chemicals are used to increase oil recovery, or the efficiency of conventional EOR processes, are considered here.

The contribution of non-conventional chemical processes to the chemically enhanced oil production in the USSR is remarkable as shown in Figure 4.

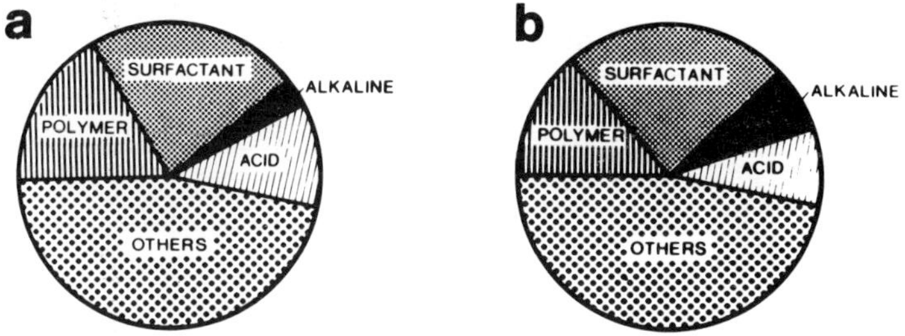

Figure 4 Chemical EOR production (a) from active projects (b) in the Soviet Union in 1989 (Total chemical EOR production 5464 t/d from 112 projects)

Sulfuric acid

Soon after the first successful field results[10] sulfuric acid injection has become a special process applied on field scale only in the Soviet Union. The sulfuric acid used is a by-product of a chemical plant close to the field where this process has actually been applied. The displacement mechanism is based on the interaction of sulfuric acid with the reservoir oil creating surfactants by in-situ sulfonation. The character of oil is naphthenic with an acid number above 0.3. The oil viscosity and density range is between 2-5 and 70-80 mPa.s and 800-900 kg/m^3, under 40-60 ^oC reservoir temperature. The porosity and permeability of the reservoir rock was 25-30% and 0.2-1.6 μm^2. Altogether 10 projects were active in 1989 resulting in a total of 585,000 t additional oil production.

The disposal of non-toxic spent hydrochloric acid into a waterflooded deep oil sand was studied in Hungary, where only a slight increase in oil recovery was observed. It was related to the insufficient amount of carbon dioxide formed due to the reaction of the acid with the cementing material.

Clay suspension

Injection of a suspension made of polyacrylamide and clay minerals is a unique procedure developed and successfully applied in the Soviet Union. The viscosity and density of oils in the reservoirs where the method has been used are 0.8-3 mPa.s and 850-890 kg/m^3 where temperature is near 60°C in sandstone and limestone formations characterized by 20-35% porosity and 0.1-1 μm^2 permeability. Results up to now prove the efficiency of this method which resulted in 465 t/y of additional oil produced.

Other chemicals

It was found that organometallic catalysts[11] promote the applicability of in-situ wet combustion with favourable air demand even in cases when low fuel content oils are involved. Large scale field tests in the Demjen oil field, Hungary, proved the efficiency of this method.

It was clearly demonstrated in the aforementioned Demjen oil field, Hungary[12], that the joint application of hot water formed from the steam injected and the surfactant dramatically increases recovery efficiency.

It has been reported in the Soviet Union that other chemicals, generally by-products (alkaline destillate liquids, trinatrium phosphate solutions etc.) have successfully been applied in various oil fields resulting in a total of 2.6 million t additional oil production.

There are many other chemicals, among them multifunctional ones, for inhibiting corrosion and scale formation, decreasing water content of gases injected and/or produced, emulsion breakers, oxygen scavengers, biocides, tracers, etc. Although they are widely applied in East European countries details about them are beyond the scope of this paper.

4. CONCLUSIONS

1. EOR techniques are widely used to increase oil production under very different reservoir conditions to increase ultimate oil recovery in the East European countries and in the USSR.

2. As a result of the application of chemicals 5.5 million tons of additional oil was produced in 1989 in the East European countries and the USSR which is nearly 80% of world oil production by chemical EOR

methods. The major part of this production comes from the Soviet Union.

3. Not only have all conventional chemical EOR techniques been tested and applied but new methods have also been developed.

4. Chemical EOR techniques are expensive and sensitive ones and there is a trend to improve the economy of these processes by improvement of technological efficiency and by using less costly chemicals.

5. An increase in the application of chemical EOR methods in the Soviet Union has been predicted but no increase in the East European countries is foreseen.

REFERENCES

1. Y.G. Mamedov, 'Implementation of EOR Technics in the World in 1989.'VNII, Moscow, 1990.
2. I. Lakatos, I. Munkácsi, S. Tröm böczky, J.Lakatos-Szabó, SPE 20996, presented at Europec 90, The Hague, 1990.
3. M.L. Surguchev, A.A. Bokserman, S.A. Zhdanov, Proc. 5th Eur. Symp. on IOR, Budapest, 1989, p. 22.
4. I. Lakatos, M. Munka, J. Tóth, M. Kristóf, S. Tröm böczky, Köolaj es Földgáz (Hungary) 1981, 12, 359.
5. A Balázs, I. Ferenczy, Gy. Gesztesi, K. Solt, S. Tröm böczky, Köolaj es Földgáz (Hungary) 1982, 5, 52.
6. R.E. Ganiev, Neft. Khoz., 1986, 1, 31.
7. N.M. Gjergji, N.N. Hamitaj, P.P. Liko, Proc. 3rd Symp. on Mining Chemistry, Siofok, 1990, p. 39.
8. I. Ferenczy, J. Pápay, M. Tóth, J. Török, A. Szittár, S. Tromboczky, Proc. 5th Eur. Symp. on IOR, Budapest, 1989, p. 13.
9. S.S. Nikolaev, A.I. Melnikov, R.E. Sofin, V.A. Popov, G.K. Cimliansky, A.G. Gorbunov, A.I. Bashurkin, S.I. Lokhmatov, Neft Khoz., 2987, 11, 48.
10. A.M. Gajnanshina, I.G. Plujan, Neftepromisl. Delo, 1983, 7, 5.
11. D. Rácz, Proc. 3rd Eur. Symp. on IOR, Roma, 1985, Vol. 2, p. 431.
12. M. Tóth, C. Gadelle, D.G. Antoniady, UNITAR/UNDP 5th Int. Conf. on Heavy Oil and Tar Sands, Caracar, 1991.

Minimising the Environmental Impact of Drilling Operations

R.C. Minton

BP PETROLEUM DEVELOPMENT CO LTD, FARBURN INDUSTRIAL ESTATE, DYCE, ABERDEEN AB2 0PB, UK

1 INTRODUCTION

Invert emulsion fluids, utilising a mineral oil continuous phase, are widely utilised in North Sea drilling operations. These fluids provide control of reactive geological formations, enhanced drilling performance, lubrication and are stable at high temperatures. Few wells fail to benefit from one, or more, of these characteristics.

In the drilling process the rock cuttings are circulated to the surface, separated from the re-circulating fluid, and discharged overboard. When invert emulsion fluids are used, the cuttings retain some 8 to 15% by weight of oil and this too is discharged. Over 12,000 MT of oil were discharged to the North Sea marine environment in 1990; from this source alone. The oily cuttings accumulate on the seabed in discreet piles distributed as a function of the prevailing currents. Smothering of the Benthic fauna is inevitable and the localised organic enrichment leads to anoxic conditions. Recolonization of the area is severely impeded and continued evidence of local damage is evident several years after drilling operations have ceased. However this is normally confined to the immediate area around the drilling site (1).

The Oil and Gas Operating companies recognise that the degree of damage being caused by the oily cuttings is unacceptable. Research effort is therefore being focused on developments to minimise the environmental impact of drilling operations.

2 ADVANTAGES OF INVERT EMULSION DRILLING FLUIDS

The upper hole sections of central and northern North Sea wells are dominated by the soft Tertiary mudstones. These react in the presence of water, softening and dispersing into the fluid.

This results in an unstable borehole wall, intrusion of the formation into the borehole behind the bit and soft sticky cuttings; all of which lead to drilling problems.

Properly formulated, invert systems prevent these reactions from taking place and significantly reduce the time, and hence cost, to drill these sections. Bailey (2) reported a 32% reduction in average tripping times through these intervals and up to an 89% improvement in reaming times, although the latter varied widely due to other problems with the invert fluids.

Deeper in the well the less reactive Cretaceous formations are often encountered. Here the use of the invert fluids leads to enhanced drilling rates particularly where the polycrystalline diamond compact (PDC) drill bits are utilised. An average drilling performance improvement of 49% was reported by Bailey (2) for the 12 1/4" hole sections in the areas he analysed, much of this through these formations.

These two characteristics of the invert fluids, better borehole stability and faster rates of penetration, combine to give a superior overall drilling performance. Consequently a reversion to water based mud use would add between 22 and 86% to the drilling costs (3) for these vertical exploration and appraisal wells, depending on the area being drilled.

For development drilling operations the cost penalties of reverting to conventional water based muds is even more severe. The borehole stability and the lubricating nature of the invert fluids permits a longer reach from the platform. This means that a single surface location is capable of draining a larger area of the reservoir, improving considerably the development economics.

Figure 1 defines the scale of these differences, indicating both the recommended maximum lateral departure, as a function of depth, for water based fluids and the theoretical limit for the invert fluids. Fields such as Bruce and Miller, now being developed, could not be drilled from a single surface platform if water based fluids were specified. This would double the development drilling costs for Miller and increase the Bruce costs by 140% (3). Under these conditions, it is questionable whether these developments would have proceeded.

The performance characteristics of invert emulsion fluids and a typical non-dispersed water based polymer fluid are summarised in figure 2. Different water based fluids will demonstrate a range of characteristics along the various axis but no water based fluid can presently compete in performance terms with the invert emulsions.

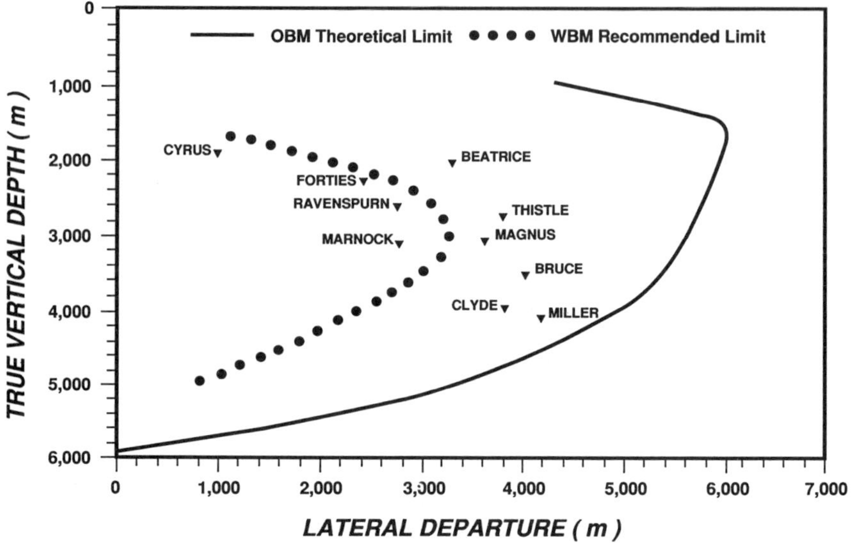

Fig 1. Recommended Departure limits for Directional Wells

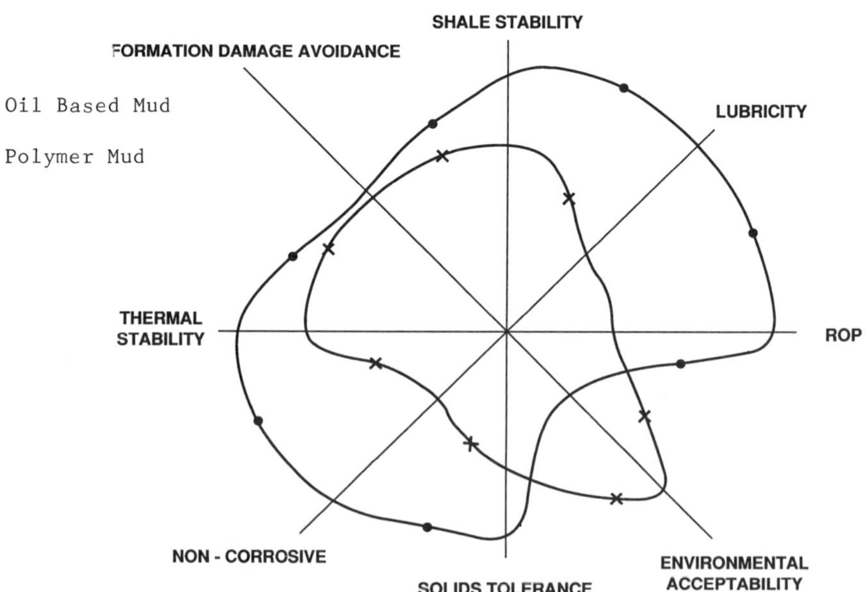

Fig 2. Performance Comparison of Different Drilling Fluids

The one area in which the invert fluids fall down is in their environmental acceptability. Davies et al. (4), reporting results of seabed surveys in the North Sea, demonstrated elevated hydrocarbon levels up to 4 kilometres from development platforms and changes in the Benthic diversity and community structure at distances of up to 2 kilometres. This is despite the change from diesel based systems to low aromatic oils, which came about due to acute toxicity concerns.

This degree of impact is unacceptable and legislative steps are in hand to control the levels of oil discharged with the drilled cuttings. The challenge for the industry is therefore to significantly reduce the environmental impact of its operations whilst retaining the operational performance advantages provided by the invert emulsion fluids.

3 DISPOSAL AND REPLACEMENT OPTIONS

Two broad options exist. Either to dispose of the oily cuttings in a more ecologically acceptable manner or to develop non-damaging fluids to replace the present invert systems. Both of these avenues are being explored with a wide variety of options under consideration.

In respect of the oily cuttings, technologies to clean them to very low residual oil levels (less than 1% by weight) and alternative disposal mechanisms are both under study.

Similarly both "pseudo oils" and novel water based mud systems are being developed and field trialed as invert fluid replacements. A combination of these technologies will ultimately be employed dependent on the relative cost benefits of the approaches.

Cuttings Cleaning

The legislative position is rapidly changing. Within the UKCS a limit of 15% weight of oil by dry weight cuttings is presently enforced for the oily cuttings and no whole fluid discharges are permitted. In Norwegian waters the comparable limit is 10% oil with an impending move to 6%.

Present discussions at the Paris Commission lead the industry to believe that the ultimate limit will be set at 1% oil on cuttings and an implementation date of January 1994 seems probable. This is the time-frame within which the industry has to act if continued use of invert emulsion fluids, as they are presently formulated, is anticipated.

Cuttings cleaning equipment, based on centrifuge or surfactant wash technology, is already utilised on platforms within the Norwegian operating area.

These, at best, achieve average oil on cuttings levels of 6 to 8%. Technology developments, aimed at further reducing the oil levels, have achieved 3% oil levels (5). This may well be the ultimate level for these technologies. Consequently both thermal distillation of the oil and solvent extraction technologies are under evaluation.

Thermal Distillation of Oily Cuttings

A number of design options have been promoted over the last few years with one, based on a rotating hammer mill (6) having undergone extensive land based trials. Recent studies (7) have shown that Rotary retort distillation equipment is theoretically capable of achieving the desired level of 1% oil on cuttings and at costs that are economic relative to a return to water based muds. However in both of these cases the plant is heavy and bulky causing problems in the practical implementation of this technology offshore.

Solvent Extraction Approaches

A viable alternative is the use of solvent extraction technology to remove the oil from the cuttings. The treatment of refinery sludge, using this approach, has already been reported (8) and a prototype unit for treating oily cuttings has recently been evaluated by BP Exploration, Amerada Hess and BP Norway working in conjunction with Thule Rigtech. The experimental programme demonstrated residual oil levels of less than 1% with total hydrocarbon levels of circa 1.5%, including residual solvent. (Pers Comm). A field unit is now under development for deployment this year.

Cuttings Disposal

A number of novel approaches to the disposal of the oily cuttings have been considered recently, ranging from concrete caissons on the seabed through to blending with cement to form bricks.

Within the Norwegian sector there is a body of opinion that favours transportation of the cuttings to shore. The mechanics of this approach have recently been published (9) but there still remains the problem of onshore disposal. An offshore solution is therefore preferred.

Annular injection of the cuttings in slurry form shows promise as a viable disposal mechanism. This approach has been successfully utilised in Alaska (10) and in the Gulf of Mexico (Pers Comm) and a Drilling Engineering Association, joint Operator project, is presently running in Europe to fully evaluate the concept. Figure 3 schematically represents the concept.

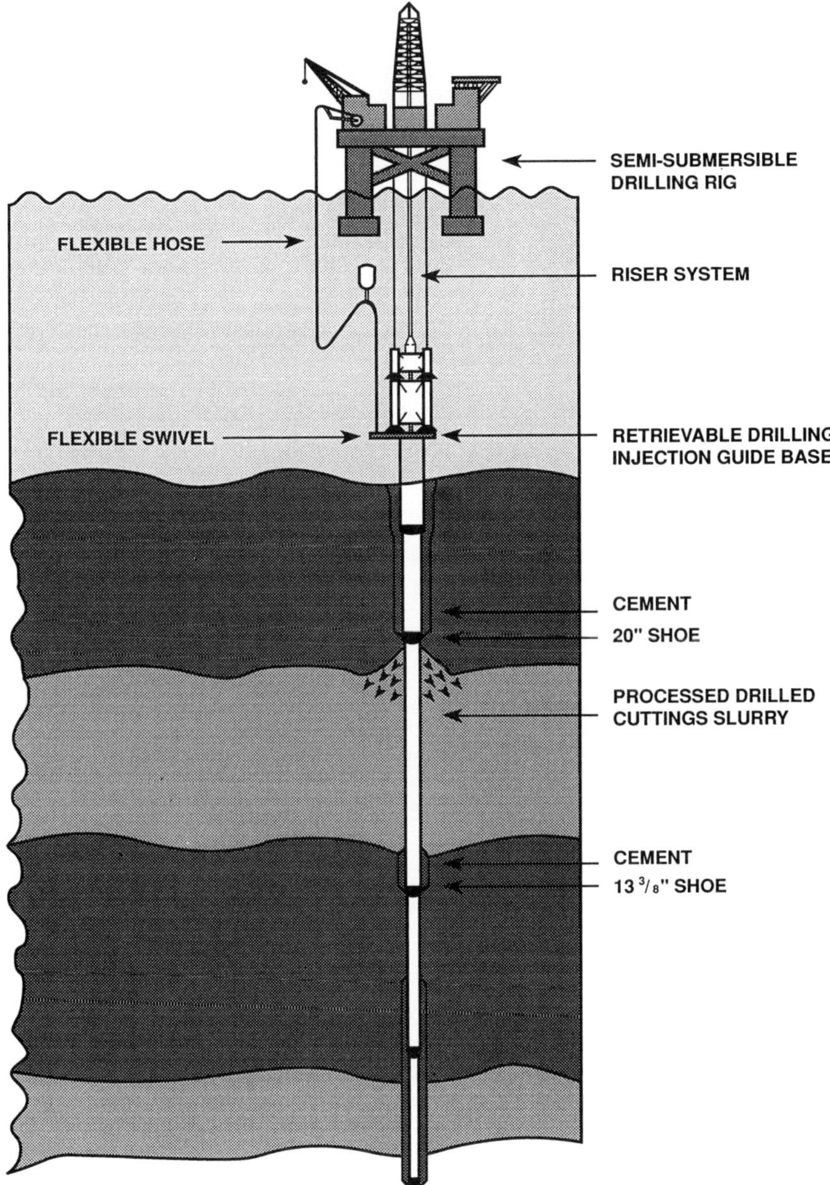

Fig 3. Schematic of the Downhole Injection of the Cuttings Slurry as a Disposal Mechanism

Application of this concept will be easiest onboard fixed platforms and jack-up rigs where annular access is easily achieved. Application for semi-submersible operations is however feasible as demonstrated in a recent patent design for a modified wellhead, permitting annular access (11).

Alternative Drilling Fluids

Replacement of the invert emulsion drilling fluids with alternative chemistries is feasible but, as previously noted (2)(3), the performance of conventional fluids is far from satisfatory.

Two primary thrusts are therefore evident in respect of new drilling fluid developments. Firstly the development of invert fluids based on readily biodegradable "pseudo" oils and secondly the design of more inhibitive water based fluids.

Pseudo oil based muds

The problem with the present invert system is the longevity of the ecological impact of the cuttings on the seabed. Consequently, oils that degrade rapidly, particularly under anaerobic conditions, should be far more acceptable to the environmental community.

Oils derived from vegetable rather than petroleum sources offer this possibility and two systems have had recent field trials. These are based respectively on Ester and Ether derivatives of vegetable oils providing stable invert emulsion systems with the desired fluid properties. Recent data (12) demonstrates that the approach is viable from a drilling standpoint and that the systems are economic when compared with conventional water based muds.

There is also evidence, although it has yet to be confirmed, that the Ester adhering to the discharged cuttings is degrading faster than was previously observed with the mineral oils.

The use of polyalcohol chemistry is also under evaluation.

Water based drilling fluid developments

In parallel with the "Pseudo oil" developments most of the drilling fluid service companies, and many operators, are evaluating novel water based drilling fluid chemistries. Most of these studies are focusing on the control of the reactive Tertiary shales since this is the area of greatest concern.

One of the present developments that appears to offer significant advantages is the use of cationic polymer species.

Much of present water based mud technology is based on anionic polymers ranging from the mildly anionic species such as Xanthan Gum through the Carboxy celluloses to the hydrolysed Polyacrylamides. However, since the surface of the clay platelets is negatively charged, cationic species can be expected to react and bind more strongly with the clay, leading to greater inhibition of swelling and dispersion. Figure 4 is a reproduction of recently published data (13) demonstrating this relationship. This test evaluates the inhibiting nature of the drilling fluid with the "best fluids" demonstrating the highest shale recovery. The first three systems (KCl/CMC, KCl/PHPA and Bentonite/Lime) can be considered as present generation water based fluids with the best recovery at less than 40% compared with the 100% recovery for the invert emulsion fluid (OBM).

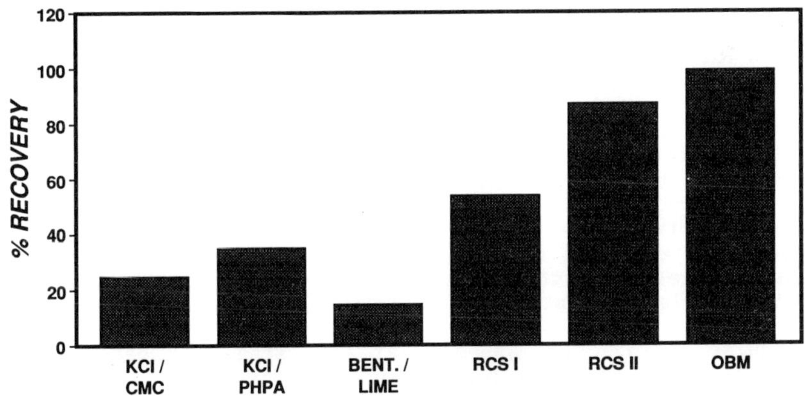

Fig 4. Comparative Shale Inhibiting Characteristics of a range of Drilling Fluids

RCS I and RCS II are fluids developed by BP Research using cationic polymer species. For RCS I a cationic starch is used whilst for RCS II the cationic starch is supplemented by the addition of a polyglycol additive. The cationic starch increases the recovery to 55% whilst the further addition of the polyglycol product increases it to 90%. Both of these systems have been field trialed and the results will be reported later.

Cationic polyamines can also be utilised in these fluids with similar levels of inhibition.

Other technologies are also showing promise and the industry can expect a range of systems offered during the coming months.

4 BP's STRATEGIC POSITION FOR THE MID 90's

A complete ban on the use of invert emulsion drilling fluids would have major repercussions for the industry and many fields presently viewed as viable would cease to be so. The problem is the environmental impact of the oily cuttings and it is this aspect that needs to be resolved. Technological developments, presently in hand, offer us this opportunity.

It is BP's stated policy to "protect the enivronment by seeking to minimise the impact of our activities" and the resolution of this present issue is a significant plank of our policy.

Development drilling operations

Continued use of invert emulsion fluids is paramount in the economics of field development. These fluids will therefore continue to be used with the generated cuttings either cleaned using solvent extraction technology or disposed of through annular re-injection. For some of the simpler geological formations water based muds may be used but this will be the exception, due to the nature of the North Sea formations.

Similarly there may be instances where the economics favour the use of water based mud for the upper hole sections. This will be in relation to the capital cost of plant installation to cope with the large quantities of cuttings generated.

Annular re-injection appears, at this stage, to offer the most exciting opportunities for the cuttings disposal.

Exploration and Appraisal Drilling Operations

As with the development drilling operations, the use of the invert emulsion fluids is indicated for the single well drilling sites.

However the logistics dictate the use of water based fluids for the upper hole sections with the cationic polymer systems appearing to offer the best solution.

The deeper hole sections will be drilled with invert fluid systems and for this the vessels will need to be fitted with cuttings cleaning equipment or be able to slurry the cuttings for re-injection. Vessels not uprated for this service will need to utilise water based muds or transport the cuttings onshore and are, therefore, likely to command a lower day rate with the Operating companies.

5 CONCLUSIONS

Continued use of the invert emulsion fluids is necessary and the industry has to address the problem of the environmental impact of the oily cuttings on the seabed.

A range of options for cuttings cleaning and disposal are presently under development. Given clear legislative guidelines the deployment of hardware could commence in 1992.

Water based mud developments will permit the substitution of these fluids for the invert emulsion fluids presently used in some instances. However this is likely to be predominantly in the upper hole sections of the wells.

The industry is committed to a better environmental image and the adoption of these technologies will minimise the environmental impact of drilling operations.

6 ACKNOWLEDGEMENTS

The author is grateful to British Petroleum Exploration Co Ltd for permission to publish the paper.

7 REFERENCES

1. J.M. Addy, J.P. Hartley, P.J.C. Tibbets, Mar. Poll. Bull, 1984, 15(12), 429

2. T.J. Bailey, J.D. Henderson, T.R. Schofield, Offshore Europe, 1987, SPE 16525/1.

3. R.C. Minton, Proc Faraday Soc, Egham, 1990 in prep.

4. J. Davies, et al., Mar. Poll. Bull, 1984, 15(10), 363.

5. M.T. McKechnie, Proc. Roy. Soc. Chem, London, 1991 in prep.

6. G.J. Potma, A.R.S. Drinkwater, Proc Europec 90, 1990, SPE 20888.

7. G.A. Young, F. Growcock, K. Talbot, Proc SPE/IADC Drill Conf, 1991, SPE/IADC 21939.

8. L.R. Poche, R.E. Derby, D.R. Wagner, Oil & Gas J. Jan 7th, 1991.

9. K.M. Arnhus, G. Slora, Proc. SPE/IADC Drill Conf, 1991, SPE/IADC 21949.

10. R. Smith, Proc. Int Artic Tech. Symp, Anchorage, 1991, in prep.

11. H.P. Hopper, UK Patent, 1990, GB 9023595.3.

12. R.L. Peresich, B.R. Burrell, G.M. Prentice, Proc SPE/IADC Drill Conf, 1991, SPE/IADC 21935.

13. P.I. Reid, P.M. Harrington, R.C. Minton, 1991, Ocean Industry, in prep.

Drilling Fluids and Wellbore Stability — Current Performance and Future Challenges

L. Bailey, J.H. Denis, and G.C. Maitland

SCHLUMBERGER CAMBRIDGE RESEARCH, PO BOX 153, CAMBRIDGE CB3 0HG, UK

1 INTRODUCTION

A high proportion of drilling problems, particularly in the North Sea, are associated with shales. Whilst sometimes these are mainly mechanical effects, there can often be a significant chemical contribution to the stresses generated in the wellbore. This arises when montmorillonite clays come into contact with aqueous drilling fluids which cause the shale to swell through uptake of water. As this occurs, the rock becomes weaker and softer, leading to either sticking problems or mechanical failure and hole enlargement. This paper is concerned with this complex interaction of chemical and mechanical factors in shale-fluid systems - shale chemomechanics: how it can help us understand the performance of existing drilling fluids in controlling wellbore stability in shales and how it might guide the future development of so-called inhibitive muds.

The most effective fluids for drilling sensitive shale formations are still invert emulsion oil-based muds[1]. However increasingly restrictive environmental legislation limiting their use in many parts of the world has led to a major effort recently to produce new, environmentally acceptable water-based fluids and OBM substitutes which can match their performance in shale stabilisation as well as meeting the many other requirements for drilling efficiency and quality. We first examine the origins of chemomechanical stability problems in shales, and then use this as a basis for exploring possible ways in which this challenge might be met.

2 OSMOTIC SWELLING

The chemical contributions to the stresses generated in a freshly drilled wellbore in shale rock are thought to be osmotic in origin[1,2]. Shales are complex heterogeneous mixtures of silt and clay minerals, the clay platelets forming highly ordered compacted local domains which give the rock a high degree of anisotropy. For both a hydrated clay phase within the shale and the drilling fluid with which it comes into contact the water chemical potential μ is given by:

$$\mu_i = \mu^o + RT \ln a_i + P_i V \quad (1)$$

where i designates the phase, μ^o is the standard chemical potential, P_i is the pressure, V is the molar volume of water, R is the gas constant and T the absolute temperature. If these chemical potentials are different, then water will diffuse from one phase to the other until a balance is achieved. Hence at equilibrium the pressure difference between the mud (m) and the shale pore water (s) is:

$$\Delta P = P_s - P_m = RT/V \ln (a_m/a_s) \quad (2)$$

This osmotic or swelling pressure will be superimposed on any mechanical stresses acting on the wellbore due to differences between the wellbore pressure and the earth stresses. Given that under typical downhole conditions RT/V is of order 1000 bar, it can be seen that even a small water activity contrast between the mud and the pore fluid can give rise to significant additional stresses.

In addition to the extra stresses generated, this chemical contrast between the mud and the shale has other consequences. The diffusion of water into the shale modifies its mechanical properties, causing it to weaken and/or soften. This increases its susceptibility to deformation or fracture in response to the rock stress field, and in some cases, if the shale can retain its integrity sufficiently to swell to a gelatinous solid, to hydraulic erosion by the drilling fluid flowing across its surface. As microfractures start to open up in the shale, convective processes will become increasingly important compared with diffusion in transporting water into the shale. In addition, there will be other transport phenomena taking place. Chemical potential imbalances between ions in the mud and in the shale, both the bound clay platelet counterions and those in the free pore fluid, lead to ionic diffusion and exchange. These in turn change the nature of the clay platelets and hence their interactions[4,5] and mechanical characteristics, and the water activities in both the shale and the mud.

There is thus a highly complex coupling between the evolving chemical exchange processes occurring at the wellbore wall and the consequent changes in stress and rock mechanical properties. In order to understand and quantify the overall process, it is helpful to try to decouple to some extent the individual contributory phenomena in controlled small-scale experiments, and to use chemically well-defined model systems, in addition to studying real mud-shale systems under wellbore conditions and geometries. The work described in the next few sections is part of an activity aimed at characterising both the extent and rate of shale swelling as a function of the major variables which will influence the shale water activity: clay and pore fluid chemistry, rock stresses and temperature. On the fluid side the major variables of interest are its chemical composition, which determines a_m, and the fluid hydraulics at the rock surface which will influence fluid convection into the shale and subsequent erosion of soft material.

The way in which these variables influence the shale water activity is first described, through both experiments and modelling. It is then demonstrated, through equilibrium swelling experiments on model clay compacts and real shale, that an osmotic stress as defined in equation (1) has the same effect on the behaviour of shales as an equivalent mechanical stress. Information on the kinetics of the swelling process is also given. Having established the basic mechanistic framework, the interaction of muds based on KCl and partially hydrolysed polyacrylamides (PHPA) with a well-characterised swelling shale (Pierre) is used as an example of how the performance of currently used inhibitive water-based muds is consistent with our physical understanding of the process.

3. WATER ACTIVITY IN SHALES

One strategy for minimising shale swelling is to attempt to match the mud water activity to that in the shale. In order to do this, it is necessary to know a_s and how this is likely to vary with the nature of the shale and the downhole conditions. In general, a_s will be a function of the ionic concentration on the clay surfaces and within the surrounding pore space. This in turn will be determined by a range of factors: the clay cation exchange capacity (CEC), the nature of these cations, the porosity ϕ, the clay fraction, the pore fluid chemistry and concentration, the amount and structure of other minerals in the shale matrix. A recent study at SCR set out to quantify these dependencies for well characterised model shales[6]: calcium montmorillonite clay compacts with a calcium chloride pore fluid.

The clay compacts were prepared by filtration of suspensions in hydraulic presses over a range of pressures up to 2000 bar. The water activity was measured at 20^oC using a Novasina Humidat-TH2 humidity meter. Water adsorption isotherms were also determined using the same apparatus, equilibrating the clay with saturated salt solutions of differing relative humidities. Figure 1 shows the variation of a_s with the concentration of fixed charges in the clay compact, A (= CEC/water content), for samples made from a deionised water suspension. The results were independent of whether the final state was reached by compaction or swelling and there is excellent agreement with the values determined from the adsorption isotherm. The activity remains close to unity until the clay molality reaches 2 mol kg^{-1}, whereafter a_s drops sharply with increasing A. This behaviour has been shown to be in line with activity coefficient behaviour which is largely determined by the clay electrical double layers at high values of A and by the co-ions in solution for less compacted systems. The large decrease in a_s for this calcium smectite has been attributed to partial removal of the outer (second) layer of adsorbed water.

The effect of increasing the salinity of the pore fluid is illustrated in Figure 2 for the case of 5M $CaCl_2$. At low values of A, where the water activity is dominated by the pore fluid, a_s levels off at a value corresponding to that of the original suspending fluid a_o. The activity starts to fall at clay molalities similar to those in the low salt case but reaches lower values at high A. Figure 2 also shows the predictions of a simple thermodynamic model, which assumes that the dominant interactions are those between the clay or the free ions and the solvent directly and ignores salt-clay cross interactions. The agreement is satisfactory. Similar behaviour was observed over a wide range of pore fluid concentrations, the major difference being the limiting activity at low A, and the value of A at which a_s started to fall from this value.

The effect of diluting the compacted laminar clay structure by inert mineral

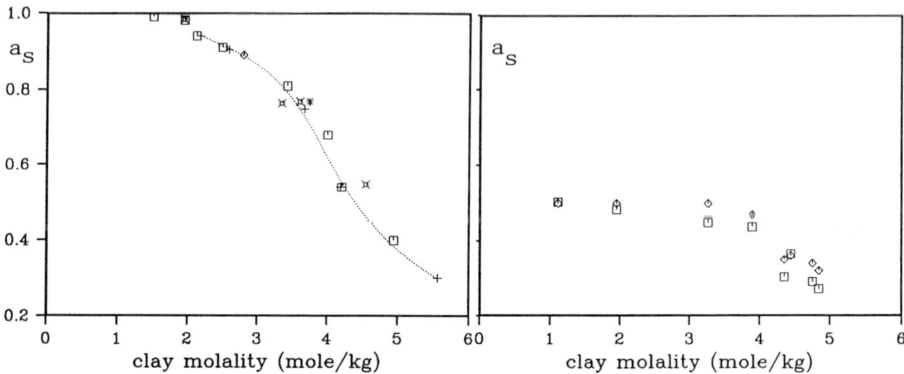

Figure 1 Variation of shale water activity a_s with clay molality A for calcium clay compact Pore fluid: water.Compaction □ swelling x; isotherm -+-; 25% $CaCO_3$ *; 50% $CaCO_3$ ◇.

Figure 2 Variation of a_s with A for same clay compact as Figure 1. Pore fluid: 5M $CaCl_2$. Measured □ ; model ◇ .

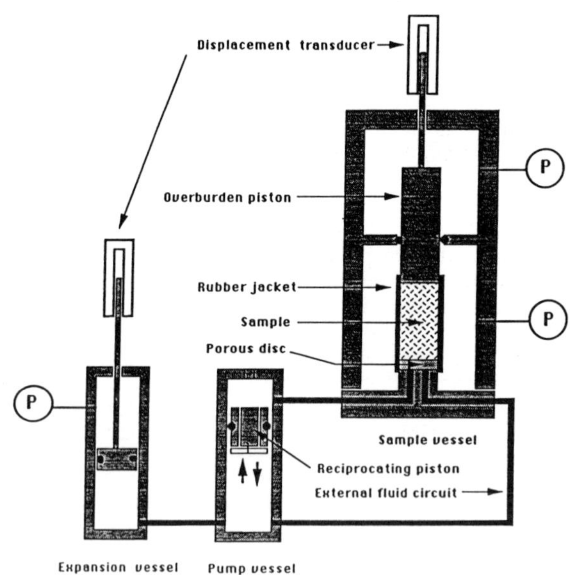

Figure 3 Schematic diagram of shale compaction-swelling cell

particles, such as occurs in real shales, was simulated by the addition of varying amounts of Analar grade calcium carbonate, up to a mass fraction of 0.5. The results are plotted on Figure 1 with A modified by the appropriate clay volume fraction; the consistency with the pure clay data indicates that no effect other than solid dilution is observed.

This study establishes a basis for understanding and predicting the water activity within a shale rock. We next look at how the swelling behaviour is related to the predictions based on osmotic behaviour.

4. SHALE COMPACTION AND SWELLING

Experiments on the compaction and swelling characteristics of clay compacts and shales in contact with a flowing external solution have been carried out[6] in the high-pressure cell illustrated in Figure 3. This enables the changes in dimensions of a cylindrical sample to be monitored as the applied axial and radial stresses, and the chemistry of the contacting fluid, are varied.

Figure 4 shows the results of two compaction/swelling cycles of a core composed of the same calcium montmorillonite clay used in the water activity study. Here the stress conditions were essentially isotropic and the contacting fluid was deionised water. The void ratio e (= liquid volume/solid volume in compact) remains relatively constant until about 40 bar, whereafter it starts to fall dramatically with increasing stress. The original sample dimensions and porosity are essentially recovered at the end of both cycles, although both swelling curves show some hysteresis compared with the compaction curve and each other. Not unexpectedly, the deformation was in fact highly anisotropic, the axial strains being about three times larger than those in the radial direction.

It is possible to relate the porosity changes, through their influence on A, to the water activity in the core at each pressure and hence to the equilibrium osmotic pressure developed within the core using the ideas described in Section 3. The values obtained are also plotted on Figure 4, where it is seen that over most of the range P_{osm} lies between the compaction and swelling pressures. Hence a large part of the behaviour of the clay core can be accounted for by osmotic effects, although for this calcium smectite specific interactions between the clay platelets also appear to restrict both compaction and swelling. The degree of hysteresis depends on the nature of the clay: those containing monovalent counterions exhibit more reversible behaviour and the swelling stress is almost entirely osmotic.

The ability of osmotic pressure to explain the swelling behaviour of clays and shales essentially quantitatively can be illustrated by two further examples. Figure 5 illustrates a compaction-swelling cycle between 10 and 800 bars for a sample of Pierre shale from N Dakota, USA. This contains 40% by weight of clay, 11% of which is montmorillonite having a CEC of 0.22 meq g^{-1} , of which 67% are Ca^{2+} and 33% Mg^{2+}. The cycle is essentially reversible and the swelling curve is well described by P_{osm}, evaluated by the above procedures. The ideas developed on model systems hence translate well to real shales.

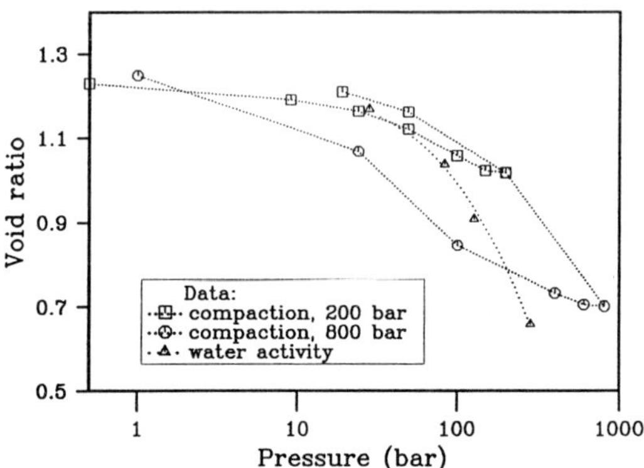

Figure 4 Compaction-swelling curves for calcium montmorillonite core. Compaction to 200 bar □; 800 bar O; Osmotic stress P_{osm} Δ

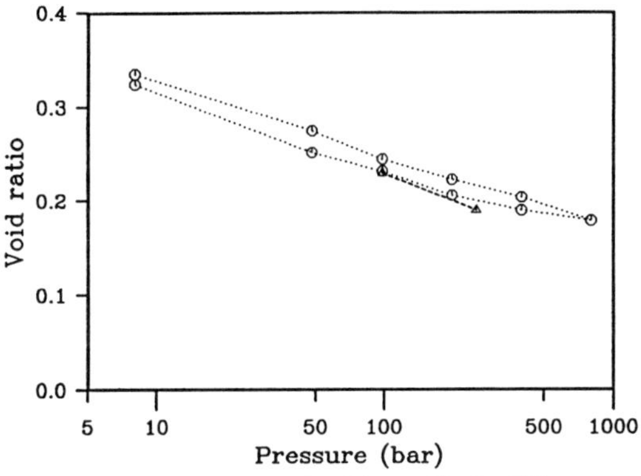

Figure 5 Compaction-swelling curves for Pierre Shale (O) compared with P_{osm} (Δ)

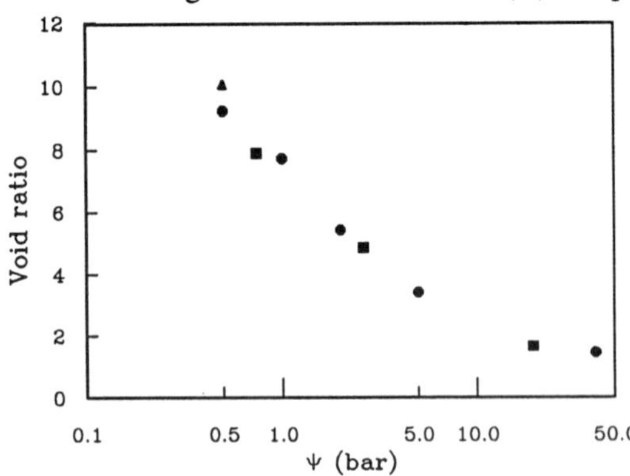

Figure 6 Void ratio e vs total stress ψ ($=P_1+P_{osm}$) for swelling of sodium montmorillonite compacts (see text). Fluid: water ●; KCl ■; PHPA ▲.

Figure 6 demonstrates the equivalence of osmotic and mechanical stresses in determining swelling behaviour in a dramatic way. It shows the void ratio of a sodium montmorillonite compact as a function of stress ψ. Three series of experiments are plotted:

* Using deionised water as the external fluid phase; here the model shale was compacted to equilibrium at 40 bar and then re-swollen to equilibrium at a pressure $P_1 = \psi$.

* With KCl solutions, ranging in concentration from 10^{-3} to 1 M, as the external phase; here every sample was compacted to 40 bar and re-swollen to P_1=0.5 bar. The final values of e on swelling varied with the salt concentration. In this case the sum of P_1 and the osmotic term $RT/V \ln(a_s^*/a_{KCl})$ is plotted as ψ, where a_s^* is the free salt contribution to the water activity in the compact and a_{KCl} is the water activity in the external KCl solution. The two sets of data lie on the same swelling curve, confirming that the osmotic stress arising from the activity contrast between the compact and the external fluid is entirely equivalent to a mechanical stress acting on the system. The inhibition of swelling by KCl solutions can be explained solely in terms of osmotic effects, without invoking specific exchange effects due to replacement of sodium counterions on the clay by potassium which have previously been suggested as the mechanism of inhibition[7].

* When a 1 gl^{-1} solution of a PHPA polymer (I) (molecular weight ~ 1.3 x 10^7, degree of hydrolysis ~ 30%) was used as the external fluid, with or without the presence of KCl, there was no additional inhibition of the equilibrium swelling of the compact compared with the base fluid.

These observations suggest that the inhibition of shale swelling by electrolytes can be fully explained by osmotic phenomena, and can be described quantitatively if one has sufficient information about the nature of the shale. They also suggest that the major inhibitive role in salt-polymer muds may be played by the salt, and that the specific nature of the cation may not be of major importance in many cases.

Before investigating further how these conclusions based on equilibrium behaviour translate to a more realistic simulation of downhole mud-shale contact, we will briefly examine the rate at which these swelling processes proceed. In the wellbore it is unlikely that equilibrium between the mud and the shale is achieved before some sort of failure process is initiated. Even if the equilibrium situation is likely to be catastrophic, it may well be that the swelling process is sufficiently slow compared with the drilling rate that it may not cause a real problem.

5 KINETICS OF SHALE SWELLING

Experiments on both clay compacts[8] and real shales indicate that the swelling follows a diffusion-type law:

$$Q = S\, t^{1/2} \qquad (3)$$

where Q is the cumulative water flux into the shale, t is the time and S is termed the *sorptivity*. The methods of soil physics[9] have been applied to the compaction and swelling of compressible soft solids like clay compacts, and enable S to be related to physical characteristics of the system such as the fluid diffusivity, D(e), and the medium permeability, k(e), both of which are functions of the void ratio e. The void ratio itself is determined by the local effective stress within the solid i.e. the difference between the external (applied) pressure and the internal (pore) pressure. For shales ($e<1$) reasonable approximations are[10]:

$$S = (2D)^{1/2}\,\Delta e \qquad (4)$$

$$k = D\,(1+e)\,\eta\,(dP/de)^{-1} \qquad (5)$$

where η is the fluid viscosity.

The linear dependence of S on the change in equilibrium void ratio on swelling implied by equation (4) is indeed observed experimentally[10]. The diffusivities inferred from such experiments for clay compacts range from 10^{-10} to 10^{-13} $m^2\,s^{-1}$, depending on the nature of the bound cations and the pore fluid. The values for shales, less anisotropic and lower in clay content, tend to be somewhat higher; Pierre shale has a value of 9×10^{-10} $m^2\,s^{-1}$ at 20°C. The permeabilities determined from swelling experiments agree well with direct measurements; for Pierre shale k ranges from 40 nD (unconfined) down to 0.7 nD at 800 bar effective stress.

Although electrolytes like KCl markedly reduce Δe in swelling experiments, they do not appear to change $S(\Delta e)$ or D significantly. Given the marked dependence of D on the nature of the bound cations, this suggests that the action of the KCl is largely osmotic and that on the timescale of the swelling experiments little ion exchange with the clay has taken place. Commonly used mud polymers appear to have little influence on Δe, S or D.

The implication for inhibitive drilling muds is that reducing the potential equilibrium degree of swelling by a factor n will have the additional benefit of reducing the rate of swelling by n^2. Given the low values of D for typical shales, it is readily understood why many shale formations can remain stable for many weeks before reaching some critical stress-strain condition leading to wellbore failure.

6 ION EXCHANGE

The experiments described in Sections 4 and 5 indicated that osmotic effects alone are capable of explaining the ability of KCl in drilling muds to inhibit the swelling of shales. This is in contrast to the normally accepted mechanism of inhibition[7,11], which involves exchange of the natural bound cations on the clay (Na^+, Ca^{2+}, Mg^{2+}) by K^+ leading to a clay fraction with a lower swelling tendency. In order to investigate the effects of ion exchange on clay swelling, both swelling-compaction tests and controlled humidity XRD measurements have been carried out on mixed Na, K montmorillonites[10]. These show that

* the degree of layer swelling increases with increasing water activity;

* about 30% K-exchange is required before swelling is significantly reduced;

* swelling is strongly inhibited by potassium fractions above 0.55;

* the water diffusivity increases by a factor of 6 as the potassium exchange fraction increases from 0 to 0.7.

These results indicate that ion exchange by potassium will ultimately reduce swelling, but that a large fraction of the clay cations must be replaced before the effect is significant. The changes in water diffusivity suggest that the rate of exchange might increase as the extent of potassium exchange increases. Thus ion specific exchange effects of this sort could reinforce the inhibitive action of reducing osmotic effects in particular cases, provided that the rates of ion diffusion and exchange into the shale are sufficiently rapid.

7 WELLBORE BEHAVIOUR OF KCl - POLYMER MUDS

The small-scale experiments described so far confirm the conclusion of much previous work[11-14] that KCl-PHPA muds can inhibit the swelling of shales, but suggest that the predominant mechanism involves general control of osmotic stresses through electrolyte concentration rather than specific ion exchange of potassium or formation of a low permeability barrier to water transport by adsorbed polymer. The relevance of these conclusions to the wellbore behaviour of mud-shale systems is being investigated at SCR by experiments designed to simulate various aspects of downhole conditions during drilling: wellbore simulator studies to examine the stability of the borehole wall, and cuttings dispersion tests[15] designed to investigate the extent to which drilled shale cuttings retain their integrity as they are transported up the annulus by the mud.

Wellbore Simulator experiments

Figure 7 illustrates the smaller of the two SCR wellbore simulators used to study mud-shale interactions under downhole conditions and geometry. It enables a cylindrical shale sample, 152 mm diameter drilled with a 25 mm hole, to be stressed up to 315 bar and contacted with a drilling fluid circulated through the wellbore at flowrates up to 15 l min^{-1} (wall shear rate ~ 800 s^{-1}). Changes in the wellbore dimensions can be monitored throughout using a mechanical caliper arm; mud samples can be taken from the flowline to monitor changes in mud composition with time. Ingress of water and ions into the shale are determined at the end of an experiment, typically 48 hours, by gravimetric analysis, ion chromatography and atomic absorption spectroscopy.

A sequence of tests was run on Pierre shale to investigate the behaviour of KCl-

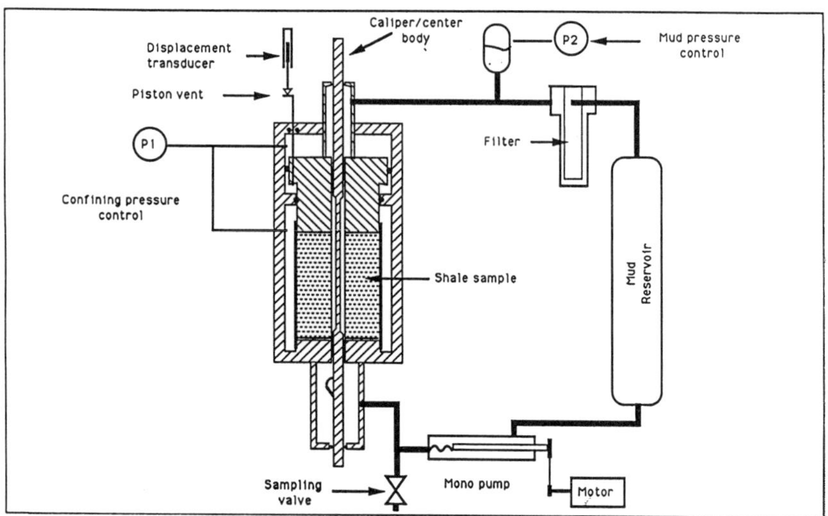

Figure 7 Schematic diagram of SCR small wellbore simulator

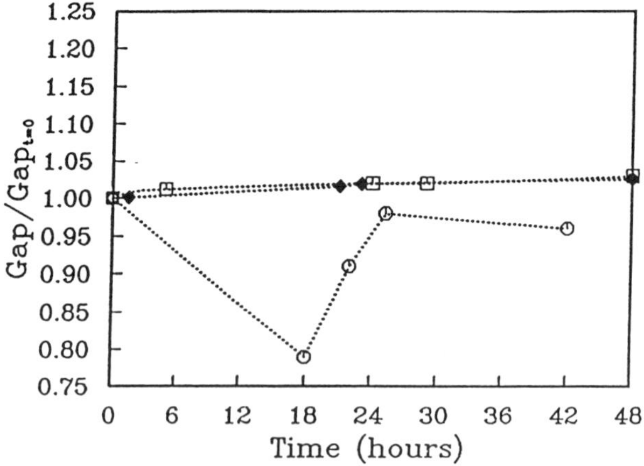

Figure 8 Relative changes in borehole diameter for Pierre Shale cores contacted with different drilling fluids in wellbore simulator. Undrained shale: FWBM □ Drained shale: FWBM O; FWBM + 5% KCl ♦

PHPA muds. As a baseline for comparison, a freshwater bentonite mud ($a_m \sim 1.0$) was used. Figure 8 shows that using an undrained shale sample ($a_s > 0.98$), essentially no change in borehole diameter was observed over a 48 hour period. This is not surprising considering the almost balanced activities of the mud and shale, but serves to illustrate the dangers of drawing conclusions about shale stability using surface equilibrated samples.

A second experiment in which the shale was pre-drained at 15 MPa for 5 days to simulate compaction at depth, reducing a_s to 0.90, showed dramatically different behaviour (Figure 8). Initially the wellbore swelled inwards, in some places closing tight on the caliper centre-body. As the shale softened and/or fractured, hydraulic erosion occurred and the wellbore diameter progressively increased until the experiment was terminated after 48 hours. Post mortem examination revealed a highly soft, sticky, irregularly eroded wellbore.

Repeating this drained experiment using a mud containing 5% w/w KCl (a_m = 0.95) gave no swelling phase and produced only a small increase in wellbore diameter over 48 hours, confirming the effectiveness of the electrolyte alone to inhibit shale swelling and weakening. Adding 1.5 g l^{-1} of PHPA I to the mud in a subsequent experiment gave essentially the same behaviour. Further tests under the same conditions using muds containing polymer, including PHPA I, but no KCl produced large swelling/erosion effects similar to those observed with the freshwater mud.

This behaviour is exactly that predicted by the small-scale model experiments described earlier. The inference is that while salt is highly effective in inhibiting shale wellbore swelling/erosion, PHPA polymer (at least the one studied extensively here, which is typical of those used in some field muds) has little additional stabilising effect at the wellbore wall. Post mortem radial concentration profiles of water and various cations (Figure 9) shows that water and ions penetrated some 4 cm (two wellbore diameters) into the shale during the experiments where there was a water activity imbalance between mud and shale, and significant swelling-erosion took place. By contrast, for the KCl muds the invaded zone was much smaller (~2 cm) and the near-wellbore region is dominated by potassium ions both in the pore fluid and on the exchangeable clay surface sites. The presence of polymer did not markedly change the ionic profiles.

Cuttings Dispersion Experiments

For comparison with the experiments on whole shale, samples of the same Pierre shale were ground to between 2.8 and 1.0 mm to simulate drilled cuttings, and rolled at various temperatures for 16 hours at 10 rpm in stainless steel cylinders of 10 cm diameter containing 2.5 g of solid in 100 ml of drilling fluid. Changes in the particle size distribution after such treatment were characterised using gravimetric sieve analysis and, for material <250 μm, low angle light scattering (Microtrac, Leeds and Northrup). Reproducibility of the results was ± 1.5%.

The major conclusions are illustrated in Figure 10, which displays simply the weight % of original cuttings retained and the weight % of fully dispersed (<250 μm)

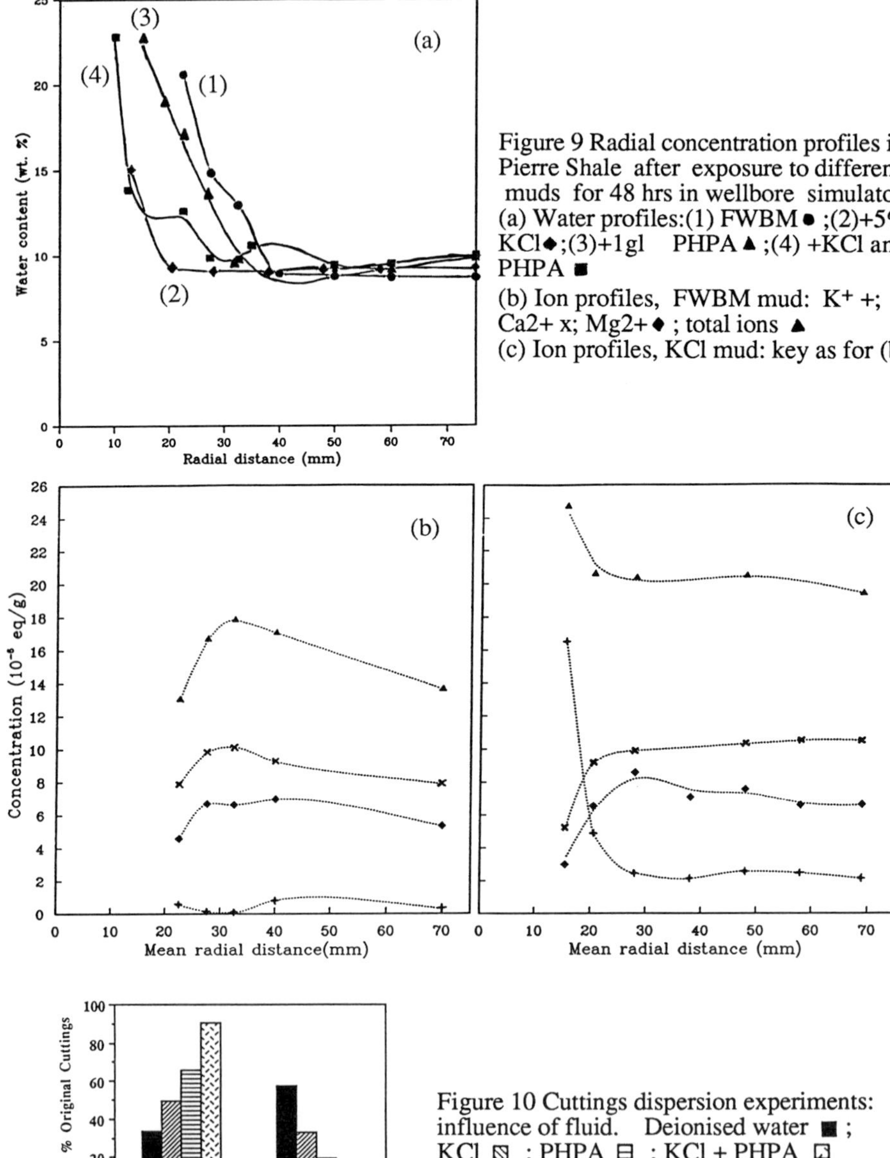

Figure 9 Radial concentration profiles in Pierre Shale after exposure to different muds for 48 hrs in wellbore simulator. (a) Water profiles:(1) FWBM ● ;(2)+5% KCl◆;(3)+1gl PHPA ▲ ;(4) +KCl and PHPA ■
(b) Ion profiles, FWBM mud: K^+ +; Ca2+ x; Mg2+ ◆ ; total ions ▲
(c) Ion profiles, KCl mud: key as for (b)

Figure 10 Cuttings dispersion experiments: influence of fluid. Deionised water ■ ; KCl ▧ ; PHPA ⊟ ; KCl + PHPA ◻

material produced:

* KCl reduces dispersion, resulting in hard, non-swollen cuttings;

* PHPA I was more effective than KCl alone at reducing dispersion, but the resulting cuttings were very soft, swollen and prone to disintegration on washing;

* KCl and PHPA I effects were additive when used together, resulting in almost complete stabilisation of the cuttings;

* Reducing the cuttings water activity markedly reduced their stability in water, but did not significantly affect the ability of both PHPA and KCl to prevent dispersion;

* Temperature has only a marginal effect over the range 25-80 ^oC.

Despite the ability of the PHPA polymer to reduce cuttings dispersion dramatically, these observations are consistent with all the previous data. Here KCl reduces dispersion of cuttings by inhibiting their swelling and subsequent disintegration. PHPA, by contrast, does not inhibit the swelling of the cuttings but limits their dispersion by adsorbing on the clay surfaces and producing a macro-floc, presumably by a bridging mechanism. The two components together are highly effective at preventing dispersion by a combination of the two mechanisms. Here the clay-binding characteristics of the polymer are able to counteract to a large extent the mechanical fragility of the unswollen cuttings in the flowing mud. PHPA I is apparently unable to perform this role with intact bulk shale at the wellbore wall.

The presence of both components in the mud is crucial to its successful operation. The KCl predominates in the minimisation of swelling, cracking and erosion of the borehole wall, whereas both components act to prevent dispersion of the cuttings and any spallings that are removed from the wall which might otherwise cause a dramatic deterioration in other mud properties such as its rheology.

8 CURRENT METHODS OF STABILISING SHALES

So far we have focused on a limited class of drilling fluids, and establishing a quantitative understanding of their behaviour in contact with swelling shales. This provides a chemomechanical framework for identifying the possible methods for minimising shale stability problems in chemically sensitive formations, which proves a useful way both to classify existing systems and to identify possible routes for the future.

A prerequisite for hole stability in any rock, of course, is that the maximum principle stress at the wellbore wall does not exceed the mud pressure by an amount set by the failure envelope for the rock. In swelling systems there will be additional criteria for the onset of hydraulic erosion. The issue here is how to minimise or control chemical effects which either cause additional contributions to the wellbore stresses or influence the failure criteria in a detrimental way. It should also be remembered that there are essentially two problems: stabilisation of the wellbore, and prevention of dispersion of drilled cuttings or material that does subsequently leave the wellbore wall. As we have seen, some mud additives may be more successful in one of these roles than the other, since the length and time scales are different for the two problems. This should be borne in mind when selecting appropriate solutions.

There are four major ways of reducing problems in swelling shales:

(a) Elimination of the driving force for ingress of water

The total driving force for hydration will be

$$P_{hyd} = P_m - P_p + RT/V \ln(a_m/a_s) \quad (6)$$

Any balancing of the mud and shale water activities should be done with this in mind. Since a_s is usually less than 1.0, the option usually used is to reduce a_m to an appropriate level; this is at the heart of most inhibitive mud systems. This is usually achieved by adding electrolytes: seawater bentonite muds, saturated salt-polymer (xanthan, guar), KCl or NaCl-polymer (PHPA, xanthan), freshwater calcium treated muds (lime, gypsum). The dispersed aqueous phase in oil-based muds is usually a calcium chloride brine at an activity $\leq a_s$. Other solutes can be used to reduce water activity; for instance water-soluble polyglycerols/glycols, which have recently been introduced with claims of good shale stability properties combined with biodegradability[16,17], have a high affinity for water and presumably reduce a_m significantly.

An alternative strategy is to increase a_s. Achieving this involves either reducing the salinity of the shale pore fluid or exchanging the clay cations for less hydratable ions. We have seen that potassium can reduce swelling through just such a mechanism; indeed one virtue it may have over other cations is this ability to act on a_m and a_s in opposite senses. Other polarisable cations, such as ammonium, can act in a similar manner. In principle, increasing both the size and charge on the cation will lead to optimum reduction in the tendency of a clay to swell or disperse, provided that this is compatible with size restrictions imposed by the overall aluminosilicate layer structure.

(b) Reduce the rate of water transport

Given that it is virtually impossible to balance water activities exactly at all points in the well (not only because it is difficult to know what a_s is, but also because it varies continuously with depth, in addition to variations with mineralogy), it is wise to take steps to reduce the rate at which water can diffuse or convect into the rock in response to any chemical potential driving force which exists. Any reduction in this driving force will in itself reduce the rate of swelling as well as the ultimate extent.

Increasing the viscosity of the aqueous phase will reduce the transport rate (see equations (3)-(5)). Here the problem is to find solutes which will increase the viscosity of water significantly, but at the same time are capable of passing through the narrow, tortuous shale pore space. Some low molecular weight polymers might act in this way, but many mud polymers are too large to enter a shale, so the fluid entering the rock has a viscosity much closer to water than the mud continuous phase.Another possibility is to reduce the permeability of the shale. We have seen that changing the clay cation can markedly influence k. Unfortunately, cations which tend to reduce intrinsic swelling, like K^+ or Ca^{2+}, also lead to a more open shale structure and so increase the permeability.

A different type of solution is to form a low permeability barrier at the shale surface or within microfractures, which may be natural or formed in the initial stages of failure. Oil-based muds do this very effectively by means of the oil continuous phase through which water must diffuse in order to reach the shale. Polymers in water-based fluids could act in this way but, because filtration rates are extremely low due to the low shale permeabilities, physically formed polymer cakes or layers of sufficient resistance are unlikely to form or survive. Consequently strongly adsorbing species, such as cationic polymers (incorporating for example quaternary ammonium groups), are likely to be more successful in this respect.

(c) Prevent water contact with shale

In some ways this is an extreme form of (b); however the approach is not simply to reduce the rate of water ingress, but to isolate the formation completely. This can be done by creating an essentially impermeable film or by rendering the exposed shale surfaces hydrophobic. Additives based on asphaltene derivatives, like gilsonite, combine both these attributes by melting under downhole temperatures and pressures onto the shale surface, including fractures, to form an essentially impermeable, hydrophobic seal.

Other systems attempt to form a hydrophobic liquid layer on the shale surface by, for instance, binding the oil droplets of oil-in-water emulsions to the clay surfaces via the charged emulsifiers. The organophilic clays in oil-based muds may well play a similar role at the shale surface.

(d) Prevent shale from disintegrating

Finally there are damage limitation solutions. If swelling has proceeded to the point where fracture or erosion are initiated, then it is necessary to preserve the mechanical integrity of the rock to limit its disintegration. This is precisely the role played by rather weakly adsorbing polymers like PHPA in limiting the dispersion of cuttings or spallings, even when they are significantly swollen, by binding the clay particles together. Performing the same function within the shale wellbore is more difficult; one reason for the binding effectiveness of the polymers on the readily accessible surface of the cuttings is their size, which can in turn prevent them from entering the bulk shale to perform a similar function there (see also (b) above).

The requirements for a polymer to reduce shale disintegration include the ability to adsorb strongly onto several clay platelet surfaces simultaneously with a total energy high enough to resist the mechanical or hydraulic forces forcing them apart. To achieve this within the bulk of the shale, rather than at the wellbore wall, requires the further ability to diffuse into microfractures, or even into the bulk shale and intercalate the clay layers. This implies short flexible chains, which may not give adequate bridging, or more rigid molecules. Again cationic polymers look promising in this respect.

Multivalent cations themselves are highly effective at binding clay platelets together. One reason why calcium-rich shales are less prone to swelling than sodium-rich shales is the greater tendency for the former to form multi-platelet tactoids[18], which take up water less readily.

9 FUTURE DIRECTIONS

Predicting the directions in which drilling fluid developments will proceed in the next few years is problematic. It is certain that the environmental pressures to replace oil-base muds will continue and intensify. Despite the large amount of activity in recent years to develop improved water-based systems, their shale stability performance is still not as reliable as the oil-based systems they seek to replace. For this reason, and because the detailed scrutiny of environmental legislation has yet to come to bear on WBM additives but surely will in time, it may well be that extensive cuttings treatment and zero-discharge systems become the norm on drilling rigs, particularly offshore. This means that OBMs and their descendents may be far from dead.

The major challenges for the future should therefore lie in two areas:

(a) Development of new fluids/additives

The relative success of oil-based muds compared with water-based, in being more forgiving and less sensitive to the details of the formation, encourages the continued evolution of fluids based on the invert emulsion principle. The trend is to replace the oil phase with organic liquids which fulfil the same roles outlined in section 8, but which are less toxic and preferably biodegradable. Apart from toxicity, there are still major challenges for these fluids to overcome, even if run with oil in closed loop systems: increasing drilling rate, solids free systems, ease of well logging.

There is much scope for developing new water-based systems and additives along the routes outlined in section 8, and there will undoubtedly continue to be a high level of activity in this area in oil, service and chemical companies. Design of appropriate polymers and surfactants will feature strongly in these developments. Here the challenge is to produce molecules having the necessary chemical features which are also resistant to degradation by the temperature and shear conditions experienced in the drilling operation. A major problem with new additives which are effective in stabilising shales is their compatability with producing other mud properties, like rheology, or operational results, like drilling rate, that are acceptable. Despite all the potential of cationic polymers outlined in section 8, their potential incompatibility with anionic species like drilled solids is a major potential problem.

It is worth considering why oil-based muds are so successful in drilling shale formations generally. It is probably significant that they feature in all three sections of section 8 which deal with prevention of swelling: they attempt to balance water activity, reduce the rate of water migration, and maybe produce a low permeability hydrophobic layer at the shale surface. Water-based muds which attempt to mimic this multifunctional approach to swelling reduction are likely to prove most robust and flexible. Components like polyglycerols which combine several functions (activity reduction, polymer-clay adsorption) look particularly promising in this respect.

As well as the requirements of conventional drilling, the advent of new and expanding techniques like slimhole exploration and horizontal drilling bring their own special challenges for shale stability and high performance inhibitive muds.

(b) Monitoring and control of drilling muds

It is unlikely that the development of better mud systems and additives alone will solve the problem of more reliable drilling in shale formations. The optimum performance of the mud is crucially dependent on the presence of components at the correct concentration and in an appropriate chemical or physical state. Despite the fact that the mud composition is being continually modified by the drilling process as it circulates around the wellbore, present practice only monitors the physical and chemical characteristics of the fluid periodically and superficially. There is great scope for introducing more frequent and detailed chemical measurements for the ionic, solid and polymeric components of drilling muds. These can be used

* to keep the mud within specification, particularly when drilling through troublesome shale formations;

* to give early indications of the onset of drilling problems, such as those involving swelling shales;

* to give information about the formation being drilled;

* as part of control procedures to optimise mud properties and overall drilling performance.

To give one example in the area of shale stability where such measurements could benefit both the specification an appropriate mud and its efficient running, consider the mud-shale water activity balance. Chemical monitoring could be used to give a better, more immediate indication of a_s, and its variation with depth. It would also enable a_m to be closely monitored and controlled by appropriate chemical changes. The fact that a_m for oil-based muds remains largely unaffected by the drilling operation is probably another feature contributing to their general effectiveness. For water-based muds this can only be achieved by better monitoring and control.

The quest for improved environmentally-acceptable water-based muds will probably lead to more complex muds. In order to run these successfully, it will almost certainly be necessary to monitor them much more closely than is done currently. On the other hand, it is possible that some of the current muds, or earlier simpler systems that have been superceded, were unreliable not because of flaws in their design, but because a higher level of monitoring is required to run them successfully in difficult shale environments. The introduction of more direct and accurate chemical measurements as an integral part of the drilling process may well therefore enable much simpler muds to be run in many cases.

9 ACKNOWLEDGEMENTS

The authors thank their colleagues in the Rock and Fluid Physics Department at SCR for stimulating discussions on many of the ideas discussed in this paper, particularly Peter Hall, Tim Jones and Paul Hammond. They also thank Marian Keall for carrying out the shale analyses presented here.

REFERENCES

1. M.E. Chenevert, J. Pet. Tech., 1970, 1309.
2. M.E. Chenevert, J. Pet. Tech., 1970, 1141.
3. M.E. Chenevert, 11th Symposium Rock Mechanics, California, 1969, 599.
4. L.M. Barclay and R.H. Ottewill, Spec. Disc. Chem. Soc., 1970, 1, 138.
5. S.D. Lubetkin, S.D. Middleton and R.H. Ottewill, Phil. Trans. Roy. Soc. Lond 1984, A311, 353.
6. J. H. Denis, Clays and Clay Minerals (in press).
7. E. A. Roehl, IADC Drilling Technology Conference, 1984, 109.
8. J.R. Philip and D.E. Smiles, Adv. Colloid and Int. Sci., 1982, 17, 83.
9. D.E. Smiles and J.M. Kirby, Separation Sci. and Tech., 1987, 22, 1405.
10. J.H. Denis, M.J. Keall, P.L. Hall and G.H. Meeten, Clay Minerals (in press).
11. R.P. Steiger, J. Pet. Tech., 1982, 1661.
12. D.E. O'Brien and M.E. Chenevert, J. Pet. Tech., 1973, 1089.
13. R.K. Clark, ACS Adv. in Chem., 1986, 213.
14. J.A. Wingrave, E. Kubena, C.F. Douty, D.L. Whitfill and D.P Cords, 45th Ann. Fall Mtg. Soc. Pet. Eng., 1987, Paper No. SPE 16687.
15. G.M. Bol, IADC/SPE Meeting, Dallas, 1986, Paper No. SPE 14802.
16. M.E. Chenevert, Oil and Gas J., 1989,
17. D. Green and T.E. Peterson, World Oil, 1989, 50.
18. I. Shainberg and A. Kaiserman, Proc. Soil Sci. Soc. Amer., 1969, 33, 547.

Chemical Based Drill Cuttings Cleaning Systems

M.T. McKechnie

BP RESEARCH CENTRE, SUNBURY-ON-THAMES, MIDDLESEX, UK

1 INTRODUCTION

Oil based drilling muds (OBM) are currently widely used in certain areas as they offer many performance advantages over alternative muds and may often be the most economic solution in drilling into sensitive geological structures. Concern over the environmental impact of discharging OBM contaminated cuttings has meant that restrictions on cuttings discharge are becoming more stringent (1,2). In the United States there is a severe restriction and indeed ban in certain States for such discharges (3) whereas the North Sea now has a tightening discharge level being imposed (4). The Norwegian State Pollution Control Authority (SFT) has, from 1st January 1991, recomended a 60g/Kg oil on cuttings level (OOC in grams oil/Kg of dry solids) on existing installations. New installations will currently have to achieve 10g/Kg. In the UK the current 150g/Kg level will be reduced to 100g/kg from January 1992 for fixed platforms, with further restrictions on appraisal, exploration and single wells by 1994. Holland currently has a 100g/Kg level which is under review. A plethora of solutions to this problem are being pursued ranging from shipping to shore, thermal treatment, annulus injection etc (5-10).

This paper will concentrate exclusively on cuttings cleaning systems and will briefly review the current technologies available and then detail the newer developments (particularly the BP Hykleen development) which are most capable of meeting the more stringent OOC levels. The mechanisms of operation will be identified for the systems.

2 CURRENTLY INSTALLED CLEANING SYSTEMS

There are two currently used systems which are based on centrifuges/screens. One is a base oil wash system (11) and the other a surfactant wash system. These are shown schematically in Figures 1 and 2 respectively. The former can typically achieve 80-90g/Kg averaged over a well and is capable of treating all well sections. The latter can achieve as low as 60g/Kg but will not routinely treat 6" sections or indeed some of the fines generated by the system itself. Both recover oil from the cuttings. The base oil wash system can be envisaged as operating in a simple manner by allowing fluidisation/dipersion of the cuttings with more low toxicity base oil and subsequent cuttings deoiling/repacking under high G forces. No chemical degradation of the cuttings occurs;some limited fines formation via mechanical attrition is apparent.

The surfactant wash system (using a few % surfactant solution) can be thought of as operating via a different mechanism. A mixture of cationic and nonionic surfactants has been found to be most suited (12). If one considers the cuttings as a model filter cake and the OBM is an oil continuous phase then the surfactant may act in several ways. It can lower the contact angle at the cutting/oil/water boundary. If $\Theta = 0$ the oil will detach spontaneously and if $0<\Theta<90°$ then mechanical agitation will release the oil. Solubilisation will also play a role in removing oil. The reduction of the base oil/water interfacial tension with the surfactant will also allow easier displacement of oil from narrow capillaries within the cuttings. This is akin to surfactant flooding for secondary oil recovery (13). The whole process is complicated by the presence of emulsifiers and wetting agents in the OBM. The use of an aqueous surfactant solution tends to lead to the generation of fines via swelling of the clays; hence the use of a very high G three phase centrifuge and the difficulty in treating the finest cuttings from the 6" hole section. The marine toxicity of any surfactant must be considered.

3 SOLVENT EXTRACTION

There are two technologies which are being considered one was originally based on a halogenated solvent (14) the other a low boiling point hydrocarbon. In both cases the process involves mixing cuttings and solvent, cuttings/liquid separation and solvent recovery via evaporation/distillation. The halogenated solvent system may well have environmental problems in light of the widening scope of the Montreal protocol on CFCs and chlorinated solvents. However it is believed that this can achieve figures

FIGURE 1

A BASE OIL WASH IN A CENTRIFUGE-CUTTINGS CLEANING SYSTEM.

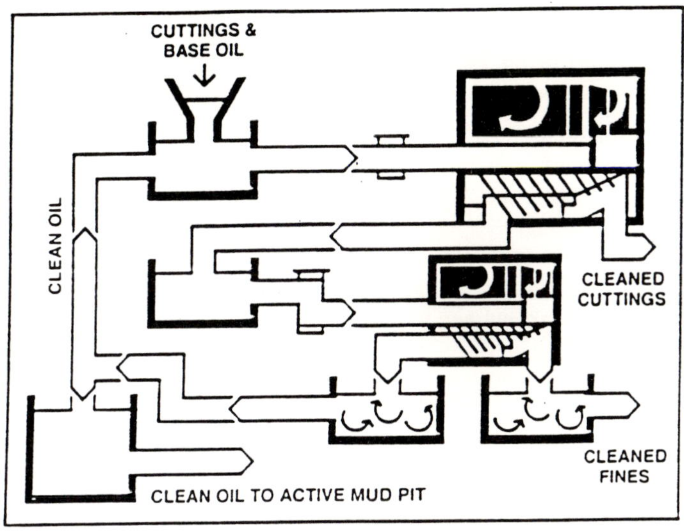

FIGURE 2

A SURFACTANT SOLUTION USED IN A WASH DRUM SYSTEM.

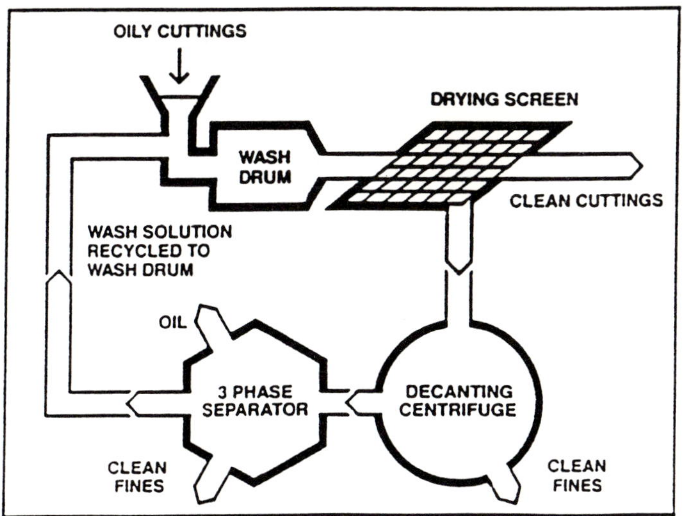

below 10g/Kg;the system being designed to accommodate full mud flow from a 12.25" section at penetration rates of 100 ft/hr. The development work on this system is currently inactive but has not been abandoned. The low boiling point hydrocarbon system can readily achieve 10g/Kg of residual base oil on cuttings with a total residual hydrocarbon level of 20g/Kg which is a consequence of solvent retention by the cuttings. The low flash point of this solvent is not considered a problem, however the final toxicity of the cuttings has yet to be determined. The designs indicate that the weight/space requirements and throughputs for this system are not outside those of the currently installed systems. Mechanisms of operation are extremely simple in that it requires selection of a solvent for the low toxicity base oils which can be readily removed.

Hildebrand parameter matching (15) can be used to model/select suitable solvents for subsequent evaluation.

4 BP SUPERWETTER (HYKLEEN)

Nature of Cleaners

This fluid works in a different manner to solvent extraction. The fluid is designed to be capable of displacing oil from cuttings surfaces by utilising oil/solid/cleaner contact angle behaviour as well as interfacial tensions. The dynamic rather than the equilibrium values are important for this fluid. The phase behaviour has been designed such that it will not solubilise large amounts of displaced oil nor indeed will it emulsify the oil. This leads to a very rapid liquid/liquid split generating cleaner for recycle and relatively pure recovered oil (98% pure) for mud make up.

The cleaner has gone through two phases; the initial prototype coded K5T and a lower toxicity launch product called Hykleen. Hykleen is different from K5T in that it can be supplied as a concentrate and used on dilution with sea water from 85% active to as low as 20% activity. It should be noted that the fluids are more hydrophilic than base oil and some fines generation could potentially occur leading to difficulties in effective solids/cleaner/oil split. This difficulty has been overcome by the careful selection of fines suppressing agents (Hykleen Additive 300) which are incorporated into the cleaner or may be dosed in separately (0.05-0.1% w/w on cleaner of active suppressant).

Marine toxicity testing of K5T has been done in extensive detail. The cuttings from full scale tests passed the relevant Norwegian test protocols however the fluid itself narrowly failed two Norwegian marine species toxicity tests (barnacle larvae and blue mussel). The 85 % active Hykleen product has easily passed the barnacle larvae test indicating that NO marine toxicity problems are anticipated (Table 1). Hykleen is readily biodegradable in aqueous environments (75% in 20 days in salt water).

The basic physical characteristics of the fluids are shown in Table 2.

Performance of K5T

K5T has undergone a series of laboratory, pilot and full scale tests. A fuller account of these has been presented elsewhere (16,17) and only the full scale data will be given in this paper.

The trials were undertaken on two installations in the Norwegian sector of the North Sea. These were chosen as they encompassed both a base oil dilution and a surfactant wash system as shown in figures 1 and 2. The currently used cleaning fluids were swapped for the K5T. In all cases the data for OOC was determined using the standard retort method. The data was spot checked with a gas chromatography technique.

Base Oil Decanting Centrifuge. Some 32 tonnes of 12.25" cuttings (1068-1270m) were treated consisting of drilled porous limestone formation. The data is shown in Table 3. The OBM was a 80/20 emulsion.

The OOC levels were well below those routinely achieved with the system running with base oil.

Wash Drum System. 118 tonnes of large Stratapax cuttings (1832-2570m) from the 12.25" section were treated. The OBM was a 80/20 emulsion. The data is shown in Figure 3 with the rig shaker data for comparison.

The cleaner showed a good split and separation of solids/cleaner/oil and fines were well controlled. All oil was returned for mud make up. The consistent OOC at 3%(30g/Kg) shows a very good performance.

In both cases some modifications to the hardware were indicated as being required for full scale routine use. The data was a significant improvement on the current systems.

Performance of Hykleen

Hykleen has undergone extensive laboratory scale testing and pilot scale testing.

Table 1 Marine Toxicity Data Hykleen (90% Active)

Test Organism = Balanus Improvisus (Barnacle Larvae)

EC 20 (mg/Kg)	EC 50 (mg/Kg)	EC 80 (mg/Kg)
1500	1900	2600
	95% Confidence (1815-2045)	

Table 2 Physical Characteristics of Cleaner

		K5T	Hykleen	Hykleen Additive 300
Flash point (°C/ASTM D93)		60	116	No Flash
Density @ 20°C (gcm^{-3})		0.96	0.99	1.00
Viscosity (cSt)	20		6.5	1100
	24	6.2		
	40	3.6		
	60	2.1		
Solubility in Water @ 20 °C		Dispersible	Complete	Complete

Table 3 Full Scale Base Oil Dilution Hardware Cuttings Cleaning Tests

Initial OOC (g/Kg)	Base Oil Wash		K5T Wash	
	Residual OOC (g/Kg)	% Oil Removed	Residual OOC (g/Kg)	% Oil Removed
173	130	25	68	61

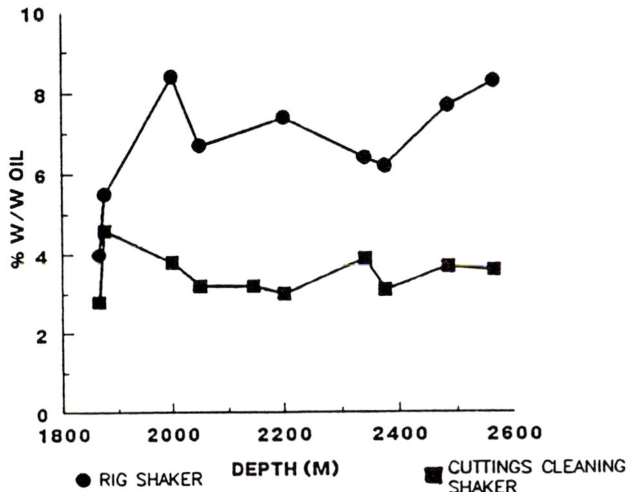

Figure 3 Full Scale (Wash Drum) K5T Performance Data

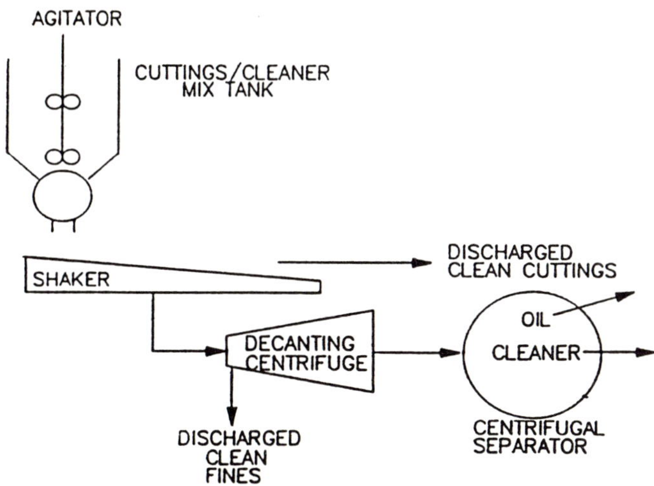

Figure 4 Schematic of Pilot Scale System Used to Evaluate Hykleen Recycle

Laboratory Testing. The procedure adopted was gentle mixing of cuttings with the cleaner for some 5 minutes with a paddle stirrer. The ratio of cleaner to cuttings was 2:1. After washing the cuttings were poured into a Buchner funnel and allowed to drain for 5 minutes under a constant flow of air through the cuttings. The analysis of oil contents was undertaken using a standard retort and gas chromatography.

The performance of the cleaner for an 80% Hykleen active formulation is shown in Table 4. The use of a 20% active with a double wash stage is also shown. This demonstrates that a range of activities of the cleaner formulation can be used dependent on the cuttings type and OBM. This is amply demonstrated in the table where one type of cuttings has been treated with a 20 and 80 % active Hykleen.

In general the cleaning is very efficient being in the range of 24-32 g/Kg for the cuttings from a single wash process. The ability to treat a range of cutting/mud types has also been demonstrated. The benefit of multiple wash has been highlighted. However the increased benefit from multiple washes may not always be significant; this must be viewed in conjunction with hardware complexity.

Pilot Scale Testing. This was undertaken using a set up as shown in Figure 4.

The drill cuttings and cleaner(approx 1:2 ratio by volume) were mixed in a large vessel for 1 minute. The mix was then fed onto a vibrating shaker by manually pouring the cuttings from the mix vessel. Top deck solids were sampled as was the shaker underflow which was then pump fed to the decanting centrifuge. The centrate was then gravity fed to a high speed disc bowl centrifuge. The cleaner was topped up as necessary and recycled.

The initial OOC before treatment was 112 g/Kg. In total some 13 recycles were undertaken and the data is shown in Table 5. Fines control (0.05% w/w in cleaner) and ease of split of cleaner was excellent throughout this test.

5 SELECTION OF LARGE SCALE HARDWARE

An expert system "p^c Select" (18) was used to give a preliminary indication of the types of hardware that may be relevant. The system requires the data input of a pressure filtration experiment or sedimentation rate to select a hardware type that will be most suitable to separate the cleaner and oil from the cuttings. Pressure filtration of a cuttings/cleaner slurry (1:2 ratio) was

Table 4 Laboratory Tests of Hykleen

Cuttings Type	OBM	OOC(g/Kg) Initial	OOC(g/Kg) After Hykleen
12.25"@ 3306m Sandstone/Lime Stringers Interbeds of Mudstone	70:30	175	25(80% active) 175(20% active)
12.25" @ 6428m Sandstone/Shale	60:40	130	26(80% active)
8.5" @ 6000m Chalk	70:30	236	32(80% active)
12.25" (Unspecified)	70:30	66	(20 % Active) 24 (1st wash) 16 (2nd Wash)

Table 5 Pilot Scale Recycle Test of Hykleen
(Initial OOC = 112g/Kg)

Recycle No	OOC (g/Kg)
1	41
2	26
3	55
4	30
5	50
6	17
7	30
8	16
9	25
10	25
11	23
12	40
13	27

Table 6 Solid/Liquid Separation Equipment Simulation & Design

DATA SHEET FOR EQUIPMENT SELECTION

Specifications

Duty	Scale:- medium (10 m^3/hr) Operation:- continuous Objective:- dewatered solids recovery
Settling	Rate:- high (greater 5cm/s) Overflow clarity:- poor Sludge proportion:- medium (2-20% vol)
Filtration	Cake growth rate:- medium (0.02-1 cm/min)

Selected Equipment Description	Selection Warnings	Particle Size (μm)	Feed Conc (%v/v)
Horizontal belt, pan or table filter	None	20-80,000	3-40
Scroll (decanter) centrifuge	1iD	1-5,000	4-40

Selected Equipment Description	F3: index	F4: index	F5: index	F6: index	f7: index
Horizontal belt, pan or table filter	7 C	7	9	8	31
Scroll (decanter) Centrifuge	4 C	4	3	3	14

F3 index:- Solid product dryness
F4 index:- Liquid product clarity
F5 index:- Washing performance
F6 index:- Crystal breakage
F7 index:- Overall performance

Equipment listed in order of overall performance rating

KEY
C = Solids in the form of a cake
1 = Alternative
i = Marginal choice for dewatering in situ
D = The settling clarity may be enhanced by pretreatment

undertaken and relevant fluid/solid physical data input. The key selection criteria for the hardware was chosen to be the dryness of the cuttings in order to ensure minimum oil retention and maximum cleaner recovery for recycle. The output from a typical programme is shown in Table 6, a high index (1-10) number indicates a good fit with the overall performance being a summation of individual indices. (The relevant abbreviations have been added on). Obviously the particle size, feed concentration, ease of use, weight and space of equipment need due consideration. This in conjunction with the full scale tests of K5T indicates that a system for Hykleen can be readily designed from currently available hardware after appropriate lab/pilot scale testing.

6 CONCLUDING REMARKS

The Hykleen product is a low toxicity water soluble drill cuttings cleaning fluid which in laboratory and pilot tests has shown good oil removal performance (typically 30 g/kg + 20). The cleaner can be diluted with seawater to give a range of activities and cleaning performances dependent on the cuttings and OBM. The use of a fines suppressant agent controls the breakdown of cuttings. The cleaner requires a hardware system to be used which will allow cleaner/oil/cuttings split. This should be achievable with the correct selection and configuration of existing filtration, centrifugation, shaker combinations without the need to specifically develop new units. It is anticipated that the product will be fully available in April (19).

7 ACKNOWLEDGEMENTS

The author would like to acknowledge the work undertaken by C Dye, B Slater, G A M Shaw of BP Chemicals, and T Smethills, G C Jeffrey of BP Research. BP Petroleum Development, Norway were instrumental in making the full scale trials possible. The extensive assistance given by Thomas Broadbent and Sons, Huddersfield is also gratefully acknowledged (S Howe and J Wright), as is assistance given by Swaco Geolograph (Aberdeen).

8 REFERENCES

1. Mar Pollut Bull, 17, NO 5, 1986, P188.

2. Dow F K, Davies J M, Raffaeili D; Mar Environ Res, 29, No 2, 1990, PP103-134.

3. Jones M, Envagelisti R; Okla Univ et al. Drilling Muds Nat Conf, Norman, Oklahoma, 29-30 May 1986, Proc PP188-211.

4. Kirby S, Noroil, 18, No 2, 1990, PP26-27

5. Mud Newsletter, No 5, 1990, PP8-9, P10.

6. Minton R C; Minimising Impact on the Environment of Drilling Operations RSC Annual Congress, London, 8-11 April 1991

7. Minton R C; Technology Development and Operational Procedures to Minimise the Environmental Impact of Drilling Operations, Drilling Fluids in the Oil Industry, Egham, 3-5 September 1990

8. Ocean Ind, 25, No 8, 1990, pp 32-33.

9. Petroleum Review, 44, No 525, p537.

10. Davies S STF/Statjford Unit Oil Based Drilling Fluids Conf, Trondheim, Norway, 24-26 February, 1986, Proc pp71-78.

11. Drilling Mud, Cuttings & Waste Pit Processing for Environmental Benefit & Reduced Costs, Thomas Broadbent & Sons, Huddersfield

12. Wiig P O, STF/Statfjord Unit Oil Based Drilling Fluids Conf, Trondheim, Norway, 24-26 February 1986, Proc pp 55-62.

13. Surface Phenomena in Enhanced Oil Recovery, Plenum Press, New York 1981, ed Shah, D O.

14. Oil & Gas Journal, Jan 30th, 1989, p8.

15. Barton A F M, Handbook of Solubility Prameters, CRC Press.

16. Shaw G A M, Slater B; Removing Oil From Drill Cuttings - An Offshore Solution; SPE 19242; Offshore Europe, Aberdeen, 5-8th September 1989.

17. Shaw G A M, Slater B; BP Superwetter - An Offshore Solution To The Cuttings Cleaner Problem; 1st International Symposium on Oil & Gas Exploration and Production Waste Management Practises, New Orleans, 10-13th September 1990.

18. p^c-Select, Separations Technology Associates, Alphington, Exeter.

19. BP Chemicals Product Brochure "Hykleen a BP Superwetter", G A M Shaw, BPC, Belgrave House, Buckingham Palace Rd, London.

Development of New Water Based Mud Formulations

R. Bland

MILPARK DRILLING FLUIDS, 7000 HOLLISTER, SUITE 300, HOUSTON, TEXAS 77040-5337, USA

1 INTRODUCTION

Progressively restrictive environmental regulations are forcing the oil and gas drilling industry to shift to more environmentally acceptable drilling fluids. Use of oil based drilling fluids has been impacted the most and in two of the most active drilling areas, offshore U.S. and the North Sea[1], discharge of any wastes from oil based drilling fluids has been either strictly banned[2] (the "zero discharge" provision) or severely restricted and moving toward zero discharge. Continued use of oil based drilling fluids under zero discharge conditions is expensive, cumbersome, totally impractical in some areas and simply transfers the disposal problem to an alternate site. An alternative approach is improved water based drilling fluids, and operators, service companies and chemical suppliers alike have devoted, and continue to devote, considerable resources toward this goal. This paper explores the design advantages of one class of polyols: polypropylene glycols (PPG).

2 THE OIL BASED DRILLING FLUID ADVANTAGE

Oil based drilling fluids offer several advantages, relative to water based drilling fluids, that the industry is reluctant to give up. The lack of water in the external phase prevents oil based drilling fluids from dissolving salts (bore-hole enlargement), hydrating and dispersing water-sensitive clays (loss of bore-hole integrity in shales and loss of permeability in producing formations) and corroding ferrous metals (loss of mechanical strength). The oil external phase also offers better lubrication (reduced torque and drag), higher boiling points (higher thermal stability) and lower freezing points (lower service temperatures).[3]

3 ALTERNATIVE WATER BASED DRILLING FLUID DESIGNS

Alternative drilling fluid designs offered as substitutes for oil based drilling fluids have, with very few exceptions, been water based drilling fluids containing one or more additives aimed at overcoming the adverse effects of the aqueous external phase and/or providing the performance improvements inherent in an oil external phase. These designs have typically been "low solids/non-dispersed" systems relying on high molecular weight polymers, such as polyacrylamide/acrylate copolymers (still referred to as PHPA), polyanionic cellulose (PAC), sodium polyacrylate (SPA) or xanthan gum. Soluble salts are often added for shale swelling and dispersion control[4,5] and addition of an immiscible liquid and/or surfactant to improve lubrication[6] is not uncommon. One of the more promising newer developments is the use of polyols to replace or supplement salts for control of activity and swelling.

4 GLYCERIN AND POLYGLYCERIN WATER BASED SYSTEMS

The first documented use of a polyol to control swelling in "heaving shale" was by G.E. Cannon in 1940[7]. Glycerin was the preferred material covered although glycols, sucrose and starch are also mentioned; concentrations of greater than 30 volume % are recommended. The mechanism is apparently osmotic control and displacement of water from the clay surface through dipolar and hydrogen bonding.[8] The extent of glycerin use by the industry is unknown, however no listing of glycerin, glycols or sucrose is found in the index of Rodgers' second, third or fourth editions.[9,10,11]

Glycerin has received increased interest in the last four years as a solution with glycerin oligomers or polyglycerins and defoamer[12]. The primary benefits claimed are a fluid which does not swell clays, is stable at high temperatures and has improved lubricating properties[13]. Low molecular weight oligomers of propylene glycol have been used similarly with comparable results[14].

5 PROPYLENE GLYCOL WATER BASED SYSTEMS

Propylene glycols are addition oligomers or polymers of propylene oxide following the general formula:

$$HO[CH_2CH(CH_3)O]_nH \quad (1)$$

where n≥1. Propylene glycols are water soluble for n<13-17 and can be used to control clay swelling, for improved lubrication, to suppress the formation of gas hydrates and to prevent differential sticking. Water insoluble polypropylene glycols (n>13-17) can be emulsified into aqueous systems to form normal or direct emulsion systems which can also be used for lubrication and to improve penetration rates, especially with PDC bits.

Water Soluble Propylene Glycols

Shale Swelling. Development of offshore fields typically requires drilling several wells from a fixed platform at large angles through hydrated and hydratable tertiary shales. Shales containing large concentrations of hydratable clays, such as tertiary Gulf Coast "gumbo", can imbibe substantial quantities of water which swells and softens the shale (the water functions as a plasticizer) resulting in a loss of mechanical integrity and bore-hole enlargement. Minimizing this swelling and maintaining the structural integrity of the bore-hole and the cuttings produced is a critical requirement for successfully drilling in many offshore fields.

One measure of clay or shale swelling is the softening (change in hardness) of the shale after exposure to a fluid and the change in penetration. The penetration (all experimental procedures are given in the Appendix) of a reconstituted sample of a typical tertiary Gulf Coast "gumbo" shale sample was found to be 0 mm after preparation, but increased outside the range of the instrument when the sample disintegrated after exposure to deionized water for 60 minutes. Table 1 lists the penetration results of identical samples of the same reconstituted tertiary Gulf Coast "gumbo" shale after exposure to 30 volume % solutions of propylene glycol oligomers for 60 minutes along with the molarity of the solutions. Note that swelling for this shale, as measured by penetration, is not a direct function of molarity or activity of the solutions (as would be expected for osmotic swelling) but is inversely related to the degree of polymerization.

Table 1 Gumbo Shale Linear Swelling Results

Glycol/ 30 Vol.% in Water	Molarity/ g-moles/L	Penetration/ mm
Propylene Glycol	3.94	>13
Dipropylene Glycol	2.24	2.5
Tripropylene Glycol	1.56	1.0
PPG 400(n=6-7)	0.75	0

These results can be interpreted as increased adsorption of the propylene glycol oligomer and displacement of water from the clay surface with increasing degree of polymerization and the resulting increase in size of the molecule; this interpretation would appear to be consistent with Bradley's[15] and MacEwan's[16] findings. This also suggests that adsorption is more important than control of osmotic potential with respect to swelling with this shale.

Lubrication. Development of offshore fields typically requires drilling several wells from a fixed platform at large angles which dictates using a drilling fluid with good lubricating properties that will minimize torque and drag. Oil based drilling fluids inherently excel in this area due to the oil external phase but water based drilling fluids can be substantially improved.[14] Table 2 (reproduced from U.S. Patent 4,963,273[14]) lists the metal to metal lubricity coefficients of 3.3 weight % dispersions of sodium montmorillonite in water after additions of a blend of predominantly dipropylene glycol and tripropylene glycol. The results for addition of 40% of the blend is comparable to values measured for oil based drilling fluids which are typically ≤0.1.

Table 2 Metal to Metal Lubricity Coefficients of Propylene Glycol Oligomer Solutions

Propylene Glycol Oligomer Solution in Water/Wt.%	Lubricity Coefficient
0	0.45
10	0.29
20	0.14
40	0.08

Gas Hydrate Formation. Deep water drilling exposes drilling fluids to conditions conducive to the formation of gas hydrates (38°F[3.3°C] and 4,000 psi[27.57 MPa])[17] which can plug subsea equipment causing serious difficulties and compromising rig safety, particularly during kill operations[18]. Salt/polymer systems such as 20 weight % (of the aqueous phase) sodium chloride/PHPA have been used extensively to suppress gas hydrate formation in deep water drilling but may be inadequate in the deepest waters[18]. Table 3 lists the effects of the same propylene glycol oligomer blend used in Table 3 on gas hydrate formation in a 20 weight % (of the aqueous phase) sodium chloride/PHPA field drilling fluid from the Green Canyon area offshore Louisiana. The equilibrium gas hydrate formation temperature is reduced approximately 10°F(5.6°C) by the addition of 30 weight % (of the aqueous phase) of the propylene glycol oligomer, and the oligomer has been used in Green Canyon for this purpose.

Differential Sticking. Differential sticking is a major problem for many operators and one of the most costly problems our industry faces. One operator has reported average annual costs due to stuck pipe of $30,000,000/year with much of it due to differential sticking[19]. Oil based drilling fluids have been the historical drilling fluid for drilling in areas prone to differential sticking due to the inherent lubricating properties of the oil external phase and the lower filtrate losses of oil based drilling fluids; water based drilling fluids, however, can be substantially improved.

Propylene glycols oligomers and polymers can be used to prevent differential sticking. Figure 1 displays the Sticking Coefficients for a 15.5 lb/gal (1857 kg/m^3) lime based field drilling fluid from North Kent Bayou,

Table 3 Equilibrium Data for Gas Hydrate Formation

20% NaCl/PHPA Drilling Fluid (Base System)		Base System + 30% Propylene Glycol Oligomer	
Temperature °F(°C)	Pressure psi(MPa)	Temperature °F(°C)	Pressure psi(MPa)
48.0(9.3)	2,730(18.82)	38.8(3.8)	3,225(22.24)
51.5(10.8)	4,120(28.40)	40.9(4.9)	4,190(28.89)
52.7(11.5)	4,875(33.61)	43.5(6.4)	5,060(34.89)

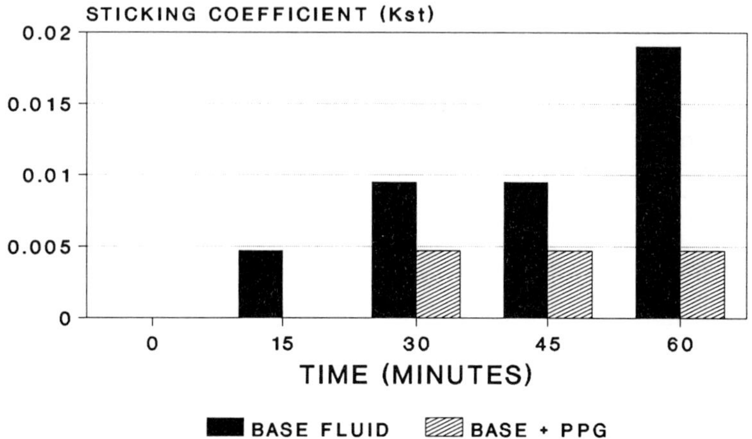

Figure 1 Effect of PPG on Sticking Coefficient

Louisiana after addition of 2 vol.% of a proprietary (patent applied for) polypropylene glycol formulation marketed by Milpark Drilling Fluids as AQUA-MAGIC™. This formulation has proved successful in preventing differential sticking in over thirty wells to date in the Gulf Coast and has proven especially helpful in drilling severely depleted formations where differential sticking is a major problem. The following two examples are representative of field performance with this formulation.

'Eugene Island Application'. An operator in Eugene Island, offshore Louisiana, experienced stuck pipe in a 34° angle hole at approximately 7,500 ft(2286.0 m). There were 11 depleted sands in this well with the pipe believed to be stuck in a sand at about 6,300 ft(1920.2 m) true vertical depth (TVD) drawn down to approximately 1,000 psi(6.89 MPa). Density of the drilling fluid at the time sticking occurred was 12.2 lb/gal(1462 kg/m^3) and API Filtrate Loss was 4.4 cm^3. Attempts to free the pipe were unsuccessful and the well was sidetracked at about 6,400 ft(1951.7 m), leaving a mud motor, steering tool,

243 ft(74.1 m) of 8 in.(203.2 mm) collars and ±17 joints of 5 in.(127.0 mm) heavy weight drill pipe in the hole. The operator at this point decided to add and maintain 2 to 2.5 volume % of the proprietary polypropylene glycol formulation described above to the system while drilling the 12.25 in.(311.2 mm) hole interval to prevent differential sticking. This section of hole was drilled to 8,872 ft(2704.2 m) with 14.6 lb/gal(1749 kg/m^3) drilling fluid and a 3.4 cm^3 API Filtrate Loss. No indication of sticking was detected while drilling or logging the interval and 9-5/8 in(244.5 mm) casing was run to bottom without difficulties. No adverse effects on drilling fluid properties were detected.

'South Marsh Island Application'. An operator on a South March Island platform offshore Louisiana was drilling an 8.5 in(215.9 mm) hole with a 16.8 lb/gal(2013 kg/m^3) sea water PHPA drilling fluid from the 11-3/4 in(298.4 mm) casing point at 8,031 ft(2447.8 m) to a programmed casing point at 9,609 ft(2928.8 m). Two depleted sands had to be drilled at approximately 9,300 ft(2834.6 m) which were drawn down to approximately 8.6 to 8.7 lb/gal(1031 to 1042 kg/m^3). The pipe stuck at 9,334 ft(2845 m) on a connection while tripping in the hole with a logging tool on the drill pipe. Drilling fluid properties at the time of sticking are presented in Table 4.

The well was sidetracked and kicked off at 8,735 ft(2662.4 m) after setting a cement plug. Approximately 3 volume % of the proprietary polypropylene glycol formulation described above was added to the system at approximately 9,285 ft(2830.1 m) and drilling proceeded to 9,646 ft(2940.1 m) without difficulty. A 75 bbl(11.92 m^3) "pill" of drilling fluid containing 10 volume % of the proprietary polypropylene glycol formulation described above was spotted in the open hole on bottom prior to running a 7 in(177.8 mm) liner to bottom and cementing without difficulty.

Toxicity. Toxicity of water soluble propylene glycol oligomers and polymers are typically low. A bioassay on the propylene glycol oligomer blend used in Table 2 and Table 3, tested at 30 volume % in a fresh water chrome lignosulfonate drilling fluid, gave an LC_{50} of 65,000 ppm using the U.S. Environmental Protection Agency's(EPA) Drilling Fluid Toxicity Test and Mysid shrimp[20]. A bioassay on the formulation tested in Figure 1 and used in the field examples, evaluated at

Table 4 South Marsh Drilling Fluid Properties at Time of Sticking

Drilling Fluid Property	Value
Density, lb/gal(kg/m^3)	16.8(2013)
Plastic Viscosity, cp(Pa·s)	39(0.039)
Yield Point, lb/100 ft^2(Pa)	19(9.10)
10 s/10 min Gel Strengths, lb/100ft^2(Pa)	5/15 (2.4/35.9)
API Filtrate, cm^3	2.0
HTHP Filtrate Loss @ 300°F(148.9°C), cm^3	8.4
Cake Thickness,32nd in(mm)	2 to 3 (1.6 to 2.4)
HTHP Cake Compressibility @ 300°F(148.9°C)[Filtrate Loss @ 500 psi(3.447 MPa)÷Filtrate Loss @ 100 psi(0.6895 MPa)]	1.4
Solids, volume %	34
Low Gravity Solids, volume %	4.2
MBT, lb/bbl(kg/m^3) bentonite equivalent	25.0(71.3)
pH	10.0
Pm,cm^3 0.02 N H_2SO_4	1.9
Pf/Mf, cm^3 0.02 N H_2SO_4	0.5/1.4
Chlorides, mg/L	10,000
Total Hardness, mg/L	160

3 volume % in a fresh water chrome lignosulfonate drilling fluid, gave an LC_{50} of >1,000,000 ppm using the EPA's Drilling Fluid Toxicity Test. Water-insoluble polypropylene glycols show comparable toxicities[21].

Water-Insoluble Polypropylene Glycols.

Polypropylene glycols of sufficient size (n≈13-17) lose water solubility but can be emulsified into water to form normal or direct emulsion drilling fluids. One system containing 3 to 5 volume % polypropylene glycol and supplemental surfactants and copolymers shows improved lubrication and creates a hydrophobic coating on metal surfaces which has been credited with improved penetration rates with PDC bits[22]. This system has been used successfully in over twenty wells to date, including two in the North Sea.

6 POLYPROPYLENE GLYCOL BASED ALTERNATIVES

Water based solutions and drilling fluids using water soluble propylene glycol oligomers and polymers have been shown to be effective in controlling clay swelling, lubrication, suppressing the formation of gas hydrates and preventing differential sticking; water insoluble polypropylene glycols (n>13-17) can be emulsified into aqueous systems to form normal or direct emulsion systems for improved lubrication and higher penetration rates. Polypropylene glycols can also be emulsified with aqueous solutions to form invert emulsion systems as well, and these systems exhibit the characteristic advantages of traditional oil based drilling fluids without the use of undesirable hydrocarbon oils. These invert emulsion systems are not strictly speaking water based systems as the external phase is not aqueous and contains relatively little if any water, but nor are they strictly speaking oil based as they contain no hydrocarbons nor any materials traditionally classified as an oil or "oily". They are instead, perhaps, a new class of drilling fluid system somewhere in between but exhibiting the performance advantages of oil based systems without the oils.

BIT BALLING

The characteristic advantages of oil based drilling fluids make them well suited to drilling shales which are prone to "balling"[23] (the build up of compacted cuttings on the bit due to inadequate removal of the drilled cuttings and the resulting decline in penetration rate). The oil external phase in oil based drilling fluids oil wets the bit and cuttings and helps prevent the adhesion of the drilled cuttings to the bit as well as the hydration of the cuttings and their cohesion into a sticky mass (prevalent in hydratable shales such as "gumbo") which adheres to the bit. Great strides have been made in recent years developing water based systems to effectively address this problem[21,22], but in the most extreme and severe cases a water-free external phase (such as an oil based or invert emulsion drilling fluid) offers real advantages where flow rates are limited.

One of the most direct means of evaluating balling tendencies is drilling a susceptible shale in a laboratory drilling device and monitoring the flow rate required to prevent balling. Water based drilling fluids are more likely to cause balling and typically require larger flow rates to keep the bit clean relative to oil based drilling fluids. Table 5 lists the minimum flow rates required to prevent bit balling while drilling Pierre I shale with a micro-PDC bit manufactured by Hughes Tool Company and tested on their microbit drilling equipment. Note that only the invert emulsion systems (the oil based drilling fluid and the PPG invert) effectively lowered the required flow rate to clean the bit and prevent bit balling with this shale.

Table 5 Minimum Flow Rates Required to Prevent Balling in Pierre Shale

Drilling Fluid System	Minimum Flow Rate/ gal/min(dm^3/s)
Fresh Water Chrome Lignosulfonate	13(0.820)
PHPA/3 wt.% KCl	>15(>0.946)
Lime	>15(>0.946)
Fresh Water SPA/5 volume % diesel oil	>13(>0.820)
Oil Based System (Diesel Oil)	<2.8(<0.177)
60/40 Invert PPG Emulsion	3.5(0.221)

7 CONCLUSIONS

Propylene glycol oligomers and polymers are useful in developing water based drilling fluids designed to replace or compete with traditional oil based drilling fluids. Water based drilling fluid systems formulated with propylene glycol oligomer and polymer systems have been shown to be effective at:

- reducing the swelling and maintaining the mechanical integrity of hydratable shales which helps prevent bore-hole enlargement, collapse of the bore-hole and the need for re-drilling.
- reducing the lubricity coefficient of water based drilling fluid systems which helps reduce

torque and drag in extended reach drilling which can lead to fewer platforms to develop offshore fields.

- reducing the equilibrium gas hydrate formation temperature in deep water drilling which allows drilling in deeper water while maintaining safe operating conditions.

- reducing or preventing differential sticking, particularly in depleted sands, which can save the industry millions in lost hole, equipment and down time.

Polypropylene glycols can also be used to form invert emulsion systems which have been shown to be comparable to oil based drilling fluids in preventing bit balling.

ACKNOWLEDGEMENTS

The author would like to thank Milpark Drilling Fluids and our customers for permission to publish these results and the many people within Milpark who contributed to this paper. Special thanks go to Dennis Clapper for the shale hardness test results, Dorothy Enright for permission to reproduce her lubricity coefficient results and her informative council on bit balling, Wade Wise for summarizing the gas hydrate test results, Gary Schlagel for providing the Sticking Coefficient data, Gary Schlagel and Bill Micho for preparation of the field recaps, and Mamie Decker for printing the manuscript.

APPENDIX

EXPERIMENTAL PROCEDURES

Reconstituted Shale Preparation. Reconstituted shale pellets were prepared similarly to Chenevert's procedure[24]. Gulf coast "gumbo" shale was dried at 180°F to 190°F and ground to pass a 30 mesh screen. Deionized water was added to bring moisture to 15 to 16 weight % and the hydrated ground shale was thoroughly blended with a Hobart mixer. Pellets were prepared using 5.3 g of re-hydrated and blended shale in a 0.5 in.(12.7 mm) die under 6,000 lbm (2700 kg) load (30,000 psi (207 MPa)) for 6 min(360 sec). Final pellets were 0.7230 in.(18.36 mm) to 0.7290 in.(18.52 mm) in length with densities of 2.13 g/cm^3 (2130 kg/m^3) to 2.16 g/cm^3 (2160 kg/m^3).

Shale Hardness. Shale hardness was evaluated with a Humboldt Penetrometer using a 0.040 in.(1.016 mm) diameter non-tapered, blunt ended needle. The reconstituted "gumbo" shale pellets described above were statically exposed to the test fluid for 60 min (3600 sec), recovered, blotted dry and the hardness determined by radial penetration with the needle for 10 seconds.

Lubricity Coefficient. Lubricity coefficients of test fluids were measured according to API Recommended Practice Standard Procedure for Testing Drilling Fluids, API RP 13B, ninth edition (1982) page 45.

Gas Hydrate Formation. Equilibrium gas hydrate formation results were obtained using a procedure developed in E.D. Sloan's group at the Colorado School of Mines[25,26]. A fixed mass, two phase "Green Canyon" gas/drilling fluid combination is cooled at a constant rate in a stirred, closed cell of constant volume (isochoric) while monitoring the temperature and pressure until gas hydrates form causing a dramatic decrease in pressure. Temperature is then increased to dissociate the gas hydrates and the equilibrium temperature is the "intersection of the hydrate dissociation trace (pressure) with the initial isochoric cooling trace."[27]

Sticking Coefficient. Differential sticking was evaluated using an OFI Differential Sticking Tester Apparatus (marketed by OFI Testing Equipment International, Houston, Texas) which measures the resistance to movement of a filter cake under differential pressure; the Sticking Coefficient is defined as the ratio of the force necessary to initiate sliding (or rotation) of a plate in contact with a filter cake under differential pressure, relative to the normal force on the plate. The test device consists of a filtration cell capable of holding 200 cm^3 of fluid, a perforated bottom capable of holding filter paper, a plate (on a plunger) which just covers the filter paper which is positioned just above the filter paper and a torque wrench. The cell, with filter paper in place, is filled with fluid, pressurized with 475 psi of nitrogen and filtrate is collected for 10 min.(600 sec). The plate (on a plunger) is then pushed down on the filter cake with 25 lbm(11.3 kg) force for 15 min.(900 sec) to stick the plate. The plate is then rotated with the torque wrench and the torque necessary to break the plate free is measured. Filtrate is then collected for another 10 min.(600 sec) and the plate is then pushed down again on the filter cake with 25 lbm(11.3 kg) force for 15 min.(900 sec) to stick the plate once more. The torque necessary to break the plate free is again measured and the process is repeated another 2 times. The Sticking Coefficient, calculated by the following equation:

$$\text{Sticking Coefficient} = \text{torque(lbf-in.)} \div 1000 \qquad (1)$$

is then typically plotted against cumulative contact time between plate and filter cake.

Micro-bit Drilling Tests. The Hughes Tool Company's microbit drilling equipment has been described in detail in previous publications[28,29]. Cylindrical Pierre I shale samples (4.0 in.[101.6 mm] by 4.0 in.[101.6 mm]) were grouted in a steel cylindrical sleeve and mounted in the drilling device and the system sealed and pressurized with drilling fluid to approximately 2000 psi(13.79 MPa) bore-hole pressure. Drilling was performed with a 1.125 in.(28.87 mm) OD 2-cutter micro-PDC bit with 2-0.094 in.(2.39 mm) ID nozzles. Rotary speed was maintained at 103 rev/min (10.79 rad/s) and weight on bit was varied automatically to maintain a constant rate of penetration (ROP) of 25 - 30 ft/hr(2.12 - 2.54 mm/s). Flow rate was varied from 2.8 - 15 gal/min(0.177 - 0.946 dm^3/s) and balling was monitored by visually inspecting the bit and bore-hole after drilling approximately 1 in.(25.4 mm) at a fixed flow rate.

The procedure developed by Hughes Tool Company to evaluate balling was to initiate testing with each drilling fluid system at maximum flow rate and monitor balling as flow rate was decreased. A clean bit responded well to weight on bit and no more than 100-200 lbm(45.4-90.7 kg) was required to maintain ROP. Onset of balling was usually accompanied by little to no response to weight on bit and exponentially increasing weight on bit (regulated by the auto-weight on bit device) to try to maintain ROP; drilling was terminated when twist-off or mechanical failure appeared imminent. Minimal to no cuttings adhering to the bit on visual inspection was defined as minimal to no balling.

REFERENCES

1. F.V. Jones and A.J.J. Leuterman, 'Alternate Environmental Testing of Drilling Fluids: An International Perspective', paper SPE 20889 presented at the 1990 European Petroleum Conference, The Hague, Oct.22-24.

2. U.S. Environmental Protection Agency [OW-4-FRL-3077-1 Permit No. GMG280000] NPDES General Permit for the Outer Continental Shelf (OCS) of the Gulf of Mexico, 51FR33130 at 33131 September 18, 1986.

3. G.R. Gray, H.C.H. Darley and W.F. Rogers, 'Composition and Properties of Oil Well Drilling Fluids', forth edition, Gulf Publishing Company, Houston, Chapter 2, p.62.

4. Ibid, Chapter 8, p.378.

5. A.G. Loomis, H.A. Ambrose and J.S. Brown, 'Drilling of Terrestrial Bores', U.S. Patent No. 1,819,646 (Aug.18, 1931).

6. Gray, Darley and Rogers, op. cit., Chapter 2, p.51.

7. G.E. Cannon, 'Drilling Fluid for Combating Heaving Shale', U.S. Patent No. 2,191,312 (Feb.20, 1940).

8. R.E. Grim, 'Clay Mineralogy', McGraw-Hill, New York, 1968, Chapter 10, p.371.

9. W.F. Rogers, 'Composition And Properties Of Oil Well Drilling Fluids', second edition, Gulf Publishing Company, Houston, 1953, p.672 & 676.

10. W.F. Rogers, 'Composition And Properties Of Oil Well Drilling Fluids', third edition, Gulf Publishing Company, Houston, 1963, Index, p.811, 816 & 817.

11. Gray, Darley and Rogers, op. cit. p. 623 & 628.

12. M.E. Chenevert, OGJ., July, 17, 1989.

13. T.E. Peterson, 'Drilling And Completion Fluid', U.S. Patent No. 4,780,220 (1988).

14. A.C. Perricone, D.K. Clapper and D.E. Enright, 'Modified Non-Polluting Liquid Phase Shale Swelling Inhibition Drilling Fluid and Method of Using Same', U.S.

Patent No. 4,963,273 (1990).

15. W.F. Bradley. J.Am.Chem.Soc., 1945, 67,975-981.

16. D.M.C. MacEwan, Trans. Faraday Soc., 1948, 44, 349-367.

17. A.H. Hale and A.K.R. Dewan, 'Inhibition of Gas Hydrates in Deepwater Drilling', paper SPE/IADC 18638 presented at the 1989 SPE/IADC Drilling Conference, New Orleans, Feb.28-March 3.

18. J.W. Barker and R.K. Gomez, 'Formation of Hydrates During Deepwater Drilling Operations', paper SPE/IADC 16130 presented at the 1987 SPE/IADC Conference, New Orleans, March 15-18.

19. W.B. Bradley, D. Jarman, R.S. Plott, R.D. Wood, T.R. Schofield, R.A. Auflick and D. Cocking, 'A Task Force Approach to Reducing Stuck Pipe Costs', paper SPE/IADC 21999 presented at the 1991 SPE/IADC Drilling Conference, Amsterdam, March 11-14.

20. U.S. Environmental Protection Agency Industrial Technology Division. May 1985. Appendix 3 - Drilling Fluid Toxicity Test Proposed Regulation for the Offshore subcategory of the Oil and Gas Extraction Point Source Category, 50 FR34592 at 34631.

21. D.P. Enright, B.M. Dye and F.M. Smith, World Oil., 1991, March, 92.

22. D.P. Enright, W.M Dye and F.M. Smith, 'An Environmentally Safe Water-Based Alternative To Oil Muds', paper SPE/IADC 21937 presented at the 1991 SPE/IADC Drilling Conference, Amsterdam, March 11-14.

23. G.R. Gray, H.C.H. Darley and W.F. Rogers, 'Composition and Properties of Oil Well Drilling Fluids', Gulf, Houston, 1980, fourth edition, Chapter 9, p.410-414.

24. S.O. Osisanya and M.E. Chenevert, 'Rigsite Shale Evaluation Techniques for Control of Shale-Related Wellbore Instability Problems', paper SPE/IADC 16054 presented at the 1987 SPE/IADC Drilling Conference, New Orleans, March 15-18.

25. T. Kotkoskie, 'Measurement and Prediction of Natural Gas Hydrates in Water Based Drilling Muds', Masters Thesis, Colorado School of Mines, Golden Colorado (1989).

26. T.S. Kotkoskie, B. Al-Ubaidi, T.R. Wildeman and E.D. Sloan, 'Inhibition of Gas Hydrates in Water-Based Drilling Muds', paper SPE 20437 presented at the 1990 SPE Annual Technical Conference and Exhibition, New Orleans, Sept.23-26.

27. E.D. Sloan, 'Clathrate Hydrates of Natural Gases', Marcel Dekker, New York, 1990, Chapter 6, p.270.

28. R.A. Cunningham and W.C. Goins, 'Laboratory Drilling of Gulf Coast Shales', paper presented at a meeting of the Study Committee on Drilling Fluids held during the 1957 meeting of the Southern District, Division of Production, Shreveport, March.

29. R.A. Cunningham, 'Laboratory Study of Effect of Overburden, Formation and Mud Column Pressures on Drilling Rate of Permeable Formations', paper presented at the 1958 SPE Annual Fall Meeting, Houston, Oct. 5-8.

Water Based Muds for Drilling High Pressure, High Temperature Wells

J.R. Plank

SKW TROSTBERG AG, PO BOX 12 62, 8223 TROSTBERG, GERMANY

1 INTRODUCTION

The drilling industry of today ventures more and more into environments of high temperatures and pressures. In the North Sea, the recent deep drilling activity in the Ekofisk area is an example. Structures at or below 5000 m are being explored now. Norway's Saga Petroleum drilled the first of these difficult wells in the Ekofisk area and experienced typical drilling problems caused by abnormally high formation pressures (70 MPa/10,000+ psi) and bottom-hole static temperatures exceeding 150°C. Reports on this Saga well [1] confirm that proper drilling fluid selection is a key to success in drilling high pressure, high temperature (HP/HT) wells.

In drilling deep, hot wells, oil-based muds have been used routinely. Thermal stability up to 260°C, high rate of penetration and easy mud maintenance were reported [2]. Since the eighties, environmental concerns have posed severe restrictions on the use of oil muds in some areas. This has prompted the development of novel polymers and advanced chemical materials which allow the formulation of water-based muds stable under HP/HT conditions.

This paper focuses on design criteria, chemical materials and field problems related to the use of water-based muds in HP/HT drilling.

2 DESIGN CRITERIA

HP/HT stable drilling fluids are required to function at temperatures from 150° to at least 200°C. Some water-based muds have been formulated to tolerate temperatures exceeding 260°C.

Pressures encountered in deep, hot wells may exceed 10,000 psi (70 MPa). This dictates high fluid densities to counteract the formation pressure. 1,8 - 2,4 g/ml are typical densities of muds used in high pressure wells.

When formulating an HP/HT stable, water-based mud system, the principle "the simpler, the better" is the best guideline. Avoid the use of "chemical cocktails". Instead, keeping the number of products to a minimum should be a rule. Disregard products with more than one primary function. Otherwise, multiple interactions between mud additives may render a system unmanageable when troubleshooting or treating the mud. Good field results have been reported when following above guidelines [3,4].

3 CHEMICAL MATERIALS

Key products for HP/HT stable water-based muds are:

Bentonite	-	for viscosity
Barite, hematite	-	for density
Fluid loss polymers	-	for filtration control
Deflocculants	-	for rheology

Bentonite

Bentonite is used in water-based muds primarily to provide viscosity which is required to suspend cuttings. In HP/HT drilling, special attention must be given to bentonite and other clays present in the drilling fluid. The reason being a high temperature induced gelation of bentonite which may result in excessive viscosities, especially when running a high density mud.

The gelation process is based on the following mechanism: When added to a drilling fluid, bentonite absorbs water. This hydration weakens the forces holding the clay layers together. A house-of-card structure with edge-to-face contact of the clay platelets is the result. When heating the fluid above 150°C, dehydration and aggregation of clay platelets leads to a state of flocculation. The fluid loses its shear-thinning characteristic. Instead, low plastic viscosity and high gel strengths are the result. This process is called the high temperature gelation of bentonite. Drilled formation clays, particularly low gravity solids, undergo a similiar gelation process.

Because of this gelation phenomena, the bentonite content in HT stable water-based muds must be kept to a minimum. As a rule, a concentration of 3 ppb (9 kg/dm3) bentonite should not be exceeded.

For HP/HT wells only Wyoming bentonite (pure sodium montmorillonite of highest quality) should be applied. The use of polymer treated (peptized) or soda ash activated calcium bentonite should be avoided. Complex reactions taking place under hostile conditions will make its performance unpredictable.

The same principle concerning bentonite applies to clayey drill solids in the mud, especially the low gravity solids (LGS). Field experience suggests that LGS should be kept below 6% [4]. Effective solids control equipment is required to achieve a mud with workable (i.e. not too high) rheology in the deeper hole sections.

Due to the limitations of bentonite and clay content, water-based muds utilized in HP/HT drilling are often referred to as "low-colloid muds".

If, out of one reason or another, the concentration of bentonite or clayey materials exceeds acceptable levels and results in viscosity problems, polymeric deflocculants such as maleic anhydride copolymers can be used as a last resort to control rheology (see "deflocculants").

Barite

Formation pressure dictates the drilling fluid density required. As a rule, fluid density must be high enough to balance formation pressure, but not so high that the formation is fractured.

The most common weighting agent used in drilling fluids is barite. In water-based muds, densities up to appr. 2,3 g/ml can be obtained. For even higher densities, hematite (iron oxide) is used instead of barite. Non-abrasive, drilling grade hematite is available in the industry.

An alternate way to increase mud density is to add soluble salts to the drilling fluid. Sodium chloride is commonly used with calcium chloride or bromide rarely applied. Brine systems allow high fluid densities at lower concentrations of suspended solids than conventional muds weighted with barite or hematite. The result is lower plastic viscosity and higher rate of penetration. Due to high cost (particularly for salt tolerant polymers), brine systems are considered only for certain environments, such as ultra-high (>2.4 g/ml) mud densities.

Fluid loss control

Adequate filtration control of a drilling fluid is essential to prevent drilling problems such as excessive torque and drag, pressure differential sticking, borehole instability and formation damage.

HP/HT filtrates of less than 20 mls are commonly specified for deep, hot wells. Recent studies indicate that the conventional method of measuring the filtrate of a mud under static conditions can produce misleading results. In one particular case, a mud showed poor fluid loss control in the static test, but performed well in the dynamic test and in the field [3]. Therefore, stirred fluid loss cells have been introduced of late [5,6]. They seem to provide more realistic data.

A variety of HT stable fluid loss polymers with disparate electrolyte tolerance are available to the industry for HP/HT wells.

Lignites and lignosulfonates. Chrome lignite has been used extensively as a fluid loss additive and thinner in deep, hot wells. However, dosage and maintenance rates are high. For example, chrome lignite dosages at 200°C of 3-4 % and at 250°C of more than 6% have been reported [7].

Chrome lignite starts to decompose at temperatures above 180°C. This is one reason behind the high dosages required. Carbon dioxide is a major degradation product. It can be detected easily by using the Garrett Gas Train. Another way to monitor lignite decomposition in a mud is to monitor the pH: a significant drop is a clear indication of lignite degradation. The decomposition products of lignites have a deleterious effect on the drilling fluid. Carbon dioxide, for example, flocculates bentonite, which results in significantly increased yield point and gel strengths of the mud. Lime is commonly used to control carbon dioxide contamination.

The accumulation of degradation by-products over time and their detrimental effect on the mud is the major reason for decreased usage of chrome lignite in recent years. A secondary concern is the environmental damage caused by the heavy metal content of chrome lignite.

Lignosulfonates, even when reacted with chrome, are temperature stable only up to 175°C and hence inferior to lignites. Their use in high temperature drilling has been limited.

Recently, monomer reacted (grafted) lignite and lignosulfonate products have been developed [8-10]. Acrylic acid and 2-acrylamido-2-methylpropane sulfonic acid (AMPS) are the most common monomers grafted onto the inexpensive lignite or lignin backbone. The grafting enhances the thermal stability and electrolyte (esp. calcium) tolerance of the products. Field experience with an acrylate grafted, so-called polyanionic lignin (PAL) fluid loss reducer is described in [3].

Synthetic polymers. The emergence of new synthetic, high-performance polymers was a major reason behind the success of water-based muds over oil-based fluids in recent years. A variety of synthetic fluid loss polymers with different molecular weights, electrolyte and temperature stability is available today. This allows tailor-made mud formulation for HP/HT wells.

Polyacrylates and polyacrylamides were the first synthetic fluid loss polymers used in drilling muds. Preparation and properties of a typical drilling grade polyacrylate fluid loss polymer is described in [11]. Products with weight-average molecular weights between 100,000 and 3 million daltons are available. They have been used successfully at temperatures exceeding 200°C. Major limitation of polyacrylates is their sensitivity to divalent cations. For example, calcium concentrations as little as 200 ppm, which are present in any lime, gyp or seawater mud, destroy their fluid loss capability. The use of polyacrylates is therefore limited to fresh water and saturated salt systems. The introduction of acrylamide polymers could not overcome this disadvantage.

The need for more calcium/magnesium tolerant fluid loss polymers prompted the development of a new generation of co- and terpolymers. Sulfonated monomers were recognized to be better building blocks than acrylic acid. Sulfonic acid is not chelated by divalent cations such as calcium/magnesium as easily as the carboxyl group in acrylic acid. Molecule engineering resulted in highly hardness tolerant fluid loss polymers using AMPS monomer as the main building block. A calcium/magnesium tolerance between 15,000 and 75,000 ppm has been achieved with these co- and terpolymers. Preparation and properties of a typical AMPS fluid loss polymer is described in [12]. Examples of commercially available co- and terpolymers are:

$$\left[-CH_2-CH(C{=}O\text{-}NH\text{-}C(CH_3)_2\text{-}CH_2\text{-}SO_3Na)-CH_2-CH(N(CH_3)\text{-}C{=}O\text{-}CH_3)-CH_2-CH(C{=}O\text{-}NH_2)- \right]_n$$

AMPS-N-methyl-N-vinylacetamide-acrylamide terpolymer

$$\left[-CH_2-CH(C{=}O\text{-}NH\text{-}C(CH_3)_2\text{-}CH_2\text{-}SO_3Na)-CH_2-CH(C{=}O\text{-}N(CH_3)_2)- \right]_n$$

AMPS-N,N,-dimethylacrylamide copolymer

$$\left[-CH_2-CH(C{=}O\text{-}NH\text{-}C(CH_3)_2\text{-}CH_2\text{-}SO_3Na)-CH_2-CH(N\text{-pyrrolidone ring}{=}O)- \right]_n$$

AMPS-N-vinyl-2-pyrrolidone copolymer

The acrylic acid-hydroxy propyl acrylate-acrylamide terpolymer described in [13] indicates that good hardness tolerance can also be achieved by inserting a bulky side-group such as the hydroxypropyl group between adjacent acrylate groups. The bulky side-group apparently prevents the calcium from exercising its chelating effect.

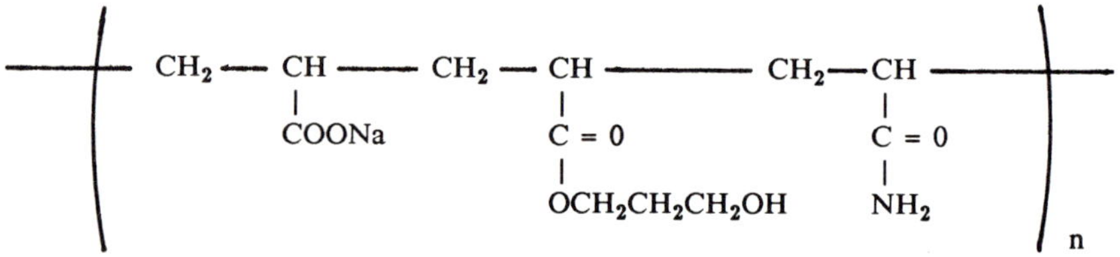

$$\left[CH_2 - \underset{\underset{COONa}{|}}{CH} - CH_2 - \underset{\underset{\underset{\underset{OCH_2CH_2CH_2OH}{|}}{C=0}}{|}}{CH} - CH_2 - \underset{\underset{\underset{\underset{NH_2}{|}}{C=0}}{|}}{CH} \right]_n$$

acrylic acid-hydroxypropyl acrylate-acrylamide terpolymer

AMPS fluid loss polymers are available at weight-average molecular weights from 400,000 to several million daltons. Because of the magnitude in molecular weight, the polymers are viscosifiers by nature. This is especially true in fresh water or low salt systems such as sea water muds. In high density muds, this secondary effect may require the addition of a polymeric thinner to offset the viscosity increase caused by the fluid loss polymer.

Advanced molecule engineering offers yet another synthetic fluid loss polymer which does not show the disadvantage of a secondary viscosifying effect. Due to an average molecular weight of only 200,000 daltons this polymer neither viscosifies nor thins the mud. Neutral behavior on mud rheology is the result. This property makes the polymer more desirable for high density, low salt systems. Thermal stability and hardness tolerance of this product are comparable to or exceed those of high molecular weight, AMPS based copolymers [14].

The conclusions made in the development of high temperature, hardness tolerant fluid loss polymers have been utilized successfully in the development of polymers for other saline environments, such as completion and workover brines or enhanced oil recovery.

Deflocculants

Short-chain, synthetic polymers are used to reduce mud viscosity at high temperature. These deflocculants are also often referred to as dispersants or thinners.

Weight-average molecular weights of deflocculants are between 1,000 and 15,000 daltons, with 2,000 - 6,000 daltons being more typical.

In calcium-free mud systems, acrylate or acrylamide based thinners are being used. Besides their limitation with hardness tolerance, they do not always respond satisfactory in muds with a high content of active solids. When using polyacrylates in such systems, a common experience is that constant treatment of the mud is required to prevent the 10 minute gel strength from continually rising. Obviously, polyacrylates do not effectively inhibit the high temperature gelation of clays.

The introduction of maleic anhydride copolymers has solved this problem [15, 16]. At small dosage (e.g. 0.2 - 1 ppb/0,6 - 3 kg/dm³), they show a lasting effect on plastic viscosity and gel strengths, indicating that they effectively inhibit the high temperature gelation of bentonite and clay. This unique property and their improved hardness tolerance distinguishes them from conventional acrylate based thinners.

Examples of commercially available maleic anhydride deflocculants are:

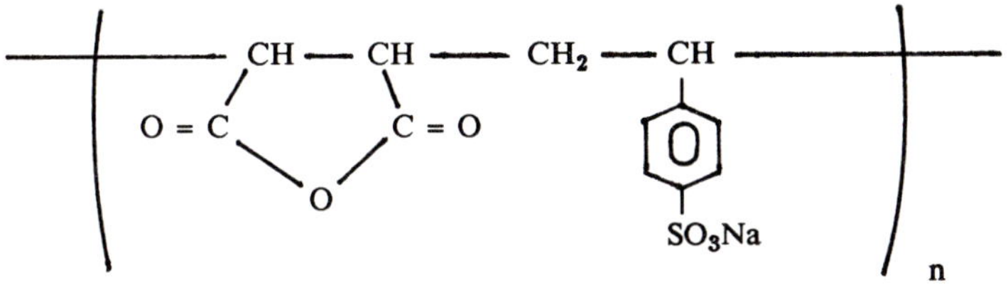

CH — CH — CH_2 — CH
O = C C = O
O
SO_3Na
n

maleic anhydride-styrenesulfonate copolymer

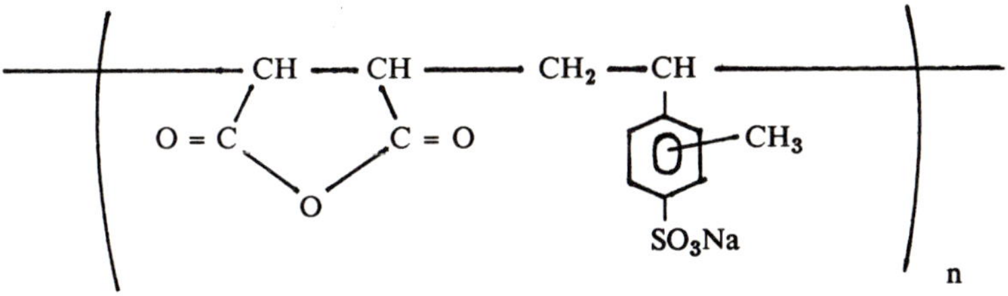

CH — CH — CH_2 — CH
O = C C = O
O
CH_3
SO_3Na
n

maleic anhydride-sulfonated vinyl toluene copolymer

Especially under environments of high active solids and hardness content, maleic anhydride copolymers have been instrumental in the success of water-based muds in deep drilling.

It should be noted that HT-deflocculants often not only reduce viscosity, but assist the high molecular weight fluid loss polymers in lowering filtrates. The dispersion of clays seems to result in a less permeable filter-cake. Improved fluid loss is observed. This effect is sometimes reported as synergism between fluid loss polymers and deflocculants. For example, in a recent study a combination of high to medium molecular weight, sulfonated fluid loss polymers together with a low molecular weight, acrylate/acrylamide thinner was found to provide a fresh water mud system with thermal stability to 290°C [17].

Polymer extenders

Thermal degradation of polymers in drilling fluids occurs by oxidation processes (initiated by oxygen radicals) and by a temperature induced oscillation of the molecule chain which ultimately results in breaching of carbon-carbon bonds. While the oscillation mechanism is determined by the molecular structure of the polymer and cannot be influenced by addition of mud chemicals, the oxidative degradation can be slowed down by certain chemicals.

Oxygen scavengers have been widely used to enhance the thermal stability of polymers. Examples of scavengers commonly used in drilling fluids are ammonium bisulfite, thiourea, 2-mercapto benzothiazole and aniline.

Recently, sodium and potassium formate also have been reported to be thermal stabilizers for organic polymers [18]. The formate seems to act as an effective free radical quencher.

The use of polymer extenders should be considered if a polymer reaches its temperature limit shortly before completing a certain well interval. In this situation, it is more economical to add the extender instead of treating the mud with a more stable (and usually much more expensive) polymer.

Shear-thinning mud system

An interesting new development in high temperature stable, water-based mud technology is the introduction of an inorganic, synthetic poly(metal hydroxy-chloride) compound, also referred to as mixed metal hydroxide [19]. A commercial product shows the following chemistry:

$$([MgAl(OH)_{4.7}]_{3.3}{}^{+}Cl^{-})_n$$

The poly(magnesium aluminum hydroxy) cation of this compound interacts with bentonite resulting in an extremely shear-thinning and fragile gel. Excellent carrying capacity (e.g. Yield Point 80-100) at low plastic viscosity (e.g. PV 5-10) is achieved. The fluid has been shown to provide effective cuttings removal, borehole stability, inhibition of sloughing and disintegrating shales and high rate of penetration [20]. Addition of mono- or divalent salts such as NaCl, KCl or $CaCl_2$, do not destroy the shear-thinning characteristic of the fluid. Because of the cationic interaction with bentonite, nonionic fluid loss polymers are best suited for this mud system. A thermal stability exceeding 180°C has been found. The poly(magnesium aluminum hydroxy chloride) seems to offer an alternate drilling fluid system for deviated HP/HT wells or geothermal drilling.

4 FIELD PROBLEMS IN HP/HT DRILLING

The following discusses typical field problems related to HP/HT drilling. Common field practice to overcome these problems is presented.

Solids control

High wellbore temperatures and drill solids buildup can result in severe mud problems. For example, low gravity solids are notorious for excessive mud viscosity. Costly mud treatments and dilution are the result.

As a rule, low gravity solids should be kept below 6%. A particle size analyzer may be used in the mud lab on the rig to monitor the efficiency of the solids control equipment.

The use of premium solids control equipment is a key to success in drilling deep, hot wells. Good results are reported from utilizing a cascading shaker arrangement (finest screens 180 mesh), a mud cleaner and two centrifuges [4]. Economical aspects of solids control are discussed in [21].

Contaminant removal

Prevalent contaminants in deep wells are gases (CO_2, H_2S) and formation brines. These influxes may contain large amounts of NaCl, $CaCl_2$ or $MgCl_2$.

CO_2 contamination results in a drop of all alkalinities plus an increase in mf and the 10 min. gel strength. Recommended treatment is to add lime to achieve the desired pm and KOH or NaOH to the desired pf. Carbonate solids in the mud are removed with fine shaker screens. It is a good idea to monitor particle size distribution of the mud solids to ensure that fine solids do not become excessive.

To eliminate the source it should be investigated whether the CO_2 is indigenous from the formation or caused by decomposition of mud chemicals such as lignite.

Apart from its toxicity, H_2S causes high mud viscosities. The Garrett Gas Train should be utilized to determine H_2S in the mud. First action is to raise the alkalinities of the mud by adding lime, KOH or NaOH. Iron or zinc based scavengers are then used to neutralize H_2S by forming insoluble sulfides. In HP/HT drilling, these scavengers must be added carefully. At high temperatures, excessive usage and rapid addition of zinc scavengers has been reported to result in high viscosities [4]. Pilot testing prior to drilling a potential H_2S interval is recommended.

Brine influxes usually increase all rheologies. If contamination is temporary, raise pH to 11-12 with NaOH or KOH to precipitate calcium/magnesium. Treatment with fluid loss polymers and deflocculants is required to maintain filtration and rheology. If severe contamination with high hardness brine is anticipated, the mud should be built around electrolyte tolerant polymers to begin with. In case the influx persists and heavily dilutes the mud, circulation of high viscosity (e.g. HEC) pills may permit drilling further. If this fails, the formation brine must be conditioned to provide some filter-cake and for fluid loss control. On a particular well, a 5 m^3/hr brine influx containing 100,000 ppm chlorides and 20,000 ppm calcium was treated with 13 ppb (37 kg/dm^3) prehydrated bentonite and 10 ppb (28 kg/dm^3) of the sulfonated polymer described in [14]. With the brine being used as the drilling fluid, more than 1000 ft (300 m) of hole was drilled successfully to the next casing point.

Lubrication

The increasing number of deep high angle wells requires drilling fluids with improved lubrication.

Among the common lubricants used in water-based muds is gilsonite, a naturally occurring bitumen.It also improves fluid loss and is non-damaging to the formation [22]. Some vendors offer specially treated, sulfonated asphalt compounds. Their advantage is better dispersability in water [23]. However, all asphalt based lubricants spoil the picture of an environmentally safe water-based fluid. In ecologically sensitive areas, their use is sometimes disputed. An interesting new development is the use of spheric copolymer beads for lubrication [24]. Products with particle sizes of 0,25 and 0,65 mm are available. Due to high cost, they are usually recovered from the mud. Recently, modified waxes, triglycerides and polyglycols also have been applied for lubrication.

Despite these existing products it is well recognized that, relative to lubricity, water-based muds still cannot compete with oil-based muds. Industry certainly needs improved, non-toxic and high temperature stable lubricants for water-based fluids. This offers opportunities for manufacturers of new materials for this purpose.

5 CONCLUSIONS

1. Design criteria for water-based muds used in HP/HT wells are:

   - use Wyoming bentonite at less than 3 ppb (9 kg/dm^3) concentration
   - prefer mud additives with only one primary function

2. Environmentally safe water-based muds functioning in excess of 260°C can be built on synthetic fluid loss polymers and polymetric deflocculants.

3. Synthetic poly (metal hydroxy chlorides) allow the formulation of highly shear-thinning, high temperature stable water-based fluids.

4. Effective solids control equipment is a key factor to avoid mud problems. Low gravity solids should be kept below 6%.

5. Industry needs improved non-toxic lubricants for water-based muds to compete with oil-based fluids.

Acknowledgement

The author would like to thank SKW Trostberg AG for permission to publish this paper. Special thanks go to many mud engineers. Their hands-on experience was a valuable help in writing this paper. E. Fröschl's patience in preparing this manuscript is also acknowledged.

REFERENCES

[1] Noroil, 1990, 1, 11-13

[2] S. Braden, 'Oil muds score high in Anadarko evaluation', OGJ., 1987, June 1, 25-30

[3] J.C. Abdon et al, 'The development of a deflocculated polymer mud for HTHP drilling', SPE 17 924, presented at the SPE Middle East Oil Technical Conference & Exhibition, Bahrain, March 11-14, 1989

[4] R.K. Mitchell et al, 'Design and application of a high-temperature mud system for hostile environments', SPE paper 20 436, presented at the 65th SPE Annual Technical Conference & Exhibition, New Orleans, September 23-26, 1990

[5] J.V. Fisk et al, 'The filterability of drilling fluids', SPE 20 438, presented at the 65th SPE Annual Technical Conference & Exhibition, New Orleans, September 23-26, 1990

[6] B.G. Chesser et al, 'Dynamic and static filtrate loss techniques for monitoring filter cake quality improves drilling performance', SPE 20 439, presented at the 65th SPE Annual Technical Conference & Exhibition, New Orleans, September 23-26, 1990

[7] H. Krause, 'Requirements of high temperature water-based drilling fluids from the view of drilling engineers', Oil Gas Europ. Mag., 1983, 1, 43-49

[8] US Pat. No. 4,678,591; assigned to Nalco Chemical Co.

[9] J. Dorman, 'Heat and electrolyte stable drilling fluids: results of development and application', 21st Petroleum Conference and Exhibition, Siofok, Hungary, Sept. 26-30, 1990

[10] W.R. Clements et al, 'Electrolyte-tolerant polymers for high-temperature drilling fluids', SPE 13 614, presented at the SPE California Regional Meeting, Bakersfield, March 27-29, 1985

[11] US Pat. No. 2,718,497, assigned to Union Oil Co. of California

[12] WO 83/02449; assigned to Casella, Hoechst and Dresser Industries

[13] US Pat. No. 4,268,400; assigned to Milchem Inc.

[14] J.P. Plank and J.V. Hamberger, 'Field experience with a novel calcium-tolerant fluid loss additive for drilling muds', SPE 18 372, presented at the SPE European Petroleum Conference, London, October 16-19, 1988

[15] US Pat No. 3,730,900; assigned to Milchem Inc.

[16] US Pat No. 4,518,510; assigned to National Starch and Chemical Co.

[17] K. Wähner, 'Combinations improve drilling fluid chemicals performance', paper presented at SKW Trostberg AG 's Drilling Fluid Symposium, Trostberg, Germany, Dec. 17-19, 1990

[18] US Pat No. 4,900,457; assigned to Shell Oil

[19] J.L. Burba et al, 'Laboratory and field evaluations of novel inorganic drilling fluid additive', IADC/SPE 17 198, presented at the 1988 IADC/SPE Drilling Conference, Dallas, Feb. 28 - March 2, 1988

[20] J.L. Burba et al, 'Field evaluations confirm superior benefits of MMLHC fluid system on hole cleaning, borehole stability, and rate of penetration', IADC/SPE 19 956, presented at the 1990 IADC/SPE Drilling Conference, Houston, Feb. 27-March 2, 1990

[21] J. Dorman and J. Gócs, 'Technological and economical aspects of mechanical solids control in the application of up to date drilling fluids', 21st Petroleum Conference and Exhibition, Siófok, Hungary, Sept. 26-30, 1990

[22] N. Davis II and C.E. Tooman, 'New laboratory tests evaluate the effectiveness of gilsonite as a borehole stabilizer', IADC/SPE 17 203, presented at the 1988 IADC/SPE Drilling Conference, Dallas, Feb. 28-March 2, 1988

[23] US Pat No. 3,028,333; assigned to Phillips Petroleum Co.

[24] US Pat No. 4,063,603; assigned to Sun Drilling Products

Correlation Between Interfacial Tension Values, Micellar Structures, and the Macroscopic Properties of Surfactant Systems

H. Hoffmann

LEHRSTUHL FÜR PHYSIKALISCHE CHEMIE I, UNIVERSITÄT BAYREUTH, UNIVERSITÄTSSTRASSE 30, W-8580 BAYREUTH, GERMANY

Abstract:

Experimental results on alkyldimethylaminoxide/ cosurfactant/water systems are used to demonstrate that the value of the interfacial tension of such solutions against a hydrocarbon reflects the packing of the surfactant in the interfacial film at the hydrocarbon water interface and at the same time at the micellar interface. The values of the interfacial tension are thus correlated with the type of micelles in the surfactant solution free of hydrocarbon. With increasing cosurfactant concentration the value of the interfacial tension decreases continuously from a few mN/m to values of the order of 10^{-3} mN/m. The high values correspond to globular micelles while the low values correspond to bilayer-type structures which are planar on a small scale. As a consequence of the variation of the radius of curvature of the micellar interface, the surfactant solutions will undergo several phase transitions with increasing cosurfactant concentration. The single phase regions which are observed with increasing cosurfactant concentration are: L_1, L_1^*, L_α and L_3. All single phase regions are

separated from each other by two phase regions.
The viscoelastic properties of the various phases differ widely. The L_1-phases can be highly viscoelastic. The shear modulus in these phases is more or less independent of the chemistry of the surfactant and scales with a power law of the concentration. The L_1*-phases can also be very viscous. They do not show the recoil effect. The shear modulus of these phases is lower than the shear modulus in an equimolar L_1-phase and follows a different scaling behavior. Both the L_1 and the L_1*-phase have a finite structural relaxation time. The L_α-phase on the other hand has a small yield stress value. The rheological properties of the L_3-phase depend very much on the chain length of the cosurfactant. For short chain cosurfactants, the phases have a viscosity which is in the range of the solvent viscosity. For intermediate chain length cosurfactants ($C_xH_{2x+1}OH$ with x > 6) the phases are of much higher viscosity. The storage and the loss moduli in these phases are very similar and independent of the frequency in a wide frequency range.

A <u>Introduction:</u>

When an aqueous micellar solution is contacted with a hydrocarbon, a surfactant film forms at the hydrocarbon/water interphase. In equilibrium the surfactant molecules at the interface and in the micelles will have the same chemical potential. The interaction energy of each surfactant molecules with its local environment must then also be the same. It is mainly for this reason that both interfaces should have very similiar properties. The packing of a surfactant film at the interface is reflected in the value of the interfacial tension. Films which are formed from surfactants which require a large interfacial area cannot be as densely packed as films from surfactants which require a smaller interfacial area. Generally it can be said that the more dense the interfacial film is packed, the lower will be the interfacial tension of that interface because a denser film exerts a higher surface pressure. A more quantitative description about this problem has recently been given by Borkovec [1]. Results which show the similarity of the packing in surfactant films and in micelles have been given by H. Hoffmann (2). The surface activity of a surfactant molecule is expressed in three parameters: the critical micellar concentration, the required area per headgroup at the interface and the value of the interfacial tension at the cmc. If the surface activity of a molecule is changed all three parameters will change. This is particularly evident when these three parameters are compared for surfactants with the same hydrocarbon chain but with different polar headgroups. The area per headgroup controls the curvature of a micelle and hence the shape of a micelle. We thus have to expect a correlation between the interfacial tension and the shape of the micelles in the binary surfactant solution without the presence of hydrocarbon.

In the present article we will further corroborate on this interrelation and will show how we can use it to make the most use of interfacial tension measurements for the prediction of properties of micellar solutions. Middle phase microemulsions with large quantities of oil and water and only a small fraction of surfactant are formed when the interfacial tension of the oil phase against the micellar solution is very low and in the range of 10^{-3} mN/m [3]. Interfacial tension measurements are therefore generally used to optimize the conditions for the formation of microemulsions [4]. If in an optimized microemulsion the surfactant concentration is increased, the microemulsion phase is generally transformed into a lamellar phase which contains the solubilized hydrocarbon sandwiched between two surfactant layers [5]. The formation of the lamellar phase is an indication that the surfactants in these situations like to form planar layers. For some surfactant systems the hydrocarbon in the lamellar phase can be completely removed without destroying the lamellar phase (6). For nonoptimized conditions the surface tension values are much higher. The micelles which are present in the aqueous phase under such conditions are usually globules. Obviously their headgroup requires a much larger area than the hydrocarbon chain in the film and the interface is therefore strongly curved. An interfacial tension and the curvature of a micellar film show the same behavior and are correlated (2,7).

B Results

1. Interfacial and surface tensions

With increasing cosurfactant concentration occurs a transition of globules to rodlike and of rodlike to disklike micelles. This is schematically shown in Fig. 1. Detailed measurements are shown in Fig. 2 for aqueous solutions of alkyldimethylaminoxide with increasing amounts of cosurfactants.

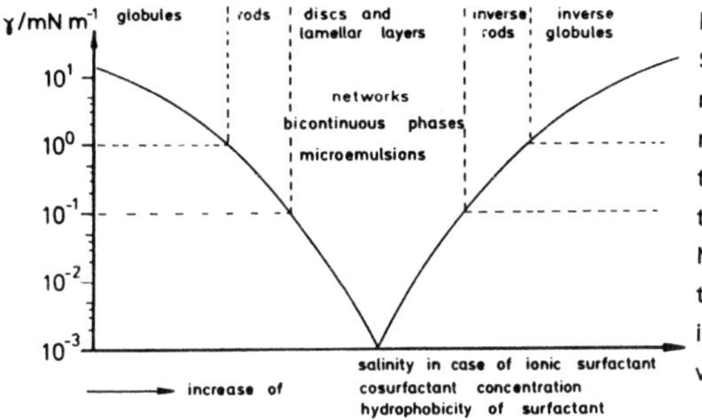

Fig. 1
Schematic drawing of the various micellar structures that exist in micellar solutions when the solution have different interfacial tensions against a hydrocarbon. Note however that the different types of micelles are present only if the surfactant is not saturated with hydrocarbon.

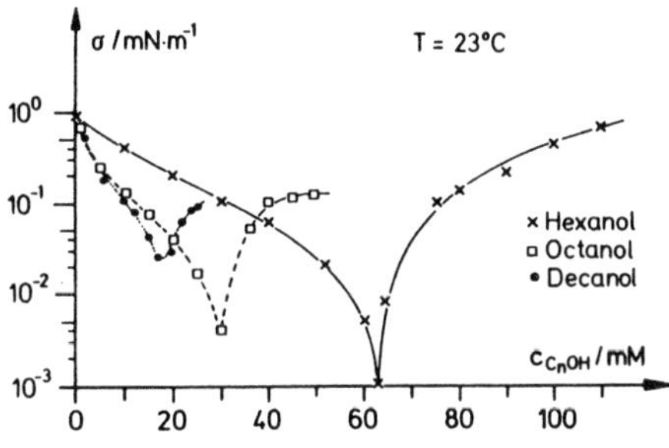

Fig. 2
The interfacial tension of a 30mM $C_{14}DMAO$ solution against the concentration of n-alcohol with a different chain length. Note that the position of the minimum is shifted to smaller concentrations with inceasing chain length of the cosurfactant while the minima of the interfacial tension are increasing in value.

Very often the surface tension of the micellar solution follows the same trend as does the interfacial tension of this solution against a hydrocarbon. There are however situations in which the changes of surface and interfacial tension do not show the same tendency and at present it is not clear why this occurs. The interfacial tension therefore seems to be a much better indicator for the state of the micellar solution and its macroscopic properties than the surface tension.

In Fig. 3 results are shown in which both interfacial tensions and surface tension measurements indicate the same tendency and clearly show the large synergism which occurs when alkylaminoxides and SDS are mixed. We obtain deep minima as a function of the mixing ratio for both parameters.

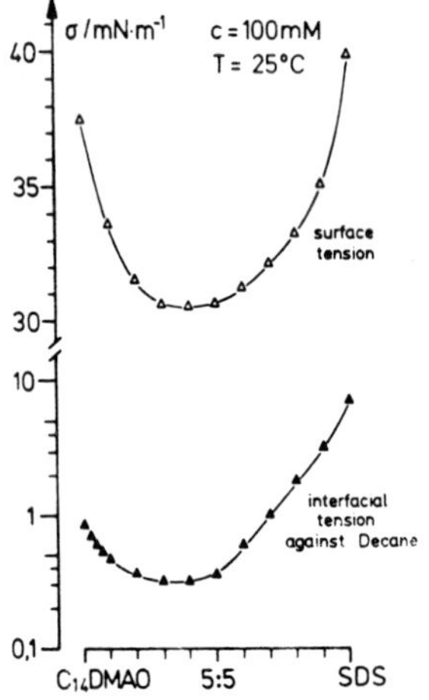

Fig. 3
Comparison of surface tension and interfacial tension measurements for surfactant mixtures of C14DMAO and SDS. Note that both measurements indicate the large synergistic effects for the mixtures.

Different applications of surfactants require different interfacial conditions. The best flooding results are obtained under optimized conditions for the interfacial tensions. This has been recognized for many years and many systems have been optimized for this application. If on the other hand we like to use surfactant systems to build up high viscosities, polyhedral foam structures or nematic phases, we cannot use surfactants which have been optimized for low interfacial tensions. Low interfacial tensions always are related to microdroplets and aggregates which can easily be deformed. High viscosities are related with hard, nondeformable particles and therefore with higher interfacial tensions. If we wish to have high viscosities in low concentrated surfactants the surfactants must form networks. Networks can be formed from entangled rodlike micelles or from entangled sheetlike structures. Rodlike structures are formed in micellar solutions which have interfacial tension values in the range of 0.2 - 2 mN/m. Disklike micelles are formed for lower surface tension values.

2. Rheological results of dilute micellar phases

With increasing cosurfactant concentration a dilute micellar solution undergoes several phase transitions in order to adjust the radius of curvature of the micelles to the packing conditions. A phase diagram for a 100 mM solution C_{14}DMAO with increasing hexanol concentration is shown in Fig. (4). For the single phase regions we find the sequence L_1, L_1^*, L_α and L_3. In the L_1-phase we generally pass with increasing cosurfactant concentration the sphere rod transition. The transition is usually indicated by a rather abrupt increase of the viscosity. The sequence of the phases will be the same when different cosurfactants are used or when some of the zwitterionic surfactant is replaced by an ionic surfactant. The four phases which we observe in Fig. (4) are thus all the various phases we can find in dilute micellar solutions. We will now proceed and look at their rheological properties in detail.

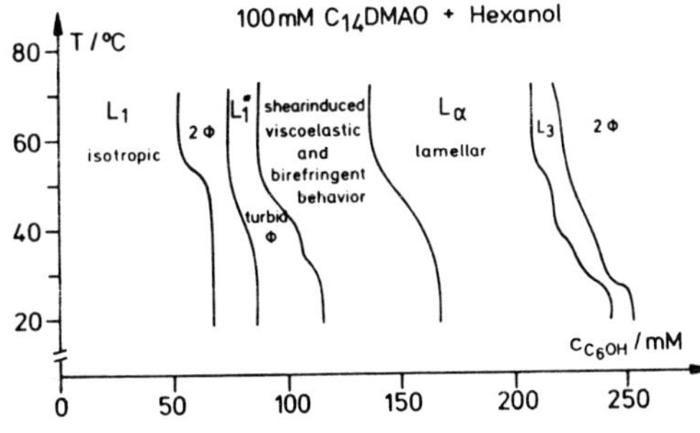

Fig. 4
Sequence of the different phases of a 100mM C_{14}DMAO solution with increasing hexanol concentration

2.1 Viscous systems of the L_1-phase

Many fluids or solutions which are used in the oil industry must have a high viscosity. In some cases it is necessary to adjust the viscosity of the pusher fluid to the level of the oil when the fluid is used to push out oil from a reservoir. In many cases polymer solutions are used for this purpose [8]. High viscosities can also be obtained with surfactant systems. Actually it is possible to adjust a zero-shear viscosity to any desired value between 1 and 10^5 mPas at a volume fraction between 1 - 2 % of surfactant. The high viscosities can be obtained at low and at high salinities and at low temperature and at temperatures which may be as high as 100 °C. High viscous solutions can be found in various positions in binary and ternary phase diagrams in surfactant systems. In the best studied systems the viscoelastic behavior comes from an entanglement network of rod-like micelles (9). Such systems can be prepared from anionic, zwitterionic and cationic surfactants. Some typical results will be presented here.

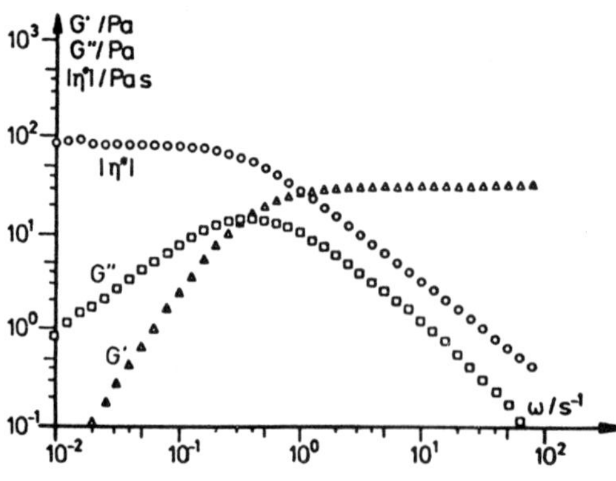

Fig. 5
Rheological results for a 100mM CPyCl solution in the presence of 60mM NaSalicylate. Oscillating measurements: complex viscosity (o); storage modulus G´(Δ) and loss modulus G´´(□) against the frequency.

In Fig. 5 the dynamic viscosity and the storage and loss modulus of a viscoelastic solution are plotted against the angular frequency. This system has the remarkable feature that it behaves like a simple Maxwell fluid, that means the rheological properties can be described for a large frequency or shear rate range with a single shear modulus and a single relaxation time. Such a simple behavior is normally not found for polymer solutions (10). The results of Fig. 5 show that the time constant of this fluid is more than seconds and that the storage modulus is much higher than the loss modulus over a wide frequency range. This means that the liquid can easily be made to recoil by a sudden rotation of the vessel containing the sample. The shear modulus in combination with the mass of the fluid results in oscillations which have a shorter period than the structural relaxation time. The system can therefore oscillate several times before the energy is lost by friction forces.

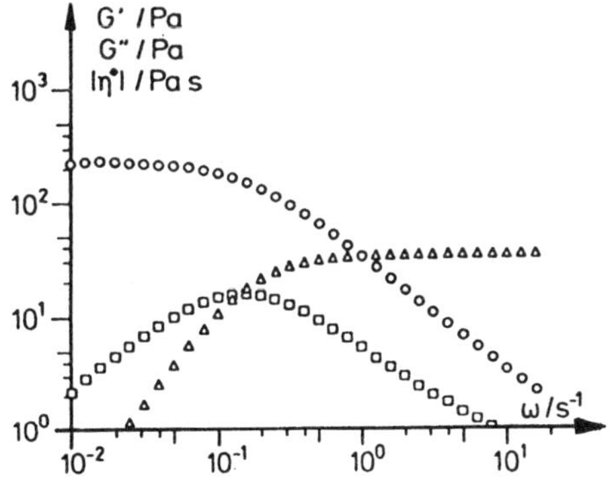

Fig. 6
Rheological results for a 100mM solution of Oleyldimethylaminoxide. Symbols are the same as in Fig.5.

Fig. 6 shows similar results for the zwitterionic system oleyldimethylaminoxide. All viscoelastic systems in the L_1-phase have a very similar flow behavior. In all systems the shear modulus scales with the same power law with the concentration (Fig. 7). There are differences however in the behavior of the structural relaxation time with the concentration (Fig. 8). The zwitterionic system shows the very remarkable behavior that the structural relaxation time is practically independent of the concentration. It is also noteworthy that the structural relaxation time for this system can even be made longer by adding a few percent of an ionic surfactant. The zero shear viscosity is then increasing further and the time constant becomes so long that it is difficult to tell whether the system might have a yield stress.

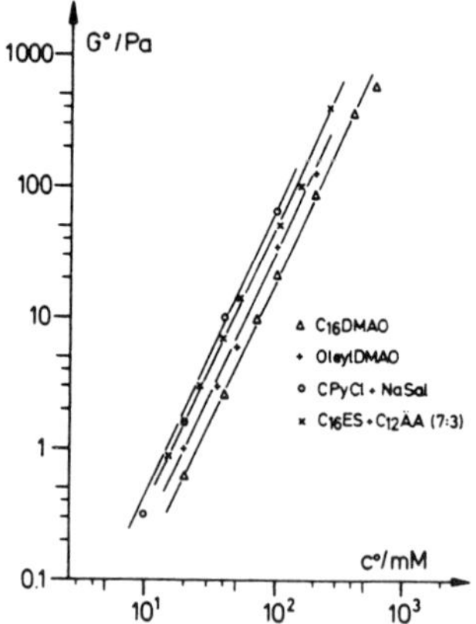

Fig. 7
A double logarithmic plot of the shear modulus for different surfactant systems against the concentration. Note that in all four systems the modulus follows the same scaling behavior.

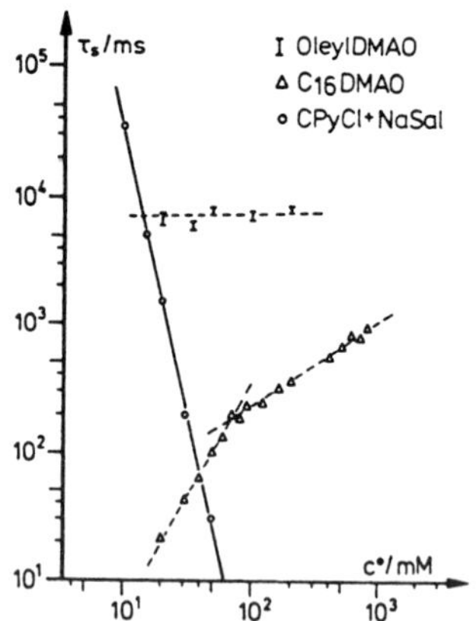

Fig. 8
A double logarithmic plot of the longest structural relaxation times for these different viscoelastic surfactant systems against the concentration.

2.2 Viscous systems in the L_1*-phase

The L_1*-phase in Fig. (9) is also highly viscous but it does not show the recoil effect. The solutions have very long structural relaxation times as is obvious from Fig. 9 and a storage and a loss modulus which are practically independent of the frequency over a large range. From these data it is not clear why the solution do not show the recoil effect of entrapped bubbles. This is easier to understand when the shear modulus in such a solution is measured as a function of the deformation. In solutions which recoil the solution can be 100 % deformed before the modulus decreases while the L_1* solutions can only be deformed by about 10 %. The shear stress due to the shear flow can then not become high enough to reverse the flow but the system rather behaves like a system with a yield stress when a shear stress is applied which is larger than the yield stress.

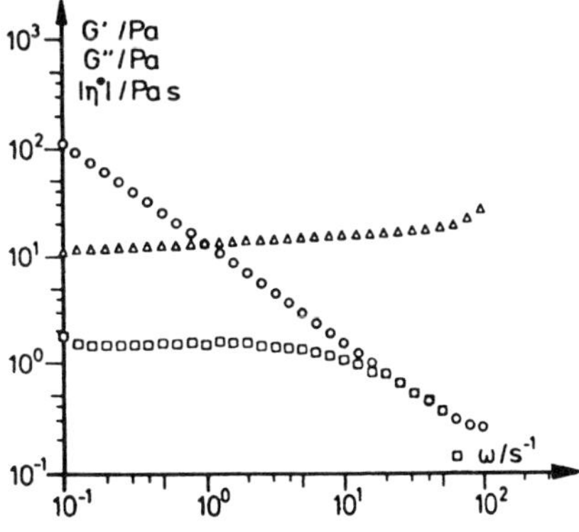

Fig. 9
Rheological data in a solution of 100 mM $C_{14}DMAO$ + 100 mM hexanol (L_1*- phase). Note that the storage modulus is practically independent of the oscillating frequency. The structural relaxation time is longer than 10 seconds. The amplitude of the deformation was 3 %.

2.3 The rheological properties of the L_α-phases

In the rheology of liquid crystalline phases one can distinguish five different viscosity coefficients. Well defined measurements can only be made on completly ordered samples. For the comparison of the properties the different phases in the discussed diagram this will not be necessary and the experiments on the L_α-phases were done in the same way as for the other phases. For the 100 mM C_{14}DMAO we could still measure a shear modulus. Its value was however much smaller than the value of the modulus of the L_1 or even the L_1*-phase. The shear modulus can however be increased to the same level as the one in the L_1*-phase by replacing 10 % of the C_{14}DMAO with a cationic surfactant. Such L_α-phases are highly elastic and perfectly clear. They show birefringence between crossed polarisers. Even for low surfactant concentrations they have a remarkably high yield stress. The systems begin to flow only if the shear stress which is applied to the system exceeds a certain level. This is shown in Figs.(10,11). The yield stress of this solution is high enough to prevent bubbles, which are trapped in the system, from rising.

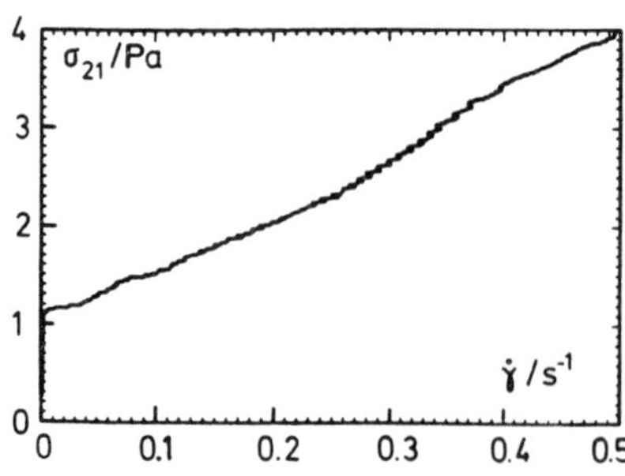

Fig. 10
Plot of the shear stress against the shear rate for the system in Fig. 11. Note that the system begins to flow when the shear stress is higher than 1 Pa.

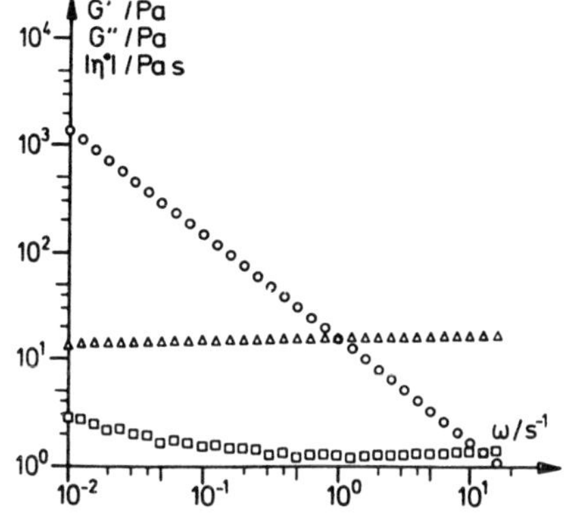

Fig. 11
Rheological data of the Lα-phase. The system consists of 90mM $C_{14}DMAO$, 50mM $C_{14}TMABr$ and 200mM C_6OH.

2.4 The rheological properties of the L_3 and L_3*-phase

For cosurfactant concentrations which are a few percent higher than in the L_α-phase one finds in the ternary $C_{14}DMAO/C_6OH/H_2O$ system an isotropic low viscous phase. This phase occurs only in a very narrow cosurfactant/surfactant ratio but over a wide surfactant concentration. In the triangular phase diagram the existence region lies directly adjacent to the L_α-phase. It is widely assumed now that the phase consists of a bicontinuous spongelike structure (11). It was recently demonstrated however that the dynamic properties of this phase can quantitatively be described on the basis of nonconnected disklike micelles (12). The disklike micelles seem to have about the same diameter as the interlamellar spacing between the bilayers in the L_α-phase. On the basis of these structures it was even possible to account in a quantitative manner for the zero-shear viscosity of the L_3-phase. These measurements appear to make it doubtful that the phase really consists of a bicontinuous structure. In any case, the L_3-phase responds to a perturbation like a phase which consists of disklike particles which can freely rotate and diffuse. If the disklike micelles should be connected, the resulting structures must then be so flexible that they can adjust so quickly on dimensions of scale of the disks as if they were not connected. The properties of the L_3-phase are remarkably insensitive to the chain length of the surfactant as is shown in Fig.(12). The viscosity data would be difficult to understand if the L_3-phase has a sponge like structure but it would make sense if it is formed from discrete disks.

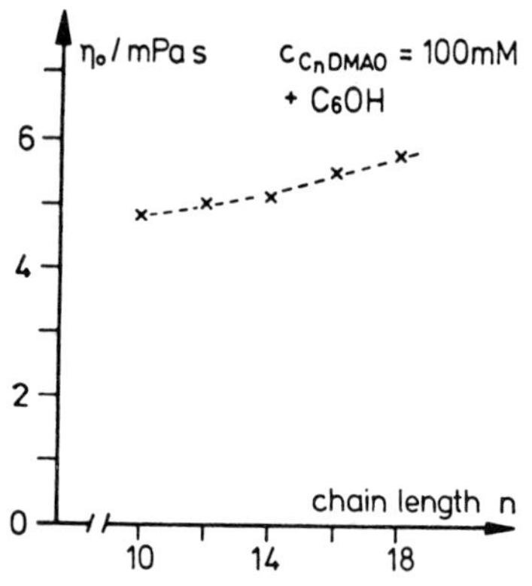

Fig. 12
The zero shear viscosity of the L3-phase for 100mM surfactant solutions with hexanol as a function of the alkyl chainlength of the surfactant at 25°C.

The position of the L_3-phase in the ternary phase diagram and their properties depend however strongly on the chain length of the cosurfactant. The L_3 and L_α-phases which were formed for a 100 mM C_{14}DMAO for different cosurfactants are shown in Fig. (13). For intermediate chain length cosurfactants one observes actually two L_α and L_3-phases. The one which occurs at the high cosurfactant/surfactant ratio has similar properties as the L_3-phase which has just been discussed. The L_3-phases which occur at the low cosurfactant/surfactant ratio are different. They have a higher turbidity and show a much stronger flow birefringence than the normal L_3-phases. They were called L_3*-phases in order to distinguish them from the L_3-phases (13).

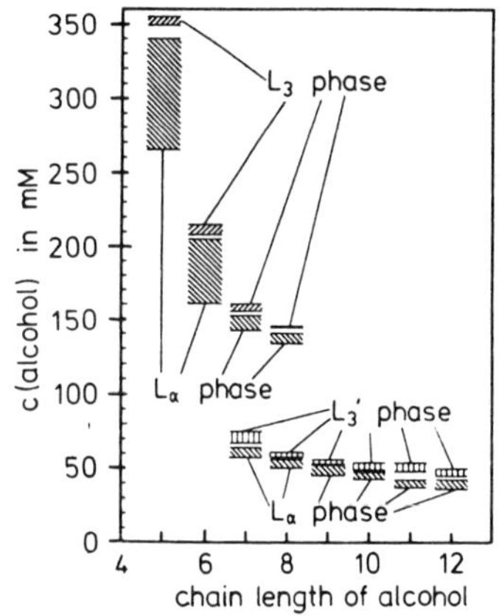

Fig. 13
Different phases of a 100mM solution with increasing cosurfactant concentration as a function of the chainlength of the alcohol at 45°C.

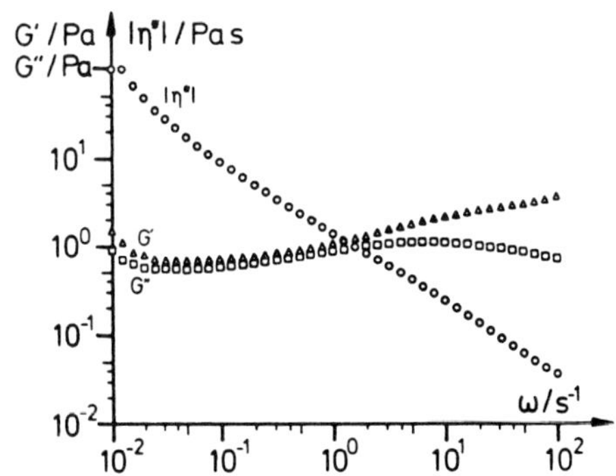

Fig. 14
Complex viscosity ($|\eta^*|$), storage modulus (G´) and loss modulus (G´´) for a solution located in the L_3´-phase at 45°C.(100mM C_{14}DMAO + 60mM 1-octanol).

These systems also have high viscosities at low frequencies or shear rates and storage and loss moduli which practically are independent of frequency Fig. (14). In addition these systems have the peculiarity that the two moduli are practically the same. Such conditions have previously been observed in gellifying systems at the Sol/Gel transitions [14]. The last two described systems have such long time constants that it becomes experimentally difficult to determine them and to decide whether the systems have a real yield stress value. Their rheological properties are very similar as the rheological properties of drilling fluids which consist of dispersions of clay platelets [15]. The microstructure of the L_3* and the L_1*-phase might actually be very similar to the microstructure of the drilling fluid . On a small scale the structures are platelike.

C. Conclusions:

The value of the interfacial tension between a surfactant solution above the CMC and an oil phase is a most sensitive parameter of the state of the interfacial film. The lower this value is, the more densely packed is the film. The packing in the film at the oil/water interface reflects also the packing of the micellar film. The value of the interfacial tension can thus be used as a scale for the curvature of micellar aggregates. A low interfacial tension around the surface tension minimum stands for a planar value while a high value indicates a strongly curved radius. The shape of the micelles in binary surfactant systems can thus be predicted from the value of the interfacial tension. Solutions with small globular particles have interfacial tension against a hydrocarbon of typically a few mN/m while optimized microemulsions are around 10^{-3} mN/m, that is three orders of magnitude lower.
At intermediate interfacial tensions between 0.1 and 1

mN/m rod- and disklike micelles do exist in the binary surfactant solutions. Such solutions can have extremly high zero shear viscosities in the range of 10^5 mPas for concentrations which contain only a few percent of surfactants.
Viscoelastic surfactant systems can be formed from anionic, cationic and from zwitterionic surfactants. The very long structural relaxation times in these systems probably comes from a hydrophobic interface of the micelles which cause adhesion between different micelles. Viscoelastic systems at low concentrations are found in the L_1, the L_1*, the L_α and the L_3-phases. The rheological properties of these systems show however marked differences. Viscoelastic L_1-phases show the recoil effect while the L_1* and the L_3-phases do not. The difference in the behavior lies in the different deformability of the network. The networks in the L_1-phase can be strongly stretched without a decrease of the modulus while the shear modulus decreases rapidly with the stretching for the other two phases. L_α-phases at low surfactant concentrations can have a yield stress which is high enough to prevent the rising of small entrapped bubbles.

References:

1. M. Borkovec, J. Chem. Phys. 91 (10), November 1989, 6268-6281.
2. H. Hoffmann, Progr. Colloid Polymer Sci. 83:16-28 (1990).
3a. D. Langevin, Colloids and Surfaces, 19 (1986) 159-170.
3b. K. Shinoda and Y. Shibata, Colloids and

Surfaces 19 (1986) 185-196

4. R. S. Schechter, Journal of Colloid and Interface Science, Vol 89, No. 1, Sept. 1982.

5a. Kahlweit, Angew. Chem. 97 (1985) 655-669.

5b. K. Shinoda and B. Lindman, Langmuir 1987, 3, 135

6. C. Thunig,H. Hoffmann and G. Platz Progr. Colloid Polymer Sci. 79: 1-11 (1989).

7. G. Oetter and H. Hoffmann, J. Dispersion Science and Technology, 9 (5 & 6), 459-492 (1988 - 89).

8. Improved oil recovery by surfactants and polymer flooding. D. O. Shah and R. S. Schechter, Academic Press, 1978, ISBN: 0-12-64/750-4

9. H. Rehage and H. Hoffmann, Journal of Physical Chemistry, 1988, 92, 4712.

10. S. J. Candau, Langmuir, 1989, 5, 1225.

11. D. Anderson, H. Wennerström, U. Olsson, J. Phys. Chem. 1989, 93, 4243

12. C. A. Miller, M. Gradzielski, H. Hoffmann, U. Krämer, C. Thunig, Coll. Polym. Sci. 268, 1066 (1990)

13. C. A. Miller, M. Gradzielski, H. Hoffmann, U. Krämer and C. Thunig, Progr. Colloid Polym. Sci., Proceedings of the ECIS conference 1990, in Press

14. N. H. Winter, Macromolecules 1986, 19, 2146

15. R. Sokus and Th. F. Tadros, J. Coll. Interf. Sci., Vol. 132, 62, 1989

Aspects of Saline and Heat Stable Polymers for Enhanced Oil Recovery

S.P. von Halasz

HOECHST AG, TECHNICAL SERVICES AND DEVELOPMENT, OILFIELD CHEMICALS, PO BOX 80 03 20, D-6230 FRANKFURT AM MAIN 80, GERMANY

1 INTRODUCTION

In many respects the chemical industry and the oil & gas industry work closely together. On the one hand oil and gas are today the most important basis for the production of organic chemicals worldwide. The chemical industry produces its current product range with its high volumes, high quality and acceptable costs only on the basis of oil and gas. On the other hand the chemical industry supplies - albeit in quite small amounts - commodities and specialities to the oil & gas industry. In other words: Chemistry lends a hand in drilling, production and refining in obtaining more oil and gas. Some of these chemicals help directly in producing more oil from the reservoirs by enhanced oil recovery methods (EOR), also known as tertiary oil recovery.

In Germany, which with an oil production of about 3.6 mty (72,000 bbl/d) is one of the world's smaller producers, EOR processes in 1989 accounted for approx. 15% of total domestic crude oil production. Some 85% of this EOR-produced crude was the result of steam and hot water injection, - the remaining 15% was recovered by means of chemical flooding processes, mainly by polymer flooding methods.

As recently published by the German Oil and Gas Association[1] the average de-oiling by primary recovery methods accounts for approx. 18% of original oil in place (OOIP) in Germany. By using secondary recovery methods the degree of de-oiling can be increased to an average of 32% of OOIP. Only by EOR methods is it possible to obtain an average de-oiling rate of 45% of

OOIP; and just a few cases are known with outstanding recovery rates of more than 55% of OOIP.

There are several different methods of EOR known, and these can be divided into four categories: thermal, gas-injection, chemical flooding and others, such as microbial methods (MEOR). The choice of the method depends on a broad range of factors; e.g. <u>technical criteria</u>, including depth, temperature, minerals, status of the reservoir, type of crude, salinity of the reservoir and/or injection water etc., - and <u>economic criteria</u>, such as attainable price for crude, total cost of production, including energy, gas or chemicals, tax conditions etc., - in total a broad range of considerations.

The newest worldwide biennial EOR survey of the Oil & Gas Journal[2] indicates that EOR methods are highly sensitive to crude oil prices. The number of active U.S. EOR projects have decreased since 1986 continuously:

<u>Table 1</u> Active U.S. EOR projects[2]

	Total EOR	Chemical only	Polymer only
1986	512	206	178
1988	366	124	111
1990	295	50	42

It should be mentioned that the total EOR production in U.S. as well as worldwide has increased due to additional carbon dioxide and hydrocarbon injection in 1990.

In Germany there were about 20 active EOR projects running last year, which can be subdivided into four hot water, seven steam, and nine polymer EOR projects. Details of a current polymer project in Northern Germany will be reported later.

2 REQUIREMENTS FOR EOR POLYMERS

In the oil field many different water soluble polymers are needed as additives, e.g. for drilling, workover and completion fluids, for cementing, fracturing, and also for EOR flooding methods. The types of polymer include celluloseethers, lignosulfonates, guar gums, starch, polyamines, polyacrylamides, biopolymers, and synthetic copolymers, which contain different monomers. Depending on the specific requirements most of the mentioned polymers have to be modified and optimized by the chemical industry before these compounds can work in the field. Some of the most important and critical requirements for EOR polymers are presented in Fig. 1.

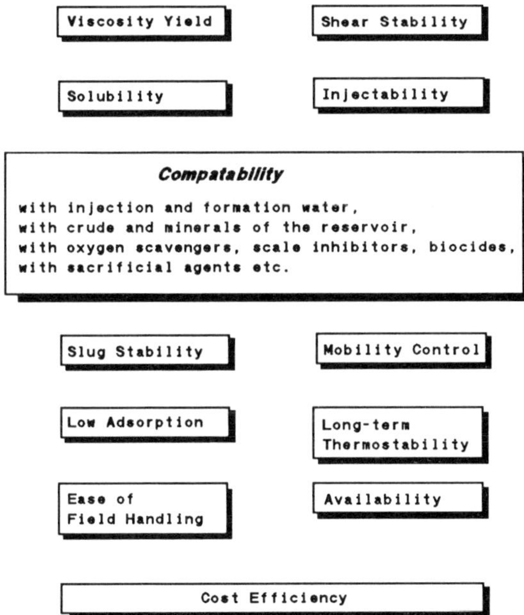

Figure 1 Requirements for EOR polymers

Polymer EOR processes involve augmenting waterflood performance and improving recovery of crudes having low to moderate viscosity. Polymers improve sweep efficiency by plugging high-permeability zones or reducing the mobility of the displacing fluid. With the knowledge of the well-studied mechanisms of polymer flooding special screening tests for suitable EOR polymer selection have been evaluated[3-8].

3 POLYMERS FOR VARIOUS EOR CONDITIONS

Depending on specific requirements for EOR only a few types of polymer succeed in showing suitable properties. In particular their EOR applicability is determinated by the salinity of both the reservoir and the injection water as well as the temperature conditions in the reservoir.

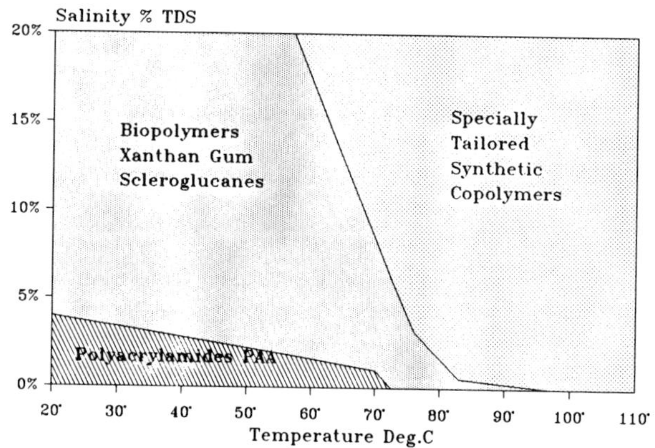

Figure 2 EOR applicability of polymer types

As shown in Figure 2, polyacrylamides (PAA) and partially hydrolysed polyacrylamides (HPAA) are suitable for low temperature and low salinity reservoir conditions. Biopolymers, such as xanthan gums and scleroglucanes, are suitable for low to moderate temperature and high salinity conditions. Copolymers, specially tailored synthetic compounds, are suitable for high temperature and high salinity reservoir conditions.

To date, most of the chemical flooding projects worldwide have been carried out by the injection of fluids which contain PAA or HPAA. These EOR applications have attained a highly successful state-of-art reputation. During the last 15 years in Germany RWE-DEA AG used this method of polymer flooding only, obtaining thereby about 700,000 t EOR crude[9-13].

Some attempts investigated by Preussag AG for the injection of biopolymers at moderate temperature conditions were made at 22°C and 56°C and high reservoir salinities[14,15].

In the literature several different copolymers suitable for EOR have been reported, - some of these are suitable for high temperature and high salinity reservoir conditions[16-18]. We have long experience with temperature and salt stable copolymers, which have been used throughout the world as fluid loss additives in drilling muds under conditions of extremely high salt concentrations and temperatures to over 200°C[19,20]. The molecular weights of these polymers at approx. 1×10^6 Dalton are too low for EOR application. Using a specially modified water-based gel-polymerization process higher molecular weights are available for producing EOR suitable copolymers[21].

Figure 3 indicates this polymerization step and the structure of this copolymer, which was developed by Cassella AG and Hoechst AG. The copolymer contains the monomers vinyl sulfonate (1), vinyl amide (2) and acrylamide (3) as building blocks; for this reason the resulting copolymer (4) is indicated as VS/VA/AM copolymer[22].

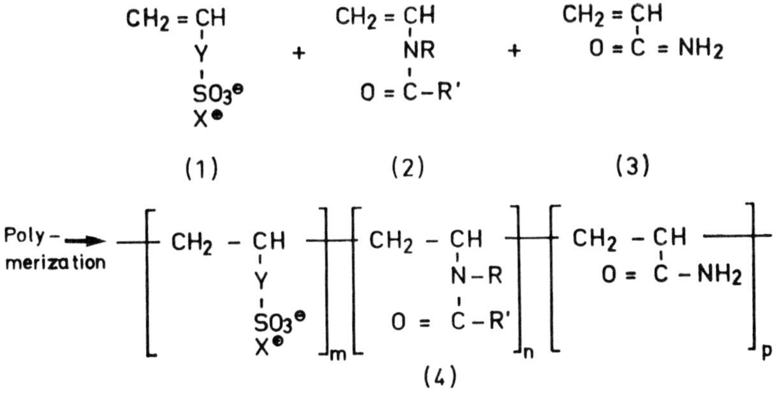

$$CH_2{=}CH{-}Y{-}SO_3^{\ominus}\,X^{\oplus}\ (1) \quad + \quad CH_2{=}CH{-}NR{-}C({=}O){-}R'\ (2) \quad + \quad CH_2{=}CH{-}C({=}O){-}NH_2\ (3)$$

Poly-merization →

$$\left[CH_2-CH(-Y-SO_3^{\ominus}X^{\oplus}) \right]_m \left[CH_2-CH(-N(R)-C({=}O)-R') \right]_n \left[CH_2-CH(-C({=}O)-NH_2) \right]_p \quad (4)$$

Figure 3 Synthesis of the VS/VA/AM copolymer

4 PROPERTIES OF THE VS/VA/AM COPOLYMER

Some properties of the VS/VA/AM copolymer, such as the molecular weight and the particle-size distribution, both of which determine the solubility of the granules, are described in Table 2.

Table 2 Some properties of the VS/VA/AM copolymer

Physical form	white granules
Particle-size distribution (approx. values)	
2000 µm: 0 %	250 µm: 41 %
1000 µm: 1.8 %	90 µm: 29 %
500 µm: 28 %	90 µm: 0.2 %
Molecular weight	7 - 9 x 10^6 Dalton
Active ingredient	80 % by weight
Monomeric AM content	0.03 % by weight
pH-value (of the 0.05 % aqueous solution)	approx. 6.5

Filtration tests of the different polymer solutions show (Figure 4) quite different flow characteristics: only the three EOR-suitable polymers have a plateau characteristic. For detailed tests of injectability the flow behaviour of the polymer solutions were examined by flooding of sand packages and original cores also. Quality control tests are made by flooding of sand packed tubes by measuring the differential pressures.

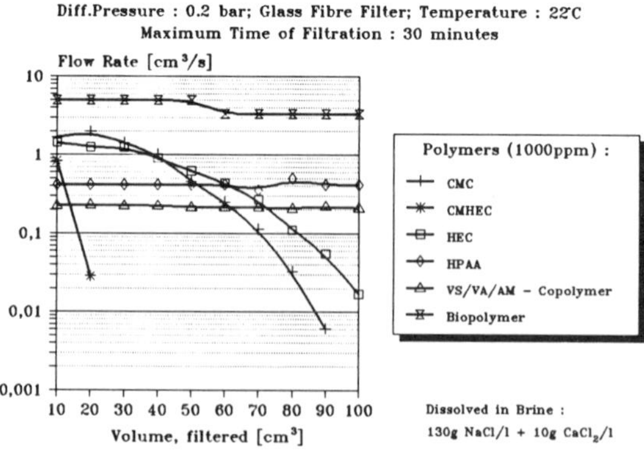

Figure 4 Filtration tests of polymer solutions

The viscosity efficiency is one of the most important criteria of polymer properties. Figure 5 gives the comparative data of different polymers: the biopolymer sample (xanthan type) yields the highest viscosities under the given saline conditions. This data has been measured by HAAKE RV 100/ CV 100 rotoviscometer.

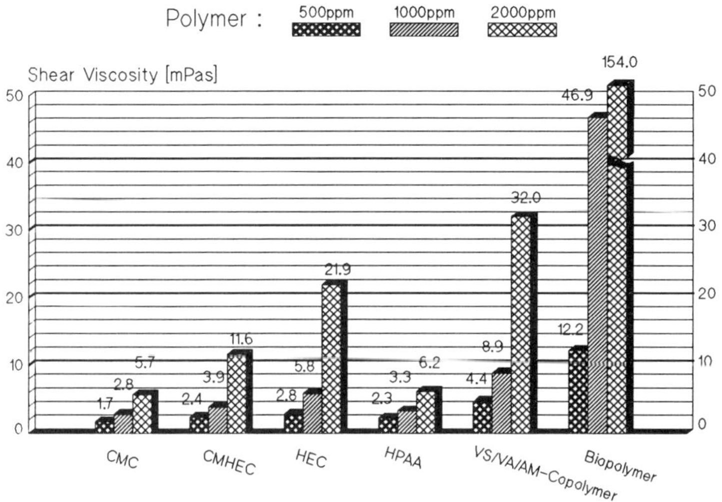

Figure 5 Shear viscosities of polymer solutions

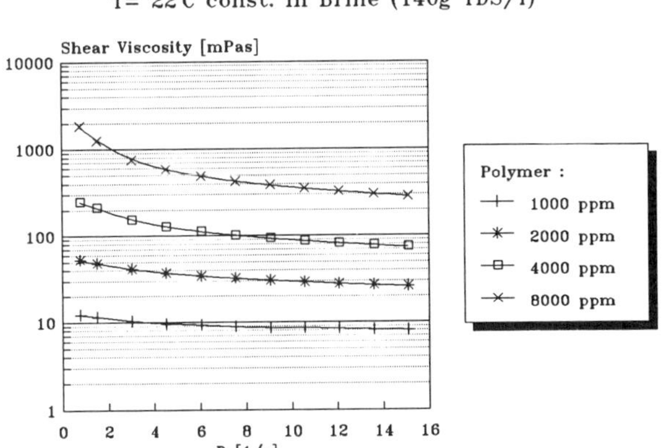

Figure 6 VS/VA/AM copolymer: viscosity vs. shear rate

Detailed viscosity data is compiled: Figure 6 shows a diagram of shear viscosity vs. shear rate and demonstrates the non-Newtonian character of these solutions. Figure 7 illustrates the viscosity behaviour of these polymer solutions at temperatures between 20 and 100°C.

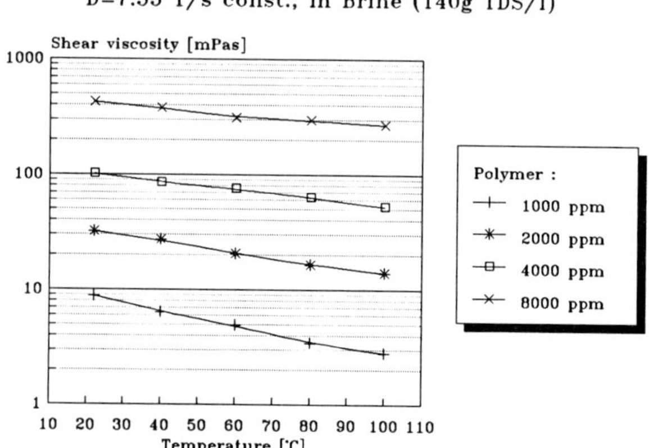

Figure 7 VS/VA/AM copolymer: viscosity vs. temperature

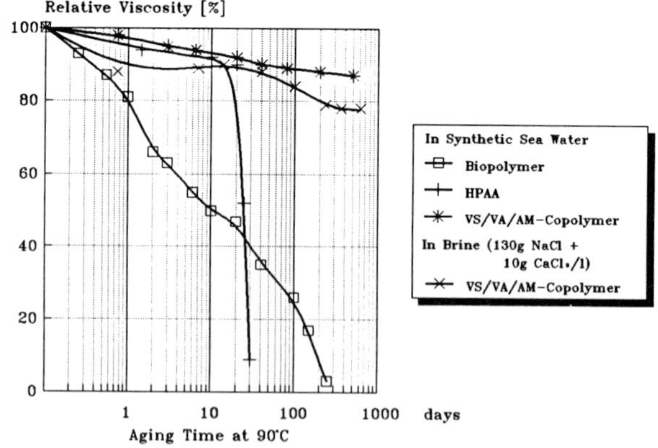

Figure 8 Thermal stability of EOR polymers

The thermal stability test of the different EOR polymer solutions at 90°C and different salinities show that the best values are found for the VS/VA/AM copolymer (Figure 8).

The efficiency of a polymer flood in a reservoir can be damaged by a high retention rate of the injected polymer. The retention rate depends on the type of minerals, on salinity, on temperature of the reservoir, on the type of the polymer and further conditions. As reported by H. Volz[23,24] the residual retention of the VS/VA/AM copolymer can be substantially reduced by using polyethylene glycols (5) as sacrificial agents.

$$HO - [CH_2 - CH_2 - O]_r - H \qquad (5)$$

Figure 9 Polyethylene glycols

The average molecular weight of the glycol has to be adjusted to obtain the lowest rates of retention of the polymer.

5 EOR APPLICATION IN THE PLOEN-OST FIELD

RWE-DEA AG, formerly Deutsche Texaco AG, started in 1989 in the Ploen-Ost field (Ost-Holstein) in Northern Germany with a polymer flooding pilot project using the VS/VA/AM copolymer. In respect of EOR aspects this oil field is characterized by outstandingly harsh environmental conditions; some of these data are compiled in Figure 10. Especially noteworthy data are the reservoir temperature of 95°C, and the extremely high salinity of 140 grams total dissolved solids per litre (TDS), including a high content of divalent cations. Additionally the reservoir matrix itself shows several complications: a relatively high content of those clays, which can swell with fresh water, and different sand layers, which have both highly permeable and low permeable zones within the same area.

Because of the sensitive clays of the matrix it is necessary to inject saline water only in all phases of reservoir flooding.

Pay zone	Dogger-Beta sand with a high content of clays which swells with fresh water
Area [acres]	30
No. wells prod.	3
No. wells inj.	1
Porosity [%]	19
Permeability [md]	200-1500
Depth [m]	2450-2550
Reservoir temperature [°C]	95
Reservoir oil	
Gravity [°API]	35
Viscosity [mPas] (under reservoir cond.)	1.3
Saturation [%] (at start of pol.proj.)	34
Reservoir water	
Salinity [TDS, g/l]	140

Figure 10 Characteristics of the pilot area for the polymer project in the Ploen-Ost field

Analytical data	Values before purification	Values recommended for flooding	
pH-value	6.4	6.4	
TDS [g/l]	130	n.c.	x)
Sodium [mg/l]	48380	n.c.	
Calcium [mg/l]	1870	n.c.	
Magnesium [mg/l]	340	n.c.	
Strontium [mg/l]	160	n.c.	
Barium [mg/l]	30	n.c.	
Anions (Cl^-, SO_4^{2-}, HCO_3^-)		n.c.	
Iron (Fe^{2+}) [mg/l]	20	0,5	y)
Oxygen, dissolved [mg/l]	0.05	0.1	y)
Crude oil, emulsified [mg/l]	30	1	y)
Solids [mg/l]	1.5	0.5	y)

x) n.c. = non-critical y) critical values

Figure 11 Quality of the injection water

The quality of the saline injection water (130 g TDS/l) had to be optimized to take account of its content of iron (Fe^{++}), molecular oxygen, traces of crude oil and insoluble solids[25]. Figure 11 shows some of these critical and non-critical values. The content of iron and dissolved oxygen had to be reduced, because higher values than recommended can decrease the viscosity of the polymer flood.

For blending the saline polymer fluid (maximum 1200 ppm polymer) in the field RWE-DEA AG built a special blender and dosage equipment. In total 300 t polyethylene glycol (PEG 1000) and 150 t VS/VA/AM copolymer have been injected up to the end of 1990. This injection phase was carried out without any serious technical problem. In the current project phase the saline waterflood (without polymer) is injected.

The targets of this polymer project are the following:

- Testing the additional de-oiling by polymer flooding method under harsh environmental reservoir conditions; temperature and salinity are extremely high for EOR chemical flooding. It has been estimated that the rate of additional oil can yield about 8-10% of OOIP, resp. approx. 10,000 to 15,000 t, - being producable in the small pilot area of the Ploen-Ost field only.
- Development of a blending technique for continuous production of a suitable polymer fluid with necessary salinity and viscosity.
- Examining the efficiency of polyethylene glycol as sacrificial agent in a field project.
- The long-term stability of the VS/VA/AM copolymer under actual reservoir conditions.

It is too early yet to estimate the success with respect to additionally produced crude. A detailed report on this pilot project will be given by RWE-DEA AG in May 1991 in Stavanger, Norway[26].

6 FURTHER ASPECTS FOR THE DEVELOPMENT OF NEW COPOLYMERS

Some of the further aspects will be discussed:

- Reducing polymer production costs
- Modification of comonomers and molecular weight; Figure 12 shows viscosities of different molecular weight VS/VA/AM copolymers. The type M (medium viscous) of the shown copolymers was used for the pilot project (Chapter 5).
- Modification of the physical form: gel, emulsion.
- Enhanced investigation in finding further efficient systems for reduction of the retention rate (sacrificial agents).
- Further activities in the Ploen-Ost field.
- EOR activities in other reservoirs with similarly harsh environmental conditions.
- Application of blends containing polymers and surfactants, such as ether sulfonates (Figure 13) which are suitable for high temperature and salinity reservoir conditions[27-31].

In total a lot of tasks and challenges for all parties concerned:the oil industry, research institutes and the chemical industry[32].

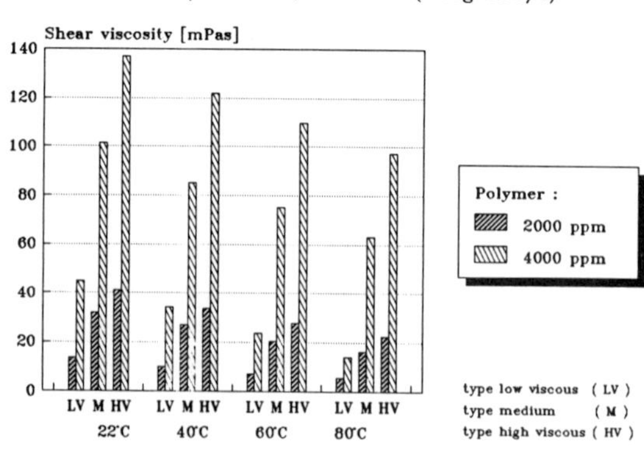

Figure 12 Viscosity of different VS/VA/AM copolymers

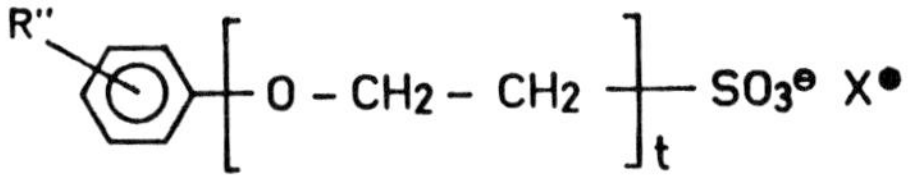

$$R''-C_6H_4-\left[O-CH_2-CH_2\right]_t-SO_3^{\ominus}\ X^{\oplus} \qquad (6)$$

Figure 13 Alkylphenol ether sulfonates

7 CONCLUSIONS

In this report the evaluation, first application, and future aspects of a highly specialized type of copolymer for EOR polymer flooding under harsh environmental reservoir conditions has been reviewed. Further activities on EOR application, especially chemical flooding methods, as well as enhanced polymer evaluation depend on several interrelated factors, e.g. on

- oil price, oil supply and demand situation,
- total costs for oil production, primary, secondary and tertiary recovery methods, - level of technical and economic risks,
- political aspects, tax conditions, environmental restrictions, crisis management,
- last, but not least, new input of improved technical methods, finding of more efficient chemicals and/or specially tailored systems of various chemicals.

Thanks are given for cooperation and discussions in the reported subject to Dr. Engelhardt, Dr. Schmitz both Cassella AG, Dr. Volz, Dr. Maitin both RWE-DEA AG, Prof. Dr. Pusch at the Institute for Petroleum Engineering, and Prof. Dr. Kessel at the Institute for Petroleum Research both in Clausthal-Zellerfeld/Germany. Part of this work was supported by the fund of the Bundesministerium für Forschung und Technologie (BMFT) as part of the project no. 032 6139 B.

REFERENCES

1. W.E.G. Wirtschaftsverband Erdöl- und Erdgasgewinnung e.V., "Erdöl-Erdgas, Entstehung, Suche, Förderung", Hannover, 1991
2. G. Moritis, Oil & Gas Journal, 1990, Apr 23, 49-82
3. G. Pusch and Th. Lötsch, Erdöl-Erdgas-Kohle, 1987, 103, 272
4. Th. Lötsch and G. Pusch, Erdöl-Erdgas-Kohle, 1988, 104, 503
5. R.B. Needham and P.H. Doe, J. Petr. Techn., Dec 1987, 1503
6. R.G. Ryles, SPE Reservoir Engng., Feb 1988, 23
7. J.G. Sothwick and C.W. Manke, SPE Reservoir Engng., Nov 1988, 1193
8. W. Littmann, "Polymer Flooding", Developments in Petroleum Science 24, Elsevier, New York, 1988
9. Erdöl-Erdgas-Kohle, 1989, 105, 388
10. W.O. Sohn et al., SPE Reservoir Engng., Nov 1990, 503
11. B. Maitin and H. Volz, Paper SPE 9794, 1981
12. B. Maitin, Proc. Third European Meeting on Improved Oil Recovery, Rom, 1985, Vol. 2, 283
13. B. Maitin et al., Paper SPE 17631, 1988
14. W. Kleinitz, W. Littmann and H. Herbst, Fifth Europ. Symp. Improved Oil Recovery, Proc., Budapest, 1989, 319
15. G. Gerken and N. Pavlik, First Int. Forum Res. Simulation, Alpach, Austria, 1988
16. G.A. Stahl, A. Moradi-Araghi and P.H. Doe, Meeting ACS Div.Polymeric Mat.,Science and Engng., Anaheim/California, 1986
17. European Patent EP 106 665 (1986)
18. P.H. Doe et al., SPE Reservoir Engng., Apr 1987, 461
19. M. Hille, Paper SPE 13558, 1985
20. A.J. Son, T.M. Ballard and R.E. Loftin, Paper SPE 13160, 1984
21. European Patent EP 044 508 (1981)
22. W. Gulden and S.P. von Halasz, Meeting ACS Div. Polymeric Mat., Science and Engng., Anaheim/California, 1986
23. German Patent DE 32 11 168 (1983)
24. H. Volz, Meeting ACS Div.Polymeric Mat., Science and Engng., Anaheim/California, 1986
25. H. Volz, 4. Vortrags- und Diskussionstagung, Chemische Produkte i.d. Erdölgewinnung, Clausthal-Zellerfeld, Germany, 1988

26. W. Schuhbauer, B. Maitin and H. Volz, Sixth Europ. Symp. Improved Oil Recovery, Stavanger, Norway, May 1991
27. G. Schneider and W. Gulden, Proc. World Surfactants Congress, Kürle-Verlag, Gelnhausen, 1984, Vol.IV, 305
28. D.G. Kessel, Erdöl-Erdgas-Kohle, 1986, 102, 504
29. S.P. von Halasz, Revue de L'Institut Français du Pétrole, 1988, 43, 545
30. A. Holst, Erdöl-Erdgas-Kohle, 1988, 104, 324
31. European Patent EP 353 469 (1989)
32. J. Combe et al., "EOR in Western Europe: Status and Outlook", Fifth Europ. Symp. Improved Oil Recovery, Budapest, 1989

Surfactant Flooding: Problems and Prospects

P.L. Bondor

SHELL RESEARCH BV, KONINKLIJKE/SHELL-LABORATORIUM, BADHUISWEG 3, 1031 CM AMSTERDAM, THE NETHERLANDS

1 ABSTRACT

The use of surfactants to improve the ultimate recovery of oil has been the subject of intensive research for more than thirty years. The body of knowledge developed allows the design of a surfactant flooding system which can efficiently recover oil under laboratory conditions, and sometimes even under field conditions. Despite technical successes, the process is still far from economically viable. In this paper the chemical aspects of the chemical flooding process are emphasised. Both the present understanding of chemical flooding and critical gaps in that understanding are discussed.

2 INTRODUCTION

The term surfactant flooding is here restricted to enhanced oil recovery methods which employ a surface-active agent injected into an oil reservoir to reduce the oil-water interfacial tension, lowering the capillary forces trapping the oil in the pore space as discrete droplets. Surfactant flooding has been a topic of active enhanced oil recovery research for the past three decades. The original concept, that of a simple introduction of very small quantities of chemical into injection water to improve recovery, was not successful. Studies in the 1960s showed that oil left in the reservoir after waterflooding was trapped in the pore space as discrete droplets by capillary forces. These analyses showed that the capillary-trapped oil could be mobilised either by increasing the viscous forces on the droplet, or by decreasing the interfacial tension between the water and the oil. The residual oil saturation after waterflood can be correlated with the capillary number, the ratio of viscous to capillary forces acting in a displacement process. To mobilise waterflood-trapped oil, it is necessary to increase the capillary number by two to four orders of magnitude. In practice, this can be achieved only by a similar decrease in the oil-water interfacial tension.

It was found that systems that achieve very low interfacial tensions generally involve a combination of surfactants and form a third, microemulsion phase. Consequently, surfactant flooding research focused on the search for surfactant systems that would exhibit the formation of such three phase systems under reservoir conditions. The development of such systems in the laboratory, and the demonstration that they could effectively recover residual oil in laboratory core floods, resulted in a high level of effort in the 1970s. The process which received research emphasis was intended to recover tertiary oil; that is, the oil remaining in the waterflooded portion of the oil reservoir. In a tertiary application, a concentrated surfactant solution (say, 10%) is injected as a small slug (say, 5 to 20% pore volume), and is followed by a mobility control drive (usually a polymer solution). The research efforts of the 1970s resulted in a variety of surfactant systems which, in laboratory experiments simulating reservoir conditions, could efficiently recover capillary-trapped oil from reservoir core material.

Unfortunately, the laboratory results were followed by a succession of disappointing field tests. In most cases, the field tests resulted in oil recoveries far below those anticipated. A variety of factors related either to the operational and/or logistical constraints of field operations, or to the uncertainties inherent in the geological description of the reservoir may have been responsible for the poor results. Even in those cases where the oil recovery was as expected, the economics of the process were such that the results were described as technical successes, but economic failures. With the drop in oil price in the mid-1980s, interest in surfactant flooding waned.

In an effort to develop economic systems for field application, the use of microemulsion systems at lower (1%) concentrations was studied for North Sea applications. Even in these low concentration systems, the operating and economic constraints were such that field applications will remain uneconomic for the foreseeable future.[1] With the history described above, it seems reasonable to re-examine the knowledge and premises of the surfactant flooding process.

In this paper, the state of the art in surfactant flooding is briefly reviewed. No attempt will be made to review the voluminous literature extant regarding the various aspects of surfactant flooding; for this purpose, the reader is referred to the paper by Kessel[2] which contains an extensive bibliography on chemical flooding. However, the aspects of the process which critically depend on knowledge of the chemistry of the systems are discussed in some detail. Finally, crucial gaps in our present knowledge are indicated, and suggestions regarding advances necessary in order to achieve economic, as well as technical successes, are made.

3 THE SURFACTANT FLOODING PROCESS

A simplified picture of the surfactant flooding process is as follows: a surfactant system, chosen by its ability to reduce the interfacial tension between a given crude oil and its reservoir brine to very low values, is injected into the reservoir as a small slug, followed by a drive fluid. The slug mobilises the trapped crude oil and pushes it ahead, forming an oil bank. The drive fluid displaces the slug and the oil bank ahead of the slug. The oil bank encounters trapped oil which is incorporated into the front end of the bank, thus the volume of mobilised oil grows both on the leading and trailing edges. The surfactant slug is displaced by the drive fluid, and the whole system moves toward the production well in a stable manner.

This picture is, by oil field standards, simplified because it does not recognise the problems introduced by the uncertainty in understanding of the reservoir geology, and by the interaction of the geology and the fluid displacement process. Unfortunately, from the physical chemistry perspective, it also greatly understates the true complexity of the process.

System Design. At present, a surfactant flooding system must be designed specifically for the oil reservoir in which it will be applied. Many of the variables which determine the behaviour of the chemical system are fixed. The reservoir pressure and temperature, the crude oil to be recovered, the composition of the reservoir brine, and the characteristics of the reservoir rock, both physical and chemical, all represent factors which must be incorporated into the chemical system design.

A surfactant system, usually a blend of a primary surfactant and a cosurfactant, is chosen which will provide the required ultra-low interfacial tensions under the above constraints. Economic constraints dictate that the surfactants chosen must be low cost; thus various commercially available petroleum sulfonates have frequently been used as the primary surfactant. Cosurfactants added, commonly either an alcohol or ethoxylated sulfated alcohol, are used to modify the solvency of the primary surfactant and add salinity tolerance to the system.

As mentioned above, residual oil saturation may be correlated with capillary number,[3] as shown in Figure 1. Conventional waterflooding is carried out at a capillary number of ca. 10^{-7}; this value is constrained by the oil-water interfacial tension (ca. 30-40 dynes/cm), the viscosity of the fluids (ca. 1-10 cp), and the flood front velocity (ca. 1 foot/day). From the figure, it can be seen that to recover significant residual oil requires a capillary number ca. 10^{-3}; i.e., an increase of some four orders of magnitude. Because an increase of this magnitude in neither the fluid viscosity nor the frontal advance rate is practical, a reduction of that order in the IFT is necessary.

Phase Behaviour. It has been shown that surfactant systems which develop the required low IFT behaviour when mixed with oils exhibit common phase behaviour characteristics: they form three phase systems, including an oleic phase, and aqueous phase, and a microemulsion phase under reservoir conditions. Extremely low IFTs have been measured in such systems, and they have demonstrated efficient oil recovery in laboratory core floods. Thus, investigation of the phase behaviour of microemulsion-forming systems, following the methods of Winsor,[4] has been the design method of choice.[5] Figure 2 illustrates the method used. The oil, water and surfactant being investigated are each treated as pseudocomponents, and the ternary

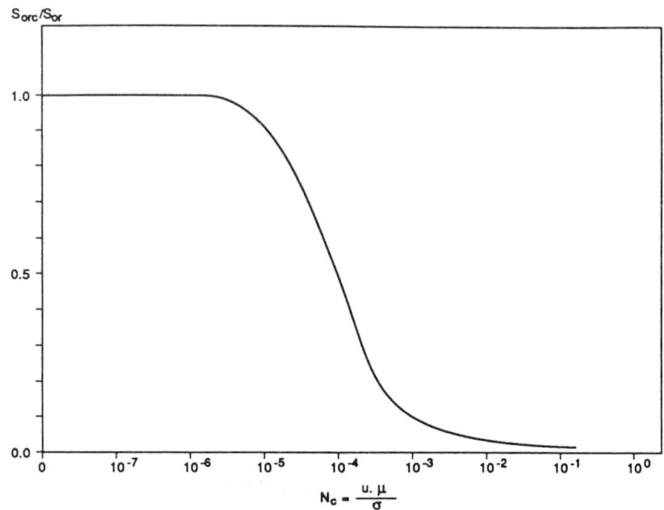

Figure 1. Remaining oil saturation as a function of capillary number. S_{or} = saturation after waterflood, S_{orc} = saturation after chemical flood.

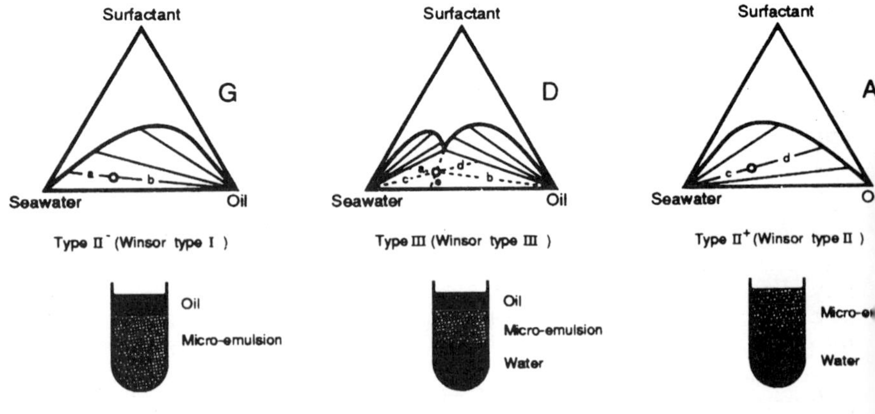

Figure 2. Pseudoternary representation of phase behaviour.

diagrams shown represent phase relationships between those pseudocomponents. As a given parameter of the system is changed (salinity or temperature), the phase behaviour of the system changes from type II-, in which most surfactant is in the aqueous phase (an "underoptimum" system), to type III, in which a third, microemulsion phase exists containing almost all the surfactant (an "optimum" system), to type II+, in which most surfactant is dissolved in the oleic phase (an "overoptimum" system). Some general observations may be made. For a given oil and brine, the phase behaviour depends on the relative solubility of surfactant in the brine and oil; greater solubility in brine tends toward type II-, greater solubility in oil towards II+. Thus, increasing the length of the alkyl chain, or decreasing the water solubility of the hydrophylic part of the molecule, will shift the system towards overoptimum. For anionic surfactants, an increase in brine salinity or hardness, or a decrease in temperature, shifts the system towards overoptimum. For nonionic surfactants, increasing the temperature shifts the system towards overoptimum.[6] When using blends of surfactants, these observations may be too simplistic.

Drive Design. Economic constraints also have dictated that the volume of the surfactant solution injected into the reservoir be small in comparison to the pore volume to be contacted by injected fluid. Thus, once in the reservoir, the chemical slug is diluted by mixing with both the formation brine at the front of the slug and the drive water at the rear. In addition, the transport of a chemical system through the reservoir results in an ion exchange process[7] taking place between the injected system and the reservoir rock (predominantly, the clay components of the rock matrix). This ion exchange also affects the chemical slug, so that the system must be able to maintain its ability to reduce oil-water interfacial tensions to the values required while being modified both in concentration and in chemical composition. Finally, the surfactants may physically adsorb on the rock surface (again, predominantly the clays), presenting another mechanism by which the injected slug is modified.

The surfactant flooding system developed from phase behaviour considerations must be tested in laboratory core floods, and modified so that the type III phase environment is maintained as the surfactant slug is displaced through the reservoir. This is accomplished using the salinity gradient concept.[8] It has been demonstrated that the type of phase behaviour a system demonstrates is dependent on surfactant concentration. This is illustrated[9] in the salinity requirement diagram shown in figure 3. As surfactant concentration decreases, the system tends toward overoptimum conditions. Because the oleic phase in overoptimum systems usually is viscous, it has a much lower mobility in the reservoir and thus has a detrimental effect on the performance of the flood. By contrast, the aqueous phase in underoptimum systems has a low viscosity and thus a high mobility in the reservoir. In the salinity gradient concept, the surfactant slug has a salinity between the reservoir brine and the drive fluid; thus, the ion exchange and dilution processes in the reservoir have a tendency to focus the surfactant slug.

The displacement conditions in the reservoir dictate that each fluid injected should have a viscosity greater than the fluid preceding it; thus the chemical slug should be more viscous than the oil and water in the reservoir, and the drive fluid following the chemical slug should be more viscous than the slug. To meet these constraints, the chemical slug has often had polymer included as a viscosifier. Similarly, the driving fluid following the chemical slug has usually been a viscous polymer solution.

4 THE PHYSICAL CHEMISTRY OF SURFACTANT FLOODING

Surfactants

For surfactant flooding, anionic surfactants, in particular the petroleum sulfonates, were used extensively in the research of the 1970s. More recently, synthetic sulfates and sulfonates have been used, and some nonionics have been studied, both singly and as blends.

Nonionic surfactants have better tolerance to divalent ions than do anionics, but in general have higher adsorption on reservoir rock, and thus have received less attention. The poor tolerance of anionic surfactants to divalents has been treated by the use of cosurfactants, as indicated above. Alkyl ethoxysulfates and alkyl ethoxysulfonates both have been used as cosurfactants. Sulfates, both as primary surfactants and cosurfactants, are cheaper, but are thermally unstable at temperatures above about 70 degrees C. Thus the sulfonated compounds are used for higher

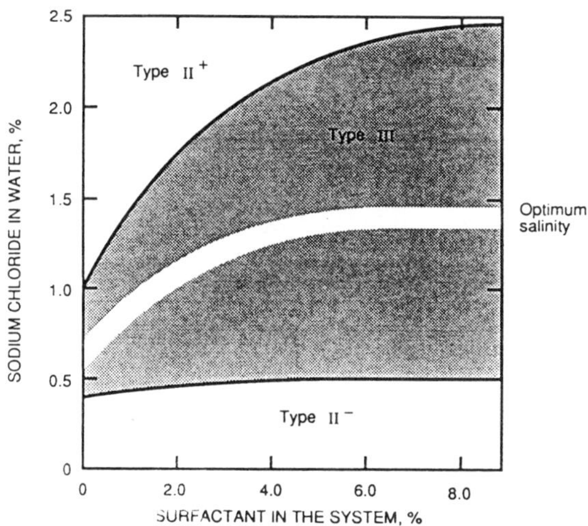

Figure 3. Salinity requirement diagram.

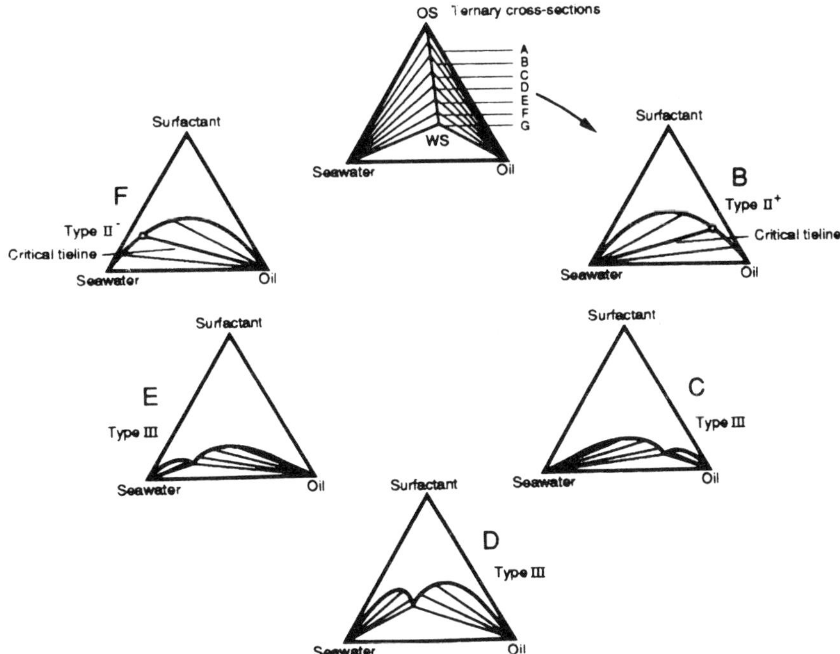

Figure 4. Effect on phase behaviour of changing the ratio between the surfactants.

temperatures. Recently, propoxy groups have been incorporated into alkyl ethoxysulfate molecules, resulting in alkyl propoxyethoxysulfates. These molecules have been used in a technically successful field test,[10] and may provide an additional method by which salinity tolerance may be achieved.

Although the above refers to individual surfactants, it is necessary to note that surfactants used for commercial operations will in general be mixtures of several, not well characterised, components. Thus both the definition of a system being used, and consistency of the supplied product on a commercial basis, can be a problem.

Phase Behaviour Studies

The use of pseudoternary diagrams as a guide to surfactant system design was outlined above. Such techniques assume that the system remains within that ternary plane during the recovery process, and that the equilibrium systems studied are appropriate models for the process in the reservoir.

Mixtures of surfactants, and mixtures of surfactants and cosurfactants, are used to provide the appropriate phase behaviour for a given crude oil/brine system. The pseudoternary diagram is used to develop a description of the system phase behaviour as the salinity of the brine is changed. Another method which may be used is an extension of the ternary diagram, the quaternary diagram. In this case, two surfactants are represented by corners of a tetrahedron. The optimum region is sought for a particular salinity by varying the ratio of the two surfactants. The ternary diagram is thus a section of the quaternary system. Figure 4 is a schematic representation of the pseudoternary diagrams resulting from a change in ratio of two surfactants in a surfactant blend.

In practice, phase behaviour studies are carried out at a limited number of water/oil ratios (usually e.g., 1/1, 3/1, 1/3). At a given water/oil ratio, samples are made up and equilibrated, and the observed phase behaviour related to positions on the model phase diagram. It is assumed that the phase behavior changes smoothly and continuously within the ternary (or quaternary) diagram, and further that the properties of the individual fluid components and IFTs are appropriately represented by this sampling. Both assumptions may not hold for a given system. Both the compositions of the phases, and the IFTs between phases, may change. In pure surfactant systems, this effect may not be so obvious; the surfactants used in surfactant flooding, however, are generally mixtures of various components as a result of the manufacturing process. Thus the phases which form at a given point within the diagram will in general be different from those at any other point. Also this phase, once formed within the reservoir, will be moved through the reservoir contacting fresh oil, reservoir brine, and reservoir rock and interacting with each. The result of this process, even if assumed to be in local equilibrium, may bear no relation to the samples and measurements taken in the static equilibrium tube samples.

Interfacial Tension

Interfacial tension measurements are carried out in association with the phase behaviour studies. As the surfactant flooding concept relies on the drastic reduction of IFT to mobilise capillary-trapped crude oil, the reduction which can be achieved is of primary importance. Usually, IFT measurements are made on oil/surfactant systems which have been allowed to equilibrate, since the assumption of local equilibrium in phase behaviour extends to an implicit assumption that such equilibrium exists throughout the course of the displacement process in the reservoir, and thus that equilibrium IFT is the controlling parameter.

Adsorption

The amount of surfactant necessary for a flood is dependent on the amount which is lost, for whatever reason, in the reservoir. A potentially very important mechanism by which surfactant may be lost from the injected slug is that of physical adsorption. Adsorption, by e.g., electrostatic interactions between charged mineral surfaces and the surfactant molecules, may be an important loss mechanism.

It is believed that adsorption may be generally described by a linear adsorption isotherm. Adsorption increases linearly with concentration until the critical micelle concentration (CMC) is reached, and then levels off to a constant value. At the CMC, individual surfactant molecules begin to aggregate. As concentration increases, the concentration of single molecules (monomers) remains constant, while the size and number of micelles increases. Thus, it is inferred that surfactant adsorption depends only on the monomer concentration. Since in surfactant flooding the surfactant concentration normally is far above the CMC, adsorption is assumed to be a constant value. It is observed that the level of adsorption of surfactants is increased drastically when clays, which have enormous surface area and many charge sites, are present, while adsorption on clean silica surfaces is very low.

The measurement of adsorption on reservoir rock thus is very important. Such measurements are carried out in core material saturated with reservoir brine; the result is taken to be the physical adsorption value.

Surfactant Retention

Other mechanisms may act to either remove surfactant from the system or render it inactive. Interaction with divalent ions, which may be present in increased concentration as a result of ion exchange during the flood, may change the phase behaviour toward the II+ regime, moving the surfactant into the oleic phase and away from the active region. In extreme cases, the calcium or magnesium salts of the surfactant may precipitate. In systems which develop three phases, the third (microemulsion) phase may have a high concentration of surfactant and a high viscosity. In this case, even though the IFT is low, displacement efficiency will be poor because the mobility of the microemulsion will be low and phase trapping of the surfactant will occur.

5 UNDERSTANDING OF THE PROCESS

While a significant effort has been devoted to the topic of surfactant flooding, in both the academic and field design and development aspects, it is fair to say that the understanding to date is phenomenological in nature. A fundamental understanding does not exist of any of the major elements of the process: surfactant molecule design, surfactant system design, phase behaviour, IFT, adsorption, or retention. Given a specific reservoir and reservoir fluids, an a priori selection of surfactants which will efficiently displace residual oil cannot be made. How the surfactant structure interacts with the other elements of the system, with the oleic and aqueous phases, and with the rock, is not understood. Further, the dynamics of the process within the reservoir cannot at present be predicted.

The above, coupled with the fact that the best systems designed (by trial and error!) to date can at best be described as technical successes and economic failures in field applications, presents a bleak picture of the prospects for the economic application of surfactant flooding.

6 NEW DIRECTIONS

If the surfactant flooding process is ever to be made economic, it will be necessary to 1) develop a scientific understanding of the elements of the process, and 2) explore new concepts in surfactant flooding which have the prospect of providing simpler, lower cost alternatives. Neither of these objectives are easy, and success is not assured: they are high risk, but potentially high reward, undertakings.

Development of Understanding

Surfactant Design. The mechanism by which a surfactant lowers IFT is not well understood at present. What is the interaction between the structure of a surfactant molecule and its ability to lower IFT between a given crude oil and its associated brine? How is that ability influenced by changing concentration of any given component? An increased understanding of the answers to these questions will allow physical chemical prediction of the properties of new surfactant molecules, and provide the ability to tailor new molecules to the problem at hand.

Surfactant Analysis. Any commercial surfactant will be a mixture of several different components. The analysis of this system is non-trivial, but the determination of how each component influences the behaviour of the system may be the key to understanding the mechanisms operating, and to detecting the components responsible for oil mobilisation. In order to accomplish these objectives, analytical tools able to identify in detail the components extant at low concentration must be developed.

Surfactant Systems. The systems which have been developed contain several components which play different roles: for example, cosurfactants may impart salinity tolerance. The interaction of the individual components in a surfactant system, and the dynamics of the process in the reservoir, are crucial to the success of the process. The assumptions inherent in the local equilibrium approach, and the pseudo-ternary and pseudo-quaternary diagrams, must be critically examined.

New Concepts

In order for surfactant flooding to become economic, it is necessary to develop systems which efficiently recover oil at very low concentrations. In this case, one cannot afford to have a microemulsion phase (with a relatively high concentration) form in the reservoir; the resultant slow propagation of surfactant through the reservoir cannot be tolerated. Thus two-phase systems which reduce IFT as efficiently as the present systems must be developed, and they must exhibit physical adsorption at least proportionally lower. While this is a very challenging set of requirements, there are indications in the literature that such systems may be developed.

Ultralow Concentration. Indications that ultralow IFT can be accomplished at lower concentrations are available in the literature. A recent report[11] gives IFTs on the order of a millidyne/cm at concentrations of 0.1% for an alkyl propoxyethoxysulfonate system. Earlier work[12] shows an extremely low IFT in a two phase system with a standard petroleum sulfonate (TRS 10-410). This is not an isolated case; at the CMC, two-phase systems do exhibit very low IFTs. This property of surfactant systems has hitherto been ignored in preference to the microemulsion systems, because the higher concentration systems can be propagated more rapidly. If, however, adsorption can be reduced significantly through intelligent design of the surfactant molecule or by better understanding of the factors influencing adsorption, then operation at the CMC may become economic.

Non-equilibrium Effects. As discussed above, previous research has developed the hypothesis that the surfactant flooding process is an equilibrium one; that the progression of the surfactant system through the reservoir can be viewed as a succession of equilibrium states. However, there is evidence that non-equilibrium IFT developed in surfactant flooding applications may be significantly lower (by one or two orders of magnitude) than the equilibrium IFT[13]. If, as a result of the changing environment and changing compositions of the phases moving through the reservoir, the system is fundamentally in a non-equilibrium condition, then such non-equilibrium IFT values may be the controlling parameters for oil mobilisation.

7 CONCLUSIONS

The surfactant flooding process, as it has been designed and understood in the past, has little prospect of becoming an economically viable recovery technique in the foreseeable future.

If surfactant flooding is ever to be anything more than a laboratory curiosity, understanding of the basic physical chemistry of surfactant systems must be developed. The dynamics of the surfactant flooding process must also be studied and the fundamentals of the process understood.

Concurrently, economic application requires that research look for new and novel chemicals, which are active and efficiently recover oil at concentrations at least an order of magnitude lower than those previously tested. In order for these systems to be viable, they must be active in the two-phase region, remain predominantly in the aqueous phase, and exhibit extremely low adsorption.
While the above are challenging objectives, there is at present no firm physical evidence that they are impossible to meet. In fact, fragmentary information from the literature may indicate potentially fruitful directions for research.

8 ACKNOWLEDGEMENTS

The author would like to thank A.M. Michels, and all the individuals in Shell's surfactant flooding research groups at both the Koningklijke/Shell Exploratie en Produktie Laboratorium in Rijswijk, the Netherlands, and Shell Development Company's Bellaire Research Center in Houston, Texas, U.S.A., for many stimulating discussions over the years. Thanks are also expressed to the managements of Shell International Research Maatschappij, B.V., and Shell International Petroleum Maatschappij, B.V., for permission to publish this paper.

9 REFERENCES

1. P.L. Bondor, "Sixth European Symposium on Improved Oil Recovery, Stavanger, Proceedings", 1991, Volume 1, Book II, p. 749.

2. D.G. Kessel, "Fourth European Symposium on Enhanced Oil Recovery, Hamburg, Proceedings", 1987, p. 1.

3. G.L. Stegemeier, Chapter 3, "Mechanisms of Entrapment and Mobilization of Oil in Porous Media", D.O. Shah and R.S. Schechter, eds., Academic Press, New York, 1977.

4. P.A. Winsor, "Solvent Properties of Amphiphilic Compounds", Butterworth's Scientific Publications, London, 1954.

5. R.C. Nelson and G.A. Pope, SPE Journal, October, 1978, p. 325, and R.C. Nelson, "Surface Phenomena in Enhanced Oil Recovery", D.O. Shah, ed., Plenum Press, New York, 1981.

6. R.C. Nelson, "Petroleum Technology into the Second Century", Proceedings of the Centennial Symposium of New Mexico Tech, 1989, p. 277.

7. G.J. Hirasaki, SPE Journal, April, 1982, p. 181.

8. G.J. Hiraski, H.R. van Domselaar, and R.C. Nelson, SPE Journal, June, 1983, p. 486.

9. R.C. Nelson, SPE Journal, April 1982, p. 259.

10. T.R. Reppert, et al., SPE/DOE 20219, presented at the SPE/DOE Seventh Symposium on Enhanced Oil Recovery, Tulsa, Oklahoma, 1990.

11. B. Kalpakci, et al., SPE/DOE 20220, presented at the SPE/DOE Seventh Symposium on Enhanced Oil Recovery, Tulsa, Oklahoma, 1990.

12. K.S. Chan, and D.O. Shah, "Surface Phenomena in Enhanced Oil Recovery", D.O. Shah, ed., Plenum Press, New York, 1981, p. 53.

13. K.C. Taylor, B.F. Hawkins, and M.R. Islam, CIM No. 89-40-44, presented at the 40th Annual Technical Meeting of the Petroleum Society of CIM, Banff, 1989.

On Minimizing Adsorption of Foam-forming Surfactants in Porous Media

Jerry J. Novosad

PETROLEUM RECOVERY INSTITUTE, 3512-33RD STREET N.W., CALGARY, ALBERTA, CANADA T2L 2A6

1 INTRODUCTION

Chemical-based enhanced oil recovery (EOR) processes that displace residual oil by lowering oil/brine interfacial tension, such as surfactant/polymer flooding, implicitly require good volumetric sweep of a reservoir for the surfactant to contact the residual oil. Their cost efficiency, however has not yet reached an attractive enough level to manifest itself in a large number of field applications. A list of reasons for this would arguably have surfactant loss caused by adsorption at the solid/liquid interface at or near the top.

Even a cursory look at the consumption of chemicals explains why. Let's consider a small section of a densely drilled reservoir: a five-spot pattern of an injector and four producers that covers an area of 10 hectares. Such a well-spacing would be typical of older oil reservoirs that are prime candidates for tertiary oil recovery processes. Assuming a thickness of the oil-bearing zone of 5 m, rock grain density of 2.6 g/ml, porosity of 0.2, and remaining oil saturation at 0.4 brings the mass of reservoir rock to $1.04*10^6$ tonnes. Even at what may appear to a chemist or a chemical engineer a nominal adsorption level of 1 mg of surfactant per gram of rock, 1,040 tonnes of surfactant would be adsorbed at the solid/liquid interface. This corresponds to consuming 26 kg of surfactant for one cubic metre of potentially recoverable crude oil. At $4.00/kg of surfactant, such an expense would represent spending $104 for each cubic metre of oil remaining in place.

The above example demonstrates that the value of the recovered oil would just cover the cost of the injected surfactant. Of course, there are additional costs for a mobility-control polymer, handling, and injection. It is thus imperative to reduce surfactant adsorption at least by one order of magnitude for surfactant flooding to have a chance of economic success.

It may be illustrative to note that the 1,040,000 tonnes of reservoir rock contained in just a small fraction of an oil field (middle-sized reservoirs would have hundreds or maybe even thousands of such areas) is equivalent to the mass content of ten thousand of the largest pressure-swing adsorption units used for gas separations in the chemical industry.[1]

Surface-active chemicals can also be used to improve volumetric sweep in gas-injection-based EOR processes by generating "foam" in porous media. Sweep improvement processes have an economic advantage over surfactant/polymer flooding in that the injected surfactant is not expected to permeate throughout the whole volume of the reservoir. It is sufficient to sweep only the zones in which the injected gas (or solvent or steam) is fingering through. Such zones are typically generated when fluids with high mobilities, such as gas, or steam, are injected, or are inherently present in highly heterogeneous or fractured reservoirs. Incremental oil recovery is not achieved by the surfactant's action on the residual oil but because of the improved volumetric sweep by the injected gas. Sweep improvement applications would also benefit from low surfactant adsorption, as the surfactant propagation is also limited by the surfactant adsorption at the solid/liquid interface.

This paper summarizes recent research results obtained in PRI laboratories that deal with various options for minimizing foam-forming surfactant adsorption. Even though the results are specific to the surfactant classes studied, the principles involved should be also applicable to surfactants being considered for surfactant/polymer flooding applications.

Three approaches to minimizing surfactant adsorption are discussed: matching surfactant type to specific reservoir rocks based upon the solid's surface charge, use of sacrificial adsorbates, and application of surfactant mixtures.

2 SURFACTANT RETENTION IN POROUS MEDIA

Surfactant adsorption is only one of several possible physico-chemical mechanisms that contribute to surfactant retention in porous media.[2-5] Phase trapping of surfactants in an immobile hydrocarbon phase and surfactant precipitation caused, for example, by changes in brine hardness, are also potential causes of surfactant losses.

The principal difference between adsorption and the other main causes of surfactant losses is that surfactant trapping and precipitation can be eliminated completely by proper surfactant screening and process design. Elimination of surfactants that may partition exclusively into an immobile hydrocarbon phase or that may precipitate under conditions they may encounter in the reservoir should be a part of every surfactant screening process. However, the enormously large solid surface present in petroleum reservoirs cannot be eliminated, and thus some adsorption of surface active materials is inevitable.

Determination of Surfactant Adsorption Isotherms

Adsorption isotherms were determined from core flooding experiments during which a slug of surfactant solution tagged with tritiated water as a brine tracer was injected into a confined core plug and displaced by several pore volumes of brine. The adsorption isotherms were evaluated by matching effluent profiles of the surfactant and the tracer with the PRI numerical model for flow of adsorbing surfactants in porous media.[6] This model takes advantage of the surface excess variable to describe adsorption at the solid/liquid interface and is specifically tailored for adsorption of surfactants from aqueous solutions. It is based upon the assumption that surfactants adsorb only from the surfactant monomer phase. As a result, sharply rising isotherms are produced that plateau at the critical micelle concentration (CMC).

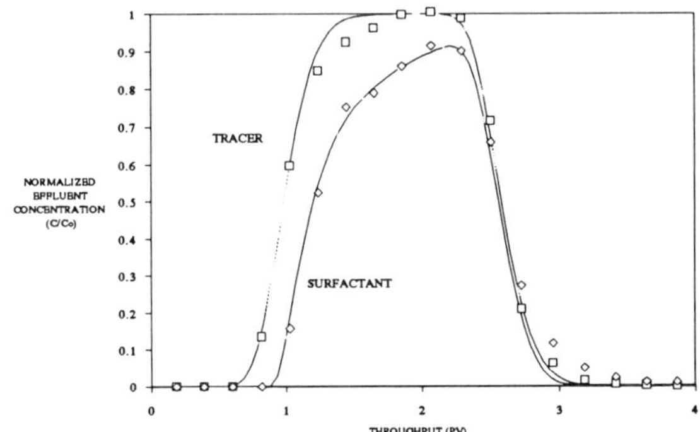

Figure 1.

Effluent profiles resulting from injection of a nonadsorbing tracer and an anionic surfactant into clean Berea sandstone. (Points are experimental data; lines represent numerical model.)

Typical tracer and surfactant effluent profiles along with the effluent pH are shown in Figure 1. The adsorption isotherm calculated from these data is shown in Figure 2, in which the x-axis is on a logarithmic scale in order to see the isotherm details below the CMC concentration.

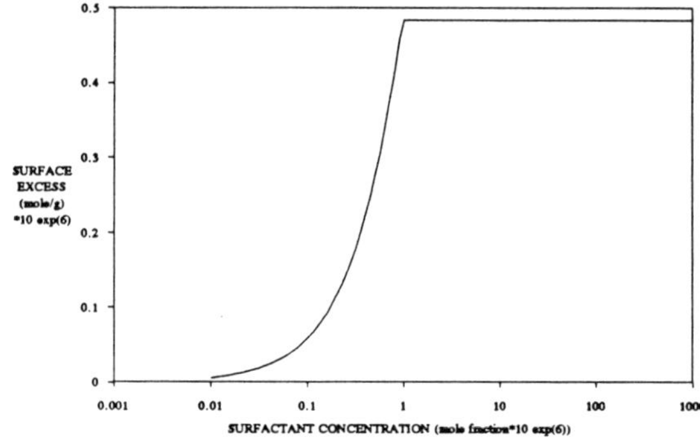

Figure 2.

Adsorption isotherm of an anionic surfactant on clean Berea sandstone.

3 MINIMIZING SURFACTANT ADSORPTION

Matching Surfactants to Specific Reservoir Solids

A survey in the *Oil & Gas Journal* showed that a large proportion of EOR field applications (63% in Canada, 16% in the U.S.A.) that utilize gas or solvent injection were conducted in carbonate reservoirs.[7] Since the vast majority of data on surfactant adsorption on reservoir solids is for sandstones, PRI initiated studies whose main objectives were to compare adsorption of two types of foam-forming surfactants (anionic and amphoteric) on two types of reservoir solids (sandstone and carbonate).[8]

In addition to comparing adsorption levels under identical conditions, an attempt was made to elucidate mechanisms that are primarily responsible for adsorption of each surfactant type. More specifically, the electrokinetic properties of both solids were measured as a function of brine pH, salinity, and hardness to assess whether the electrostatic interactions dominate the adsorption process.[9] Since these measurements were made in realistic reservoir brines containing up to 23,200 ppm total dissolved solids (TDS), special experimental techniques had to be utilized. The results showed significant variations from those obtained in distilled water or brines of nominal electrolyte content.

Two classes of commercially available foam-forming surfactants that generated significant mobility reductions in cores containing reservoir brines of high salinity and hardness under reservoir conditions were chosen for the comparative adsorption study.[10] The anionic surfactant was a 44% active mixture of a diphenyl ether disulfonate and an alpha-olefin sulfonate (sold under trade name Dow XS84321.05 by Dow Chemical Canada), and the amphoteric surfactant was a 30% active alkyl amido betaine (sold under the trade name Empigen BT by Albright and Wilson).

Berea sandstone and Indiana limestone were selected as representative solids. The main mineral components of these two rocks and specific surface areas are listed in Table 1.

Table 1 Mineral Composition and Specific Surface Areas of the Solids

Berea Sandstone	%	Indiana Limestone	%
Quartz	87.8	Calcite	99
Kaolinite	6.3	Quartz	1
Feldspar	3.8	Illite	trace
Dolomite	1.1		
Illite	1.0		
Chlorite	trace		
Specific Surface Area	$0.91 m^2/g$	Specific Surface Area	$0.37 m^2/g$

Adsorption plateaus for all twelve isotherms (two surfactants in three brines on two solids) are shown in Figure 3.

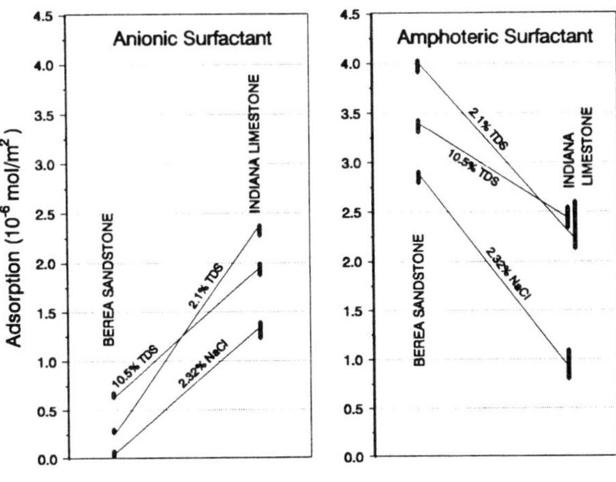

Figure 3.

Adsorption plateaus for two surfactant types on two solids in three brines.

Adsorption trends are opposite for the two solids. The anionic surfactant adsorbs less on the Berea sandstone from all three brines, while the opposite is true for the amphoteric surfactant. The differences in adsorption are substantial and their impact on the surfactant propagation in radial coordinates is schematically illustrated in Figure 4.

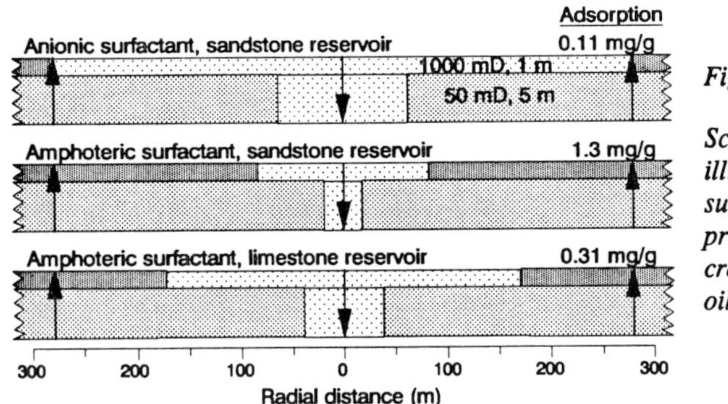

Figure 4.

Schematic illustration of surfactant propagation into a cross section of an oil reservoir.

The above comparison is for the propagation of anionic and amphoteric surfactants in sandstone and limestone reservoirs containing 2.1% TDS brine. In the sandstone, the amphoteric surfactant would propagate only 83 m from the wellbore in the more permeable zone, while the less-adsorbing anionic surfactant would reach all the way to the producing well, 281 m away. The amphoteric surfactant in the limestone reservoir would have a contact radius of 171 m in the more permeable zone. These are substantial differences in propagation considering the radial geometry, and the foam performance would be significantly different for each reservoir/surfactant combination.

Such potential differences in foam performance caused by different limits of foam-forming surfactant propagation require that adsorption information be made an integral part of the surfactant selection process. However, adsorption measurements are time consuming and sometimes difficult to make at all, especially when complex commercial mixtures are considered. Thus even approximate a priori assessment of surfactant adsorption would be useful.

There are several possible physico-chemical mechanisms affecting adsorption of surfactants at the solid/liquid interface: electrostatic attractions/repulsions, van der Waals interactions among adsorbed molecules, covalent bonding, non-polar chain/solid interactions, hydrogen bonding, and solvation/desolvation of the surfactant and the solid. Additional mechanisms have been suggested for adsorption on salt-like minerals such as calcite.[11] Electrostatic interactions are often significant or probably even dominant for charged surfactants. Whether these interactions enhance or reduce adsorption depends on the ionic charges of the surfactant and the surface charge of the reservoir minerals in reservoir brines.

The sensitivity of electrophoretic mobilities of Berea sandstone and Indiana limestone to brine pH, salinity, and hardness is shown in Figures 5 and 6.[9] These data show that the assumption of negative sandstone and positive limestone surfaces may not be correct in brines of certain pH and, especially, hardness.

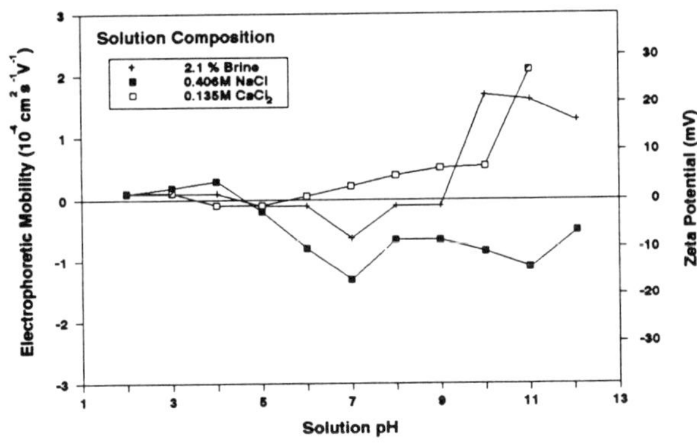

Figure 5.

Dependence of electrophoretic mobilities of Berea sandstone particles on brine pH, salinity, and hardness.

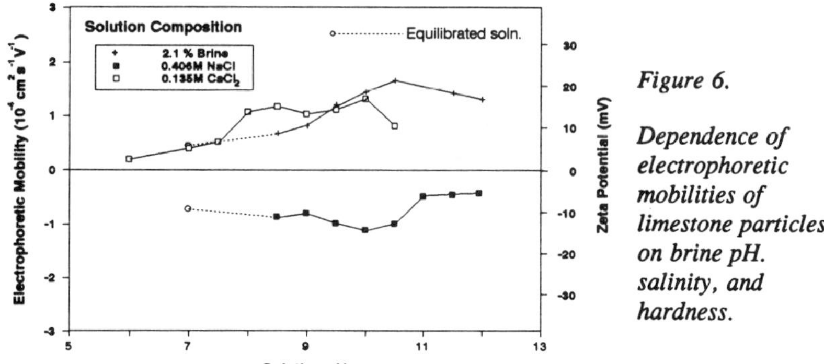

Figure 6.

Dependence of electrophoretic mobilities of limestone particles on brine pH. salinity, and hardness.

Detailed examination of all twelve adsorption isotherms and the trends in the electrophoretic mobilities of both solids in the brines of different salinity and hardness show that adsorption of the anionic surfactant is consistent with the electrostatic interactions in all reported experiments.[8] For the amphoteric surfactant, the relationship between electrophoretic mobility and adsorption is not as straightforward. This indicates that physico-chemical mechanisms other than electrostatic interactions also contribute to the adsorption process or that the dominant charge on the amphoteric surfactant molecule may be affected by the brine pH and composition.

Sacrificial Adsorbates

The idea of preflushing a reservoir with cheaper chemicals, such as lignosulfonates, to cover the adsorption sites and thus reduce subsequent surfactant adsorption has been extensively studied. However, there is an inherent problem with all reservoir preflushes in that it is extremely difficult to place the sacrificial chemicals in the same zones as the surfactants that follow them. This problem can be traced to the changed mobility during surfactant injection that can be caused either by higher viscosity of the surfactant solution or by the fact that residual oil is being mobilized and moved ahead of the surfactant bank.[12] In foam flooding, the desired foam action of reducing fluid mobilities ensures development of different flow patterns, thus defeating the purpose of the sacrificial adsorbates.

Proper functioning of sacrificial adsorbates is based upon two basic assumptions. The first is that no significant desorption of the sacrificial agent takes place when the surfactant solution reaches the surface with the preadsorbed materials. The second is that sacrificial agents adsorb on the same sites that would later attract the surfactant.

There is very limited information on adsorption/desorption kinetics of sacrificial materials,since they typically are complex mixtures, such as lignosulfonates. Complexities of adsorption/desorption behavior of surfactant mixtures are examined next, with the implication that similar behavior may be exhibited by materials considered for sacrificial adsorption.

Adsorption and desorption kinetics can also be estimated from effluent data used for determining the equilibrium surface excess isotherms.[5,6] For example, the anionic surfactant on Berea sandstone (Figures 1 and 2) adsorbs at one reciprocal hour, while its desorption rate constant is 0.001 reciprocal hours (first order rate equations are assumed for both adsorption and desorption). Similar values, 0.2 - 1.0 hr^{-1} for adsorption and 0.01 - 0.001 hr^{-1} for desorption, were obtained in all experiments reported in this paper.

Figure 7 shows normalized surfactant adsorption and desorption as a function of time based on rate constants for the anionic surfactant on Berea sandstone. These calculations show that there would be significant surfactant desorption from the solid surface within several weeks. If sacrificial materials desorb anywhere near these rates, their usefulness would be severely limited in field applications that typically require protection from surfactant adsorption for months, or even years.

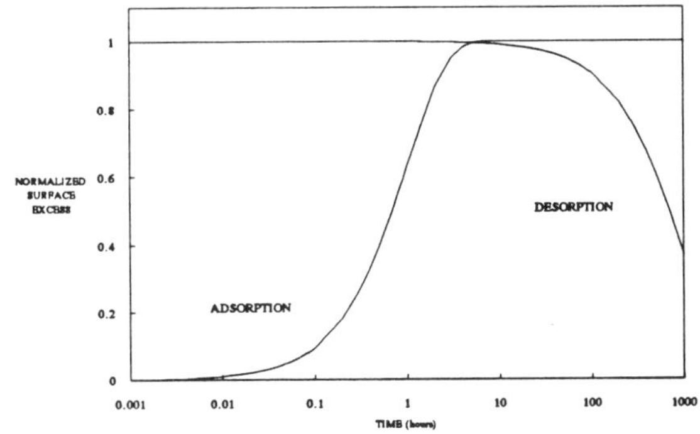

Figure 7.

Normalized adsorption and desorption of an anionic surfactant in 2.1% TDS brine in Berea sandstone.

The second uncertainty generally associated with sacrificial adsorbates is demonstrated next in experiments with sequential injection of foam-forming surfactants into Berea cores. In the first experiment, a slug of anionic surfactant was followed by a brine injection, and then by an injection of the same surfactant. In a way, the first surfactant was being used as a sacrificial adsorbate for the second injection of the same surfactant. The results, shown in Figure 8, confirm that the surfactant injected in the second slug flows out of the core outlet at the same time as the nonadsorbing tracer, indicating that there is no additional adsorption. This is exactly what one would hope to achieve with a sacrificial adsorption agent.

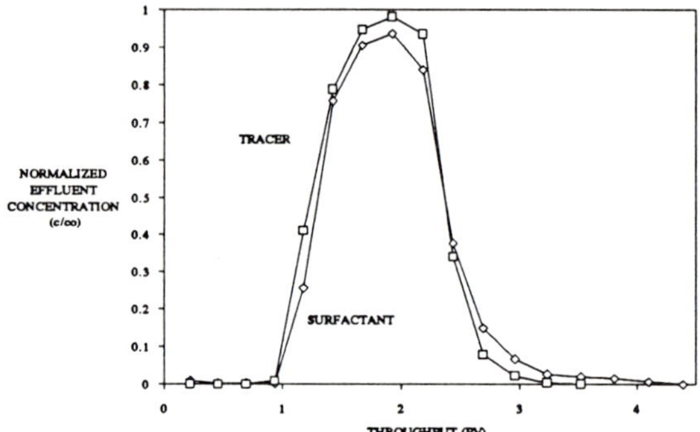

Figure 8.

Effluent profiles of a nonadsorbing tracer and an anionic surfactant injected after prior injection of the same surfactant.

If the surfactant in the second slug is different from that in the first slug, there may be some adsorption of the second surfactant, as indicated by its delay relative to the nonadsorbing tracer. Figure 9 shows an example of such behavior in the case of an amphoteric surfactant following an anionic surfactant. This would be an indication that the second surfactant may be adsorbing on sites different from those on which the first surfactant was preadsorbed. This interpretation is confirmed by the fact that no apparent desorption of the first surfactant took place, since no anionic surfactant was found in the effluent during and after the second surfactant injection.

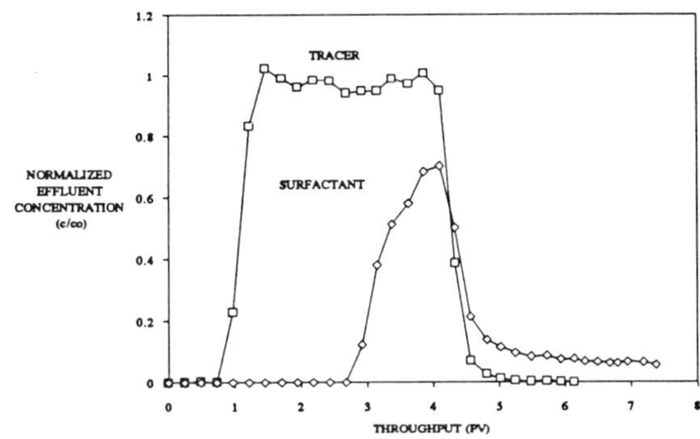

Figure 9.

Effluent profiles of a nonadsorbing tracer and an amphoteric surfactant injected after the earlier injection of a slug of an anionic surfactant.

Finally, the worst scenario is demonstrated in Figure 10. Effluent profiles shown there resulted from the reverse order of surfactant injection from the previous example. That is, the amphoteric surfactant was injected in the first slug as a sacrificial adsorbate and was followed with a slug of brine and a slug of the anionic surfactant. The effluent analyses showed that more anionic surfactant was adsorbed on Berea covered with some amphoteric surfactant than was adsorbed on clean Berea sandstone surface (Figure 1). In addition, significant amounts of amphoteric surfactant were found in the effluent, indicating strong desorption from the surface (Figure 10). Results such as these lead to speculation that there is some interaction between the adsorbed amphoteric and the anionic surfactant molecules. Obviously, such a scenario would be disastrous for sacrificial adsorbates.

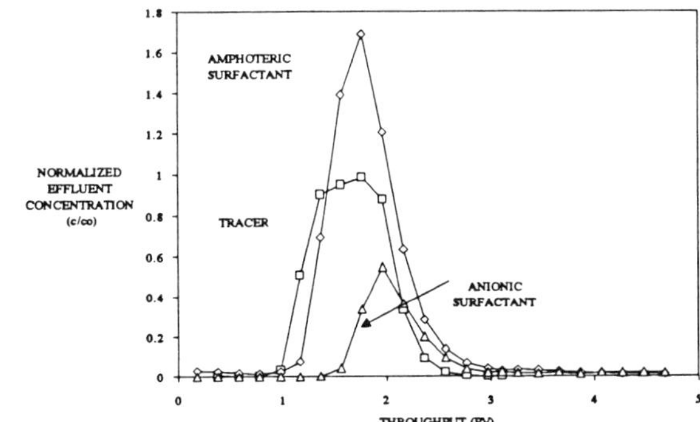

Figure 10.

Effluent profiles of a nonadsorbing tracer, and anionic and amphoteric surfactants resulting from the injection of an anionic surfactant into Berea sandstone previously flooded with an amphoteric surfactant.

In summary, preflushes with sacrificial adsorbates frequently show some improvements in the chemical flooding processes,[13] but it remains to be demonstrated that the additional costs related to sacrificial materials can be fully compensated for by lowering the requirements for the primary surfactant.

Surfactant Mixtures

Another promising approach to minimizing surfactant adsorption is through the utilization of surfactant mixtures. It was shown previously that surfactants adsorb from the monomer phase and that adsorption from surfactant mixtures can be significantly affected by competition for the surfactant monomer species by the micelles.[14,15]

This competition could be taken advantage of by "hiding" the surfactant molecules in the micelles and thus preventing them from adsorbing at the solid/liquid interface. An example of two individual surfactant adsorption isotherms and their mixture that obeys the ideal micelle mixing rule is shown Figure 11.

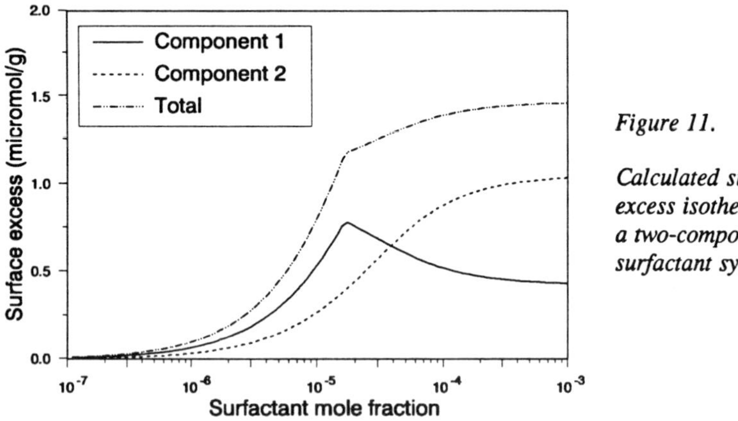

Figure 11.

Calculated surface excess isotherms of a two-component surfactant system.

The isotherm for one surfactant exhibits a well-defined maximum and its adsorption at concentrations above its CMC is relatively low. Ideally, the properties of surfactants forming the mixture would be such that the sum of adsorption of both surfactants is lower than the sum of adsorption of both components from single-component solutions. Two examples of adsorption from commercially available surfactant mixtures are shown next to demonstrate the potential of this idea.

Effluent profiles resulting from an injection of a mixture of two components (C12 and C18 fractions) extracted from a commercially available sulfobetaine (sold under trade name Varion CAS by Sherex Chemicals) are shown in Figure 12.

One would expect that the heavier C18 fraction would elute later than the lighter C12 fraction, since the higher molecular weight component should adsorb more strongly than the lighter component.[14] The results in Figure 12 show just the opposite. The C18 component has a lower CMC which forces the C18 molecules into the micelles, preventing them from adsorption at the solid/liquid interface. This reversal of component order in the effluent is correctly predicted by the PRI adsorption model that includes the ideal mixed-micelle theory for calculating compositions of the mixed micelles. The model matches the trends shown in Figure 11.[6]

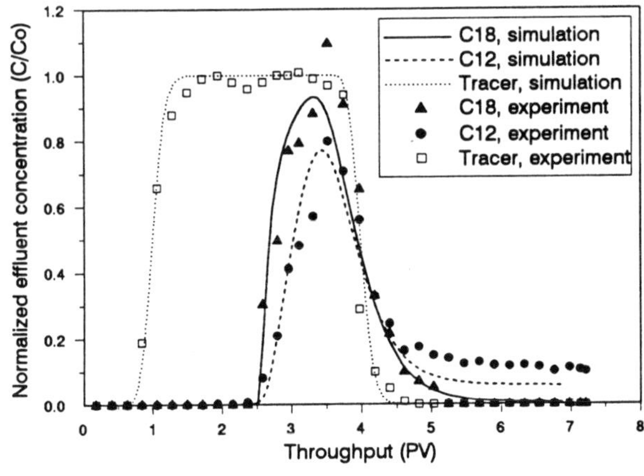

Figure 12.

Effluent profiles of a two-component surfactant mixture.

The next example illustrates adsorption from a mixture of different classes of surfactants for which the ideal mixed-micelle theory probably does not apply. The results, summarized in Table 2, show that the betaine adsorption from a mixture is almost one order of magnitude lower than the adsorption from a single component solution.

Table 2 Comparison of Adsorption from Single-Component Surfactant Solutions and Their Mixture

	Anionic	Betaine
Maximum adsorption on Berea sandstone from a single-component solution (mg/g)	0	1.7
Maximum adsorption on Berea sandstone from a 50/50 mole mixture (mg/g)	0	.2

4 SUMMARY

The objective of this paper is to summarize some of the approaches that can be considered to lower adsorption of EOR surfactants at the solid/liquid interface. It is shown that

- Matching the ionic surfactants and reservoir solids based upon the solid surface charge properties may yield significant benefits in lower adsorption.

- Determination of surface charges of reservoir minerals should be made in reservoir brines, since the magnitude and even the sign of the charge depends strongly on brine pH, salinity, and especially hardness.

- Sacrificial adsorbates may not be effective unless they adsorb on the same sites that would later attract the surfactant.

- Sacrificial adsorbates may not be beneficial in field applications unless their kinetics of desorption (in the presence of surfactants) is not several orders of magnitude slower than the desorption rate of surfactants studied in this work.

- Application of binary surfactant mixtures may offer significant benefits through lowered adsorption of both components.

5 REFERENCES

1. S. Sircar, Private Communication, 1991.
2. J. Novosad, '1981 European Symposium on EOR', Elsevier Sequoia S.A., Lausanne, Switzerland, 1981, 101.
3. G. Hirasaki, SPEJ, 1982, 22, 181.
4. J. Novosad, SPEJ, 1982, 22, 962.
5. K. Mannhardt and J.J. Novosad, Rev. Inst.Français Pétrole, 1988, 43, 659.
6. K. Mannhardt and J.J. Novosad, J.Petrol.Sci.Eng., 1991, 5, 89.
7. L.R. Aalund, Oil & Gas J., 1988, 86, 33.
8. K. Mannhardt, L.L. Schramm, J.J. Novosad, 'Proceedings of the 65th SPE Annual Technical Meeting held in New Orleans', SPE 20463, 1990, 31.
9. L.L. Schramm, K. Mannhardt, and J.J. Novosad, Colloids and Surfaces, 1991, in print.
10. J.J. Novosad and E.F. Ionescu, 'Proceedings of the 38th Annual Technical Meeting of CIM held in Calgary', CIM 87-38-80, 1987, 1345.
11. P. Somasundaran and H.S. Hanna, SPEJ, 1985, 25, 343.
12. R.N. Healy, R.L. Reed, C.Carpenter, SPEJ, 1975, 15, 87.
13. J. Novosad, JCPT, 1984, 23, 24.
14. F.J. Trogus, R.S. Schechter, and W.H. Wade, J. Coll. Interf. Sci., 1979, 70, 293.
15. K. Mannhardt and J.J. Novosad, Chem. Eng. Sci., 1991, 46, 75.

6 ACKNOWLEDGEMENT

The author would like to thank Karin Mannhardt, whose help and many valuable discussions and suggestions are gratefully acknowledged.

An Assessment of the Suitability of Acid Stimulation for Clyde Production Wells

C.A. Cade, J.A. Simmons, P.S. Smith,

BP RESEARCH, CHERTSEY ROAD, SUNBURY-ON-THAMES, MIDDLESEX, UK

and G.L. Wood

BP EXPLORATION, FARBURN INDUSTRIAL ESTATE, DYCE, ABERDEEN, UK

ABSTRACT

Hydrochloric acid (HCl) matrix stimulations are amongst the most commonly used methods of improving the productivity of damaged wells. Such treatments are particularly effective in limestone formations, carbonate cemented sandstones or where damage was caused by acid soluble species (for example, calcium carbonate lost circulation material).

This paper describes a study to assess the suitability of an HCl stimulation treatment for damaged Clyde production wells. Acid solubility studies on Clyde core showed very low weight losses with 15% HCl; this was consistent with mineralogical data which showed little carbonate cement to be present. No significant increase in mineral dissolution was observed with mud acid.

Further testing identified a serious incompatibility; viscous emulsions formed when HCl was mixed with Clyde crude oil. The presence of soluble iron exacerbated this incompatibility. In coreflood tests, reductions in permeability were observed, rather than the anticipated stimulation (permeability increase). This incompatibility was believed to be due to the precipitation of asphaltenes from the crude oil upon reaction with the acid. The use of a mutual solvent, to remove residual oil from the core, failed to prevent the permeability reduction. In contrast, it was shown that a scale dissolver exhibited no incompatibilities and produced a small increase in permeability in a coreflood test.

The study clearly demonstrates that prior to using any chemical stimulation treatment (especially acids), testing needs to be performed to assess the potential benefits, or damage, that is likely to result.

1 INTRODUCTION

Hydrochloric acid (HCl) matrix stimulations are amongst the most commonly used methods of improving the productivity of damaged wells. Such treatments are particularly effective in limestone formations, carbonate cemented sandstones or where damage was caused by acid soluble species (for example, calcium carbonate lost circulation material).

During matrix stimulation treatments, formation damage can occur due to reaction of the crude oil with the injected acid (or additives in the acid), producing an oily "sludge". Nearly all crude oils contain constituents, such as asphaltenes, resins, paraffin waxes and other high molecular weight hydrocarbons, which can cause such sludge formation[1]. However, heavy crude oils (less than 30° API gravity), in particular, have a tendency to produce sludge when mixed with acid[1-5]. The asphaltic components are thought to be the main cause of these adverse reactions with acid[6].

Asphaltenes found in crude oils have high molecular weights (10,000 to 100,000), a density of about 1.2 g/cm^3 and occur as colloidal suspensions (typical particle diameter between 0.003 and 0.0065 μm)[7,8]. The nature of asphaltic materials in crude oils is complex, the chemical structures comprise many aromatic/ heterocyclic rings. These structures are susceptible to protonation by acids at, for example, the basic nitrogen groups in the structure. Such protonation results in a large increase in polarity, resulting in precipitation from the low polarity oil.

The precipitated sludges are only slightly soluble in their original crude oil, consequently they are not easily re-dissolved during hydrocarbon production. This type of damage can often be identified due to the production of asphalt-like material in the returned acid[1]. Ultimately, however, the damage results in the slow "clean-up" of the well.

It has been estimated that 30%, or more, of US crude oils form sludge when mixed with acid[9]. For a randomly selected set of Michigan crude oils, 85% of the samples precipitated sludge[10]. The majority of these gave 1-3% w/v of the sludge but in extreme cases as much as 17% w/v was found.

In addition to the adverse reaction between acid and crude oils, reactions between acid and the rock matrix can also be problematic. Aluminium is extracted from aluminosilicates by reaction with HCl[11-18]. This weakens the structure of clay minerals, which in simple terms comprise layers of octahedral sheets of aluminium oxide alternating with tetrahedral sheets of silicon oxide, resulting in fragmentation[19-22]. Such physical degradation of rock constituents into mineral fragments and amorphous silica particles has important implications for stimulation treatments. The injection of HCl into clay-rich or feldspathic sandstones can create fines which, in turn, can cause formation damage.

Owing to the potential for matrix acid treatments to reduce well productivity it is important that their suitability is assessed. This paper describes one such assessment for damaged Clyde production wells.

2 BACKGROUND

The Clyde field, situated in the central North Sea, produces crude oil from the Jurassic sands, which has a high propensity for asphaltene precipitation.

Following drilling, coring and completion, well A exhibited a poor productivity index, especially when compared to previous wells in the field. A chemical stimulation treatment was identified as a potentially cost effective method to attempt to remove any acid soluble damage, without having to workover the well. The specific aim of this study was to identify whether acid stimulation was a suitable treatment for the improvement of the productivity index of this well. At this stage, however, the exact nature of the damage was still unknown.

3 MATERIALS

Simulated Formation Water (SFW). A formulation obtained from the analysis of a water sample from well A was used in this study, and is given below.

		g/litre
Calcium Chloride	$CaCl_2.6H_2O$	45.0951
Magnesium Chloride	$MgCl_2.6H_2O$	6.4370
Potassium Chloride	KCl	6.4828
Strontium Chloride	$SrCl_2.6H_2O$	1.7040
Barium Chloride	$BaCl_2.2H_2O$	0.0445
Sodium Chloride	NaCl	144.4311
Sodium Sulphate	Na_2SO_4	0.3401
Sodium Hydrogen Carbonate	$NaHCO_3$	0.3167

pH = 6.8 (at surface)

15% Hydrochloric Acid (Live Acid). Hydrochloric acid (15%) was prepared from Analar Grade hydrochloric acid (35%) and deionised water. A proprietary corrosion inhibitor was added to the acid. Iron$^{(III)}$ chloride (1% w/w) was added to a portion of the acid, in order to simulate iron that may be dissolved by the acid during the stimulation treatment.

Spent 15% Hydrochloric Acid. 15% hydrochloric acid was prepared as above and then neutralised by addition of sodium hydrogen carbonate. The free liquid was then decanted off and used. Iron$^{(III)}$ chloride (1% w/w) was also added to a portion of the acid, for the reason described above.

Mud Acid. Standard strength mud acid (12% hydrochloric acid/ 3% hydrofluoric acid) was prepared by diluting concentrated Analar Grade hydrochloric acid (HCl, 35%) and Analar Grade hydrofluoric acid (HF, 60%) with deionised water to the required concentration.

Asphaltene Dissolver. A commercially available asphaltene dissolver was used.

Mutual Solvent. A proprietary mutual solvent was tested.

4 METHODS

Fluid Compatibility Tests. Fluid-fluid compatibility tests were conducted using a procedure based upon an API procedure[23], but modified by BP Sunbury. Equal volumes of each test fluid were mixed together and then heated in an oil bath at 95°C. The temperature of these tests was chosen to simulate reservoir temperature but was limited by the boiling point of the aqueous phase. The combinations of fluids tested were based upon the probable inter-mixing that could occur in the formation. For all of the tests, any interaction which occurred immediately upon mixing was recorded. Thereafter, the mixtures were observed at regular intervals and any incompatibilities noted.

One percent of iron chloride, by weight, was added to several of the tests to simulate the presence of soluble iron (from the tubulars and/or from iron bearing minerals in the formation).

Acid Dissolution of Reservoir Rock from Well A. Samples of rock from well A were tested to assess their general suitability for acidisation. All of the samples were treated with 15% HCl. For each test, 100ml of the acid was added to a known mass of rock and the mixture heated to 95°C. On the completion of reaction the mixture was filtered and any residue dried and weighed.

Preparation of Preserved Core and Core Flood Testing. The preserved core obtained for this study was cut, under simulated formation water, into 1.5 inch diameter plugs. The plugs were visually inspected and those suitable were selected for coreflood testing. These plugs were then solvent cleaned using toluene and methanol. For each coreflood test, the prepared plug was mounted in the permeameter and pressurised (see Figure 1). A confining pressure of 7000 psi was used to simulate reservoir overburden stress and a pore pressure of 5000 psi was used to simulate reservoir pore pressure.

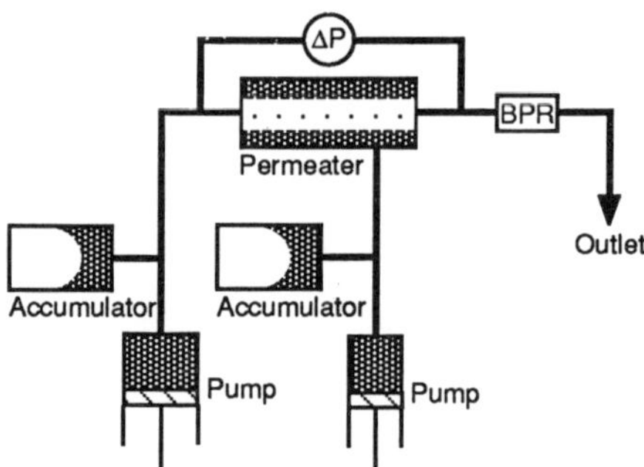

Figure 1 Core Flood Apparatus : Schematic Layout

The permeameter temperature was raised to 120°C. These conditions were chosen to best reflect downhole conditions, within the capability of the core flood apparatus. The core plug was then flooded with SFW to displace the methanol and the base permeability of the plug, to SFW, was determined in both the forward direction (simulating flow into the reservoir) and the reverse direction (simulating flow from the reservoir). After achieving acceptable base permeability measurements, the coreflood test was undertaken.

5 RESULTS

Fluid compatibility tests. A total of 11 compatibility tests were performed, of which 8 tests showed some form of incompatibility :

Fully Compatible	Incompatibility Observed
Crude/SFW	SFW/Spent 15% HCl
SFW/15% HCl	SFW/Spent 15% HCl + 1% $FeCl_3$
SFW/15% HCl +1% $FeCl_3$	Crude/15% HCl
	Crude/Spent 15% HCl
	Crude/15% HCl + 1% $FeCl_3$
	Crude/Spent 15% HCl + 1% $FeCl_3$
	Crude/Mud acid
	Crude/Mud acid +1% $FeCl_3$

SFW/Spent 15% HCl. A clear solution resulted from the mixing of these fluids, a slight incompatibility was noted with a small amount of precipitate being formed (≃ 1% by volume). The precipitate was formed by reaction of the excess sodium hydrogen carbonate, in the spent acid, with the soluble calcium in the formation water.

<u>SFW/Spent 15% HCl + 1% Iron Chloride.</u> The mixing of the two fluids produced a clear solution. However, on the addition of iron$^{(III)}$ chloride a large quantity of precipitate (>10% v/v) was produced.

<u>Crude/15% HCl.</u> The initial mixing of the two fluids produced a totally emulsified mixture. Over a period of 24 hours the two fluids began to separate and a solid plug of emulsion (≃ 10%) was left at the interface.

<u>Crude/Spent 15% HCl.</u> An emulsified fluid was formed on mixing the two fluids. This emulsion gradually separated out to leave two immiscible fluids with 4% emulsion remaining at the interface. The incompatibility observed was slight.

<u>Crude/15% HCl + 1% Iron chloride.</u> An extremely viscous emulsion resulted on the mixing of these fluids. Over a period of 24 hours only a slight separation occurred, with over 40% of the fluids remaining emulsified. The amount of emulsion produced, and its viscosity, increased significantly due to the presence of iron.

<u>Crude/Spent 15% HCl + 1% Iron chloride.</u> Despite a totally emulsified phase resulting from the mixing of these fluids, a rapid partial separation was achieved with the fluids separating into two immiscible liquids after 15 minutes. However there was a large amount of orange solid produced, both at the interface of the two fluids and at the base of the acid (10% v/v).

<u>Crude/Mud Acid.</u> The mixing of these fluids produced an opaque emulsion that partially separated into three phases in a short time. The phases consisted of 20% clear acid at the base, 45-50% mobile crude oil at the top, with 30-35% of a totally solid viscous emulsion as the mid phase.

<u>Crude/Mud Acid + 1% iron chloride.</u> The addition of iron chloride to the mud acid/crude mix increased the amount of emulsion formed and its viscosity. After a period of 24 hours the phases consisted of 10% clear acid, 35% mobile crude oil and 55% remained as an immobile viscous emulsion.

<u>Acid Dissolution of Reservoir Rock.</u> The acidisation of Clyde reservoir core samples, from various depth intervals of Well A, was carried out to assess the general suitability of Clyde reservoir rock for acid stimulation. The percentage of rock dissolved after acidisation with 15% HCl is shown in Table 1.

<u>Corrosion.</u> Corrosion in the presence of acid was considered due to the use of a 22% chrome liner. Using a 5 inch length of this duplex stainless steel, the formulation 15% w/w hydrochloric acid formulation (including corrosion inhibitor) was tested. A corrosion weight loss of 0.016 lb/sq.ft. was recorded when the steel was aged

Table 1

INTERVAL	% WEIGHT LOSS
a	1
b	2
c	30
d	0.5
e	2
f	6
g	3
h	3
i	3

in the fluid at 290°F for 6 hours. It was concluded that the proposed corrosion inhibition treatment gave adequate protection.

Geological Background and XRD Analysis. The rock samples used in this study were from well A. This reservoir zone, in common with two other Clyde wells, exhibited natural fractures which showed cementation with silica, dolomite and calcite. The samples were all of similar grainsize, very fine or very fine-fine sand grade (80 to 120μm), with a moderately compacted grain fabric. The samples contained approximately 25% v/v matrix clay and were poorly cemented. It is likely that the presence of significant matrix clay prevented the nucleation of silica as grain overgrowths. Chalcedonic or microcrystalline silica cement was locally significant. Dolomite and ankerite, the HCl soluble components, were only patchily developed, as pore-filling rhombic crystals.

Table 2

	SAMPLE DEPTH I	II
WHOLE ROCK DATA	%	%
Illite	5	2
Chlorite	1	–
Quartz	17	76
Potassium Feldspar	9	11
Albite	10	5
Calcite	54	–
Ankerite	3	5
Haematite	–	–
Pyrite	Trace	1
CLAY FRACTIONS		
Illite	91	90
Chlorite	9	10

Acid solubility tests, and X-ray diffraction analysis (see Table 2), showed that calcite was only well developed at a few horizons, where it was extensive (up to 54% of solid rock volume). The acid solubility tests showed predominantly small weight losses, except in these calcite-rich bands. Sample I was not considered representative due to the very high calcite content.

Core Flood Tests. Three coreflood tests were undertaken to assess the effect of acid stimulation on Clyde core. A further series of coreflood tests were undertaken in order to compare the performance of a scale dissolver treatment with acid stimulation.

Core flood test No 1. This test was carried out on a plug that had been solvent cleaned. Simulated formation water was initially flowed through the core until a stable differential pressure (ΔP) had been achieved (70 psi at a flow rate 0.92 ml/min). Clyde crude was then flooded through the core in both directions until a stable ΔP was obtained. The test fluids were then injected through the plug and backflowed. These comprised 1 pore volume of asphaltene dissolver followed by 1 pore volume of a mixture of concentrated hydrochloric acid and mutual solvent (Ratio 1:4). Clyde crude was then passed through the core again, to assess the performance of the test fluid. This treatment gave rise to a significant reduction in permeability (see Table 3), in contrast to the anticipated stimulation (usually associated with acidisation).

The core plug used in this test had a low permeability compared to the typical permeability of the Clyde reservoir. This could have had a significant impact on the results obtained. Assuming the damage was due to incompatibility between the Clyde crude and acid, then the permeability reduction observed could be explained by ineffective displacement of the crude by the pre-flush (likely to occur due to low permeability core). A higher permeability core might allow a better displacement and thus minimise interaction between the crude and the acid.

Coreflood Test No 2. This test was carried out on a plug of higher permeability, that had also been solvent cleaned. SFW was initially flowed through the core until a stable ΔP was observed

Table 3

	Differential Pressure Before	Differential Pressure After	% Change In Permeability
Forward	72 psi	97 psi	-26%
Reverse	80 psi	102 psi	-22%
	(Flow Rate Throughout Test = 0.185 ml/min)		

Table 4

	Differential Pressure Before	After	% Change In Permeability
Forward	48 psi	48 psi	0%
Reverse	44 psi	48 psi	-10%
	(Flow Rate Throughout Test = 0.91 ml/min)		

(9.7 psi at flow rate 0.91 ml/min). Clyde crude was then flooded through the core in both directions until a stable ΔP had been achieved. The test fluid was then injected through the plug and backflowed. The test fluid comprised 1 pore volume of asphaltene dissolver followed by 1 pore volume of a mixture of concentrated hydrochloric acid and mutual solvent (Ratio 1:4). Clyde crude was then passed through the core again to assess the performance of the test fluid.

No significant change in permeability was observed after acid stimulation (see Table 4). The small permeability reduction in the reverse direction could indicate a small amount of damage; however this was within the experimental error of the method. The asphaltene dissolver/mutual solvent treatment virtually eliminated any adverse reactions between the acid and the crude in this higher permeability core. However, as no increase in permeability was observed, the acid stimulation treatment does not offer any benefit.

Coreflood Test No3. - HCl/Mud Acid treatments. Clyde crude was flooded through the core in both directions until a stable ΔP had been achieved. The test fluid was then injected through the plug and backflowed. The test fluid comprised 2/3 pore volume of asphaltene dissolver followed by 2/3 pore volume of a mixture of concentrated hydrochloric acid and mutual solvent (Ratio 1:4), followed by 2/3 pore volume of a mixture of mud acid and mutual

Table 5

	Differential Pressure Before	After	% Change In Permeability
Forward	32 psi	38 psi	-16%
Reverse	30 psi	39 psi	-23%
	(Flow Rate Throughout Test = 0.91 ml/min)		

Table 6

	Differential Pressure Before	After	% Change In Permeability
Forward	40 psi	37 psi	+ 8%
Reverse	45 psi	40 psi	+12%
	(Flow Rate Throughout Test = 0.50 ml/min)		

solvent (Ratio 1:4). Clyde crude was then passed through the core again to assess the performance of the test fluid. The results are given in Table 5.

A reduction in permeability was observed whereas an increase in permeability would be anticipated from an acid stimulation treatment. It is possible that the asphaltene dissolver/mutual solvent treatment failed to eliminate adverse reactions between the acid and the crude in this test. It is therefore concluded that mud acid stimulation treatment is not appropriate.

Coreflood Test No 4. A coreflood test was performed in order to assess the potential for a scale dissolver treatment to increase permeability. Clyde crude was flooded through the core plug in both directions until a stable ΔP had been achieved, prior to the injection of the scale dissolver. The treatment resulted in a small improvement in permeability (see Table 6). The magnitude of this improvement could be due to a) the limited volume of treatment fluid used, b) the short contact time between the treatment fluid and the core plug or c) limited quantities of "soluble" material in the core plug.

Coreflood Test No 5. An additional coreflood test was performed to assess whether the asphaltene dissolver caused any reduction in permeability. No detrimental effects were observed (see Table 7).

Table 7

	Differential Pressure Before	After	% Change In Permeability
Forward	32 psi	33 psi	- 3%
Reverse	34 psi	34 psi	0%
	(Flow Rate Throughout Test = 0.45 ml/min)		

6 CONCLUSIONS

1. This study demonstrates the potential downside of conducting acid stimulation treatments on wells without first assessing the suitability of such treatments.

2. Clyde core exhibited low solubility in both hydrochloric acid and mud acid. Mineralogical analysis of core samples showed little acid-soluble carbonate cement. Hence, matrix acidization of "undamaged" wells in the Clyde field would not be beneficial. However, damage may have occurred after the coring operation, thus acidisation could still be of benefit on such wells.

3. A serious incompatibility between Clyde crude and acid was identified, thus it could be expected that acidisation could cause formation damage instead of stimulation. This damage could not be avoided, even with the use of an asphaltene dissolver and a mutual solvent to minimise contact between the crude and the acid.

4. An EDTA type scale dissolver gave a small degree of stimulation rather than damage, confirming that such a treatment would be more appropriate than acid stimulation.

REFERENCES

1. E.W.Moore, C.W.Crowe and A.R.Hendrickson; J.Pet.Tech.,1985,17, 1023.

2. T.S.Knobloch, A.Farouq and M.J.Trevino Diaz; SPE 7627,1976.

3. P.M.Lichaa; Can.Inst.Mining Met.Spec.,1977,17,609.

4. M.E.Newberry and K.M.Barker; SPE of AIME Prod.Oper.Symp.Proc., 1985,53,(SPE 13796).

5. I.C.Jacobs and M.A.Thorne; SPE 14823,1986.

6. P.M.Lichaa and L.Herrera; SPE 5304,1975.

7. P.A.Witherspoon and Z.A.Munir; Prod.Monthly,1960,Aug,20.

8. J.Briant and G.Hotier; Revue Inst.Français Pet.,1983,38,83.

9. R.F.Krueger; J.Pet.Tech.,1986,38,131.

10. L.Blodgett; Northeast Oil Reporter,1983,3,37.

11. S.Hulbert and D.Huff; Clay Minerals,1970,8,337.

12. S.Ziegenbalg and G.Haake; Light Met,1983,1119.

13. R.Olsen, W.Gruzensky, S.Bullard, R.Beyer and J.Henry; US Bureau of Mines Report of Investigations 8834.

14. S.Gajam and S.Raghavan; Hydrometallurgy,1985,15,143.

15. J.Rupert, W.Granquist and T.Pinnavaia; "Catalytic Properties of Clay Minerals," in Chemistry of Clays and Clay Minerals, Ed. A. Newman, Mineralogical Society Monograph No. 6, Longman Scientific and Technical (1987), 275-319.

16. C.Thomas, J.Hickey and G.Stecker; Ind.Eng.Chem.,1950,42,866.

17. K.H.Kanzler; Ger.Pat.Off.,1069583,1959.

18. J.Hodgkiss; UK Patent Spec.,1028475,1966.

19. R.Miller; Soil Sci.,1968,105,166.

20. G.Ross; Clays and Clay Minerals,1969,17,347.

21. M.Gastuche and L.Vielvoye; Bull.Soc.Chim.Fr.,1960,6,1216.

22. M.Gastuche, B.Delmon and L.Vielvoye; Bull.Soc.Chim.Fr.,1960,1, 60.

23. Laboratory testing of surface active agents for well stimulation; API RP42, January 1977.

The Use of Adsorption Thermodynamics in the Development of Scale Inhibitors for High Barium Content Oilfield Brines

P.J. Breen, H.H. Downs,

BAKER PERFORMANCE CHEMICALS, ERISMUS HOUSE, QUEENS SQUARE, MIDDLESBOROUGH, CLEVELAND TS2 1AA, UK

and B.N. Diel

CHEMICAL WASTE MANAGEMENT, 150 WEST 137TH ST., RIVERDALE, ILLINOIS 60627, USA

1 INTRODUCTION

To date, problems with barium sulfate scale inhibition in various North Sea oil fields have not yielded to conventional solutions[1-4]. A combination of low pH and high sulfate to barium ratios renders most commercially available scale inhibitors ineffective. Our primary objective has been to develop new and more effective inhibitors for such conditions. We sought to accomplish this with measurements of performance and adsorption thermodynamics as functions of variation in inhibitor structure.

The focus of this study is a mechanistic one; attention is centered on the interaction of scale inhibitors with growing barium sulfate embryos or nuclei in solution. Thus, our experiments are aimed at measuring the adsorption thermodynamics of scale inhibitors adsorbing to barium sulfate, rather than to reservoir rock.

The data obtained in this study provide new insight into the mechanisms of scale inhibition. Important questions dealing with the difficulty of scale inhibition at acidic pH, and key structural characteristics for performance are addressed.

Homologous series of phosphonate-type inhibitors were prepared in which molecular weight and charge density were systematically varied. The performance of these inhibitors under both seeded and nonseeded conditions was evaluated, and correlated with measurements of adsorption thermodynamics on barium sulfate. Improved insight into the mechanisms of scale inhibition has resulted in the development of new, proprietary inhibitors which outperform all other inhibitors tested under typical acidic North Sea reservoir conditions.

2 EXPERIMENTAL

All inhibitors used were prepared in house and were checked for completeness of reaction with H^1 and C^{13} NMR methods. Barium sulfate powder was purchased from Fisher Scientific Products and was used as received.

Static and Seeded Bottle Test

Inhibitors were tested in static and seeded bottle tests to evaluated relative performance. Synthetic brines were used and were based on analyses of actual North Sea formation brines and seawater brine. The compositions are given in Table 1.

The test was conducted by placing the desired amount of inhibitor in 85 mL of the barium-containing synthetic formation brine, followed by adjustment of the pH to the desired value (usually 4.0) with HCl. The sample was equilibrated in a water bath, along with a pH-adjusted sample of synthetic seawater, to 85°C, whereupon 15 mL of the seawater were added to the formation brine sample. The 85:15 ratio yields a $Ba^{2+}:SO_4^{2-}$ ratio of 1:1. After approximately 17 h the bottle was sampled and analyzed for barium content using atomic absorption spectroscopy. Filtration of the samples through a 0.45 μm filter was observed to have no effect on measured barium ion. 100% inhibition was taken to be 621 mg/L barium ion. Percent inhibition was calculated according to equation 1:

$$\% \text{ inhibition} = \frac{(\text{sample} - \text{blank})}{(621 - \text{blank})} \times 100 \tag{1}$$

Seeded test conditions were obtained by simply placing 0.5 g of barium sulfate powder in the 85 mL of formation brine prior to pH adjustment.

Table 1 - Synthetic Brine Compositions

Formation Brine		Seawater	
Salt	mg/L	Salt	mg/L
NaCl	65,520	NaCl	28,483
KCl	2,060	KCl	671
$CaCl_2 \cdot 2H_2O$	1,213	$CaCl_2 \cdot 2H_2O$	1,470
$MgCl_2 . 6H_2O$	360	$MgCl_2 . 6H_2O$	11,835
$BaCl_2 . 2H_2O$	1,300	Na_2SO_4	3,919

Adsorption Thermodynamics

Adsorption of scale inhibitors on barium sulfate was measured at two temperatures, 25°C and 85°C. 20 mL of a 500 ppm solution of an inhibitor in deionized water was pipetted into a test tube containing a weighed amount (5-20 g) of barium sulfate. The surface area of the barium sulfate was found from a BET analysis to be 2.994 x 10^4 cm^2/g. The test tube was capped and vigorously shaken over the course of a day and allowed to stand overnight. The tubes were then centrifuged. Samples of clear supernatant were taken and analyzed for inhibitor content using a Zr(IV)-arsenazo titration developed for this purpose.

Following the methods of Naono[5], ΔG_{ads} was calculated from equation 2,

$$\Delta G_{ads} = -RT\ln[(C_i-C_f)V/gC_fAt] \quad (2)$$

where C_i=500 ppm, C_f=final inhibitor concentration, g=grams of barium sulfate, A=barium sulfate surface area per gram, t=monolayer thickness, and V=volume of solution. ΔH_{ads} was calculated from equation 3,

$$\Delta H_{ads} = R\,\frac{T_1T_2}{T_1-T_2}\,\ln\left[\frac{C_{T2}C_i-C_{T2}C_{T1}}{C_{T1}C_i-C_{T2}C_{T1}}\right] \quad (3)$$

where C_{T1} and C_{T2} are the final inhibitor concentrations at temperatures T_1 and T_2. ΔS_{ads} was calculated from ΔG_{ads} and ΔH_{ads} using equation 4,

$$\Delta S_{ads} = (\Delta H_{ads}-\Delta G_{ads})/T. \quad (4)$$

Since no information concerning t, the monolayer thickness, was available, we have used for the purpose of our calculations the value given by Naono for sodium tripolyphosphate[5], which is 1×10^{-7} cm. The accuracy of this value should have minimal impact on conclusions arising from comparisons of ΔG_{ads} for similar types of inhibitors, since t is not expected to vary significantly within an homologous series of products.

Zeta Potential Measurements

The zeta potential of suspended barium sulfate particles in the presence and absence of scale inhibitors at varying pH was measured with a PenKem Model 501 Lazer Zee Meter. The procedure consisted of placing 1 g of barium sulfate powder in 4 L of deionized water followed by vigorous stirring. The suspension was allowed to stand for several minutes, to allow large particles to settle to the bottom. The suspension was then decanted into another vessel, allowed to settle for 10-15 minutes more and then decanted again. 100 mL of suspension containing 500 ppm of inhibitor were used for the measurement. The solution was adjusted to the desired pH with HCl, and used to rinse the potential cell several times prior to making the actual measurement.

3 RESULTS AND DISCUSSION

The phosphonate inhibitors A-H represent an homologous series in molecular weight, ranging from 436 to 159,000 g/mole. Figure 1 illustrates percent inhibition obtained for an unseeded bottle test vs. log molecular weight for this series of inhibitors. The data, also listed in Table 2, demonstrate a performance maximum corresponding to a molecular weight of 1920 g/mole. The trend in Figure 1 could simply reflect changing adsorption characteristics as molecular weight is varied. Measurement of ΔG_{ads} for the A-H series of inhibitors should be a useful way of examining such a statement.

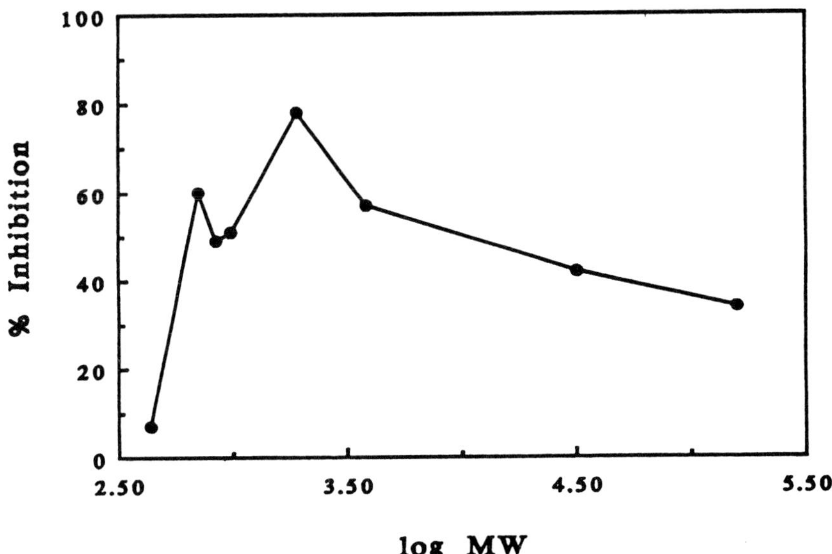

Figure 1. Percent inhibition as a function of molecular weight for unseeded conditions at 25 ppm inhibitor, 85°C and pH=4.0.

Table 2 - Static Bottle Test Results

$BaSO_4$ Inhibitor[a]	MW	% Inhibition Unseeded	Seeded
A	436	7	4
B	710	60	47
C	847	49	37
D	984	51	39
E	1,920	78	64
F	3,840	57	79
G	31,800	42	84
H	159,000	34	35
I	1,660	71	-
J	1,390	49	-
K	492	0	2
L	206	0	2
M	205	0	-
N	275	0	-

[a]A-H; homologous series, increasing molecular weigth. E,I,J; homologous series, decreasing number of PO_3H_2 groups (same polymer backbone). K,L; typical commercial phosphonate inhibitors. M,N; other phosphonate inhibitors.

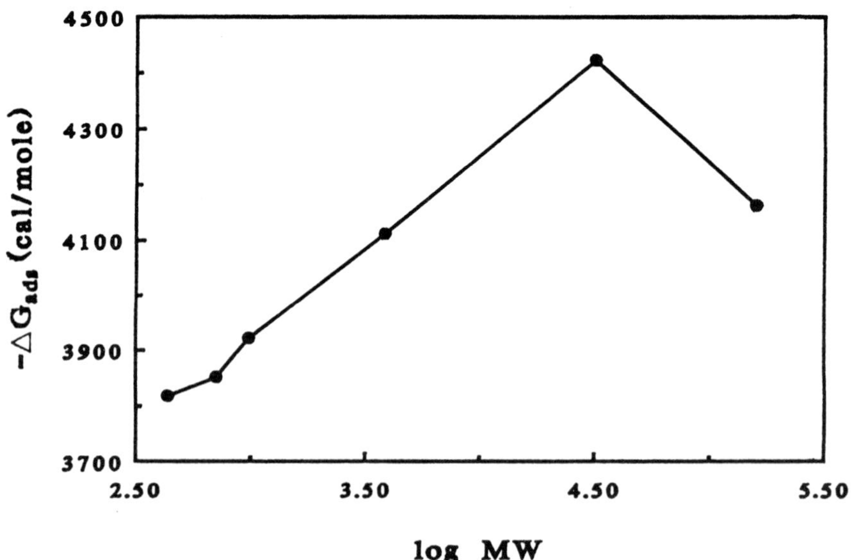

Figure 2. $-\Delta G_{ads}$ on $BaSO_4$ as a function of molecular weight. T=298 K.

Figure 2 is a plot of ΔG_{ads} as a function of log molecular weight for inhibitors A through H. This figure shows a maximum in ΔG_{ads} at a molecular weight of approximately 31,800 g/mole. The maximum in Figure 1 occurs at a different molecular weight than in Figure 2, thus implying that the trend in inhibition performance observed in Figure 1 is not soley due to changes in adsorption characteristics.

A closer correlation with observed trends in ΔG_{ads} can be seen in Figure 3, which is a plot of the percent inhibition vs. log molecular weight for inhibitors A through H under seeded bottle test conditions. As in Figure 2, Figure 3 shows a maximum at a molecular weight of 31,800 g/mole.

The data in Figure 2 illustrate that as molecular weight increases, ΔG_{ads} becomes increasingly negative, suggesting that adsorption does increase with increasing molecular weight, up to a point. Such a result is consistent with multi-segmental attachment: as molecular weight increases, the number of adsorbed functional groups increases and the reversibility of the adsorption decreases. As more points of attachment become established, it becomes statistically less probable that all will become detached at the same

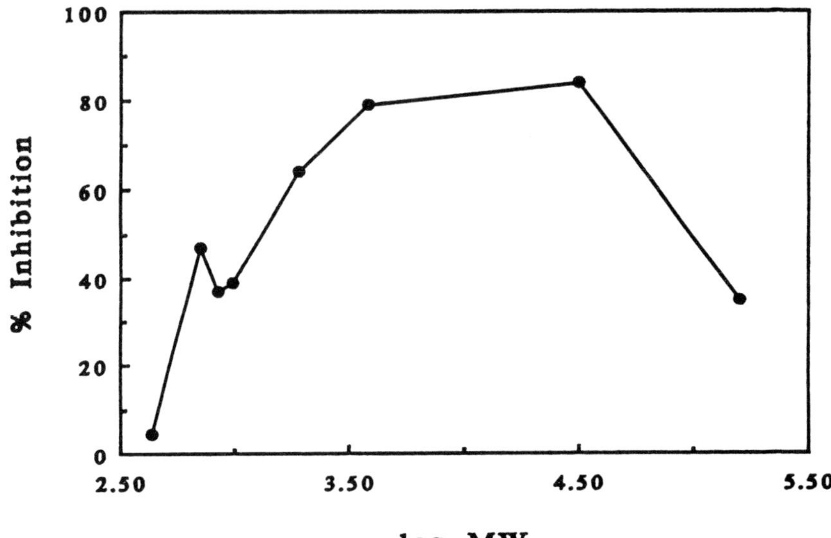

Figure 3. Percent inhibition as a function of molecular weight for seeded conditions, at 25 ppm inhibitor, 85°C and pH=4.0.

time. Countering this trend is the decreasing diffusion rate and concomittant drop in collision frequency with increasing molecular weight. The result is that at high enough molecular weight, inhibition efficiency drops off. It seems likely that the maxima observed in both Figures 2 and 3, at 31,800 g/mole, correspond to maximum absorption yielding maximum inhibition, presumably by the blockage of active growth sites on the crystal surface. A different mechanism of inhibition is needed, however, to explain the location of the maximum in Figure 1, for nonseeded bottle test conditions.

In a supersaturated environment, crystals need to exceed a critical size (critical nucleus) before they are thermodynamically stable due to the unfavorable chemical potential associated with surface sites relative to solvated ions[6]. A crystal which is less than critical nucleus size (an embryo) has many surface sites relative to bulk sites and will face an energy barrier to continued growth. Naono[5] postulated a mechanism of inhibition based on the idea that endothermic adsorption is capable of raising the energy barrier which exists in order for a barium sulfate "embryo" to achieve thermodynamically stable growth.

Data listed in Table 3 for ΔH_{ads} show that a positive value was obtained for every inhibitor tested. This endothermicity presumably arises from the electrostatic repulsion between the negatively charged barium sulfate surface (isoelectronic point of $BaSO_4$=3.8[7]) and the negatively charged functional groups of the scale inhibitor molecule.

Endothermic adsorption of an inhibitor to barium sulfate embryos will increase the probability that they will redissolve. This mechanism of inhibition can only

Table 3 Thermodynamic Data for Adsorption Onto $BaSO_4$

Inhibitor	pH	ΔH_{ads} (cal/mole)	ΔS_{ads} (cal/mole°K)	ΔG_{ads} (cal/mole)
A	4	2061 ± 200	19.9 ± 0.9	-3818 ± 240
A	8	3139 ± 7	24.8 ± 0.1	-4214 ± 48
B	4	1243 ± 663	17.2 ± 2.1	-3852 ± 61
D	4	951 ± 162	16.5 ± 0.4	-3921 ± 57
F	4	1224 ± 385	18.0 ± 1.5	-4112 ± 252
G	4	492 ± 280	16.6 ± 0.8	-4423 ± 151
H	4	1031 ± 467	17.5 ± 1.6	-4163 ± 62
K	4	1180 ± 366	15.7 ± 1.0	-3468 ± 66
L	4	1866 ± 247	19.4 ± 2.5	-3876 ± 199
M	4	1085 ± 247	15.1 ± 1.0	-3400 ± 335
N	4	2025 ± 240	15.8 ± 0.1	-2656 ± 250

be effective in an environment where crystal growth has not proceeded beyond the critical nucleus stage, i.e., thermodynamically stable crystal growth sites do not exist. This mechanism cannot therefore be effective under seeded growth conditions. It can, however, be used to explain the results observed in Figure 1 for nonseeded bottle test conditions.

A positive value for ΔH_{ads} would require a large value of ΔS_{ads} for adsorption to be spontaneous. Naono[5] theorized that when inhibitor molecules adsorb to the surface of the barium sulfate embryo, water molecules which are hydrogen-bonded to the inhibitor are released resulting in a positive value of ΔS_{ads}. It is interesting to note that scale inhibitors with $\Delta S_{ads} < 10$ cal/mole-K give poor bottle test performance (Tables 2 and 3). A low value for ΔS_{ads} is presumably related to smaller numbers of water molecules coordinated to the inhibitor prior to adsorption. This implies that successful inhibitors operating by this mechanism will possess atoms and functional groups capable of hydrogen bonding. This is consistent with the observation that many commercial scale inhibitors contain heteroatoms. Hydrophobic stuctures should be unfavorable for scale inhibition. This conclusion was also reached by van der Leeden et al.[8] in a study evaluating polycarboxylate scale inhibitors.

The effect of varying pH on the performance of various scale inhibitors was also investigated. Figure 4 is a plot of percent inhibition at 25 ppm of inhibitor (unseeded bottle test) as a function of pH and molecular weight. Low molecular weight inhibitors in this series show steadily improved performance with increasing pH. This type of behavior is not unusual and has commonly been attributed[3,9] to the increased degree of ionization of phosphonate or carboxylate functionalities on the inhibitor, resulting in improved chelating abilities and adsorption.

Higher pH conditions will also result in a more negatively charged barium sulfate surface, thus making inhibitor adsorption more endothermic. This larger ΔH_{ads} should be offset by a larger value for ΔS_{ads} arising from a larger hydration sphere generated by the increased ionization of the inhibitor. Table 3 illustrates this for inhibitor A at pH 4.0 and 8.0. At pH 8.0, ΔH_{ads} and ΔS_{ads} are considerably larger than at pH 4.0. Thus the improved performance at higher pH may be a reflection of improved efficiency at destabilizing barium sulfate embryos, through a more highly endothermic adsorption, as well as improved efficiency at simply adsorbing and blocking active growth sites.

Popular arguments[3,4,9] contend that inhibition is

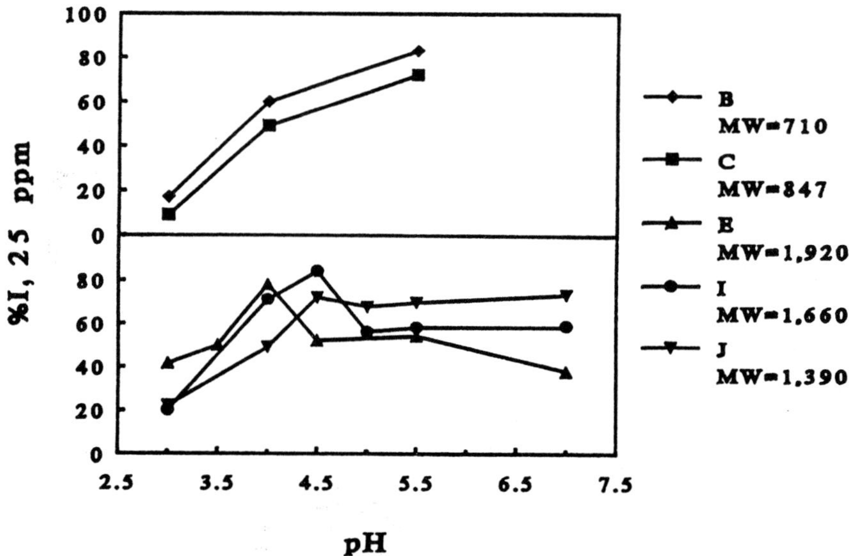

Figure 4. Percent inhibition as a function of pH, unseeded conditions, for 25 ppm inhibitor, 85°C.

usually worse at acidic pH because of reduced binding capability due to lack of ionization. Such arguments miss the important added factor of the (typically) reduced endothermicity of adsorption at acidic pH, and the resultant lessened ability to destabilize barium sulfate embryos.

Figure 4 illustrates that at high enough molecular weights, performance ultimately begins to fall as pH is steadily increased. Polymeric inhibitors thus demonstrate substantially different behavior from their lower molecular weight homologues. Figure 5 is a plot of ΔG_{ads} as a function of pH and molecular weight for the homologous series of inhibitors A-H. ΔG_{ads} becomes increasingly favorable as pH is increased for low molecular weight inhibitors (A and D), but goes through a maximum for the higher molecular weight members of the series. The same trends in inhibition as a function of molecular weight are observed for these inhibitors in Figure 4.

Literature precedent exists to explain the maxima observed for polymers in Figures 4 and 5. Williams, et al.,[7] observed a decrease in the adsorption of

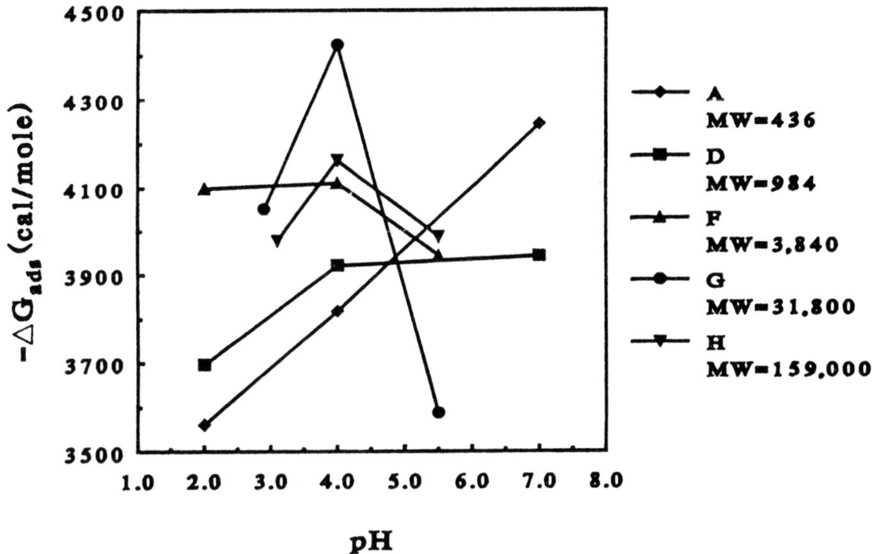

Figure 5. $-\Delta G_{ads}$ as a function of pH, T=298 K.

sodium carboxycellulose on barium sulfate as the pH was raised, and attributed the effect to increased charge density on the polymer. Unadsorbed polymer segments extending out from the surface could thus exert a shielding effect and prevent more polymer from adsorbing. This explanation is supported by data presented in Figure 4 for inhibitors E, I and J. Inhibitors E, I and J form a homologous series in which charge density is varied. I and J have 20% and 40% fewer phosphonate groups, respectively, substituted onto the same polymer backbone as E. As the degree of substitution decreases (charge density decreases, E > I > J), maximum adsorption ($-\Delta G_{ads}$), and maximum performance shift to higher pH. Thus, as pH increases, the degree of ionization increases until charge shielding effects become evident. As the density of phosphonate groups is reduced, a greater degree of ionization is needed (higher pH) to produce charge shielding effects.

Figure 6 is a plot of the zeta potential of barium sulfate particles, in the presence of inhibitors E, I and J, as a function of pH. As expected, increasing pH results in increasingly negative zeta potentials, corresponding to a higher negative charge density on the scale particles. This is due to the increased degree of ionization of the phosphonate groups of the

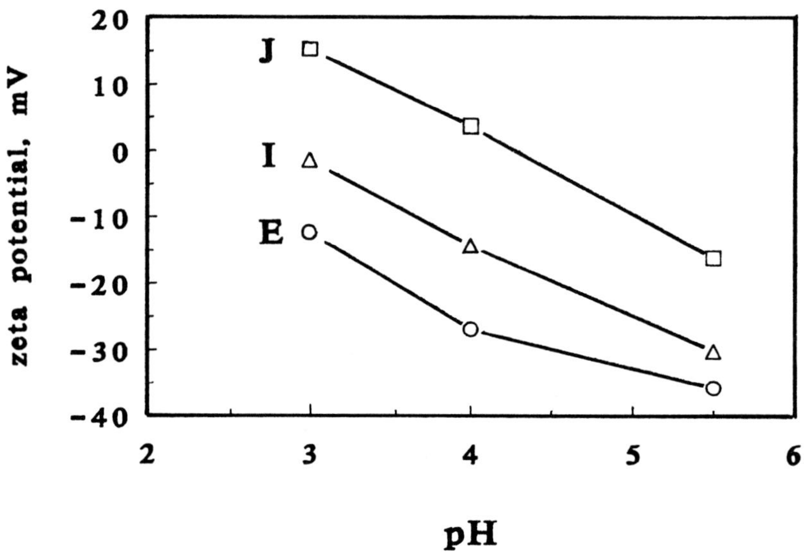

Figure 6. Zeta potential of $BaSO_4$ as a function of pH and charge density on the scale inhibitor, at 25°C and 500 ppm inhibitor.

inhibitors. The decreased number of phosphonate groups in going from inhibitor E to I to J results in the curves for these inhibitors being increasingly displaced towards higher pH. The data illustrate that the total charge of the barium sulfate surface can be affected by the charge density, or number of phosphonate functionalities, on the scale inhibitor molecule.

The data in Figure 4 illustrate that as charge density is reduced, the maximum shifts to higher pH and becomes less pronounced. It thus appears from the data in Figures 4 and 5 that there is an optimal, pH-dependent charge density for polymeric inhibitors which allows for the greatest amount of adsorption.

The data presented here support the hypothesis that two mechanisms of scale inhibition are operative: one involving endothermic adsorption resulting in dissolution of barium sulfate embryos, and another involving relatively irreversible polymer adsorption at the active growth sites of barium sulfate crystals, resulting in the blockage of such sites. The first mechanism is expected to be more important for low molecular weight inhibitors since such molecules would have difficulty blocking growth sites due to adsorption which is anticipated to be highly reversible. Endothermic adsorption is not as important for polymeric inhibitors since irreversible adsorption to crystal growth sites should be relatively irreversible. The thermodynamic data in Table 3 support this conclusion: the largest values for ΔG_{abs} were generally obtained for the highest molecular weight inhibitors, ie., inhibitors F, G and H. Also, polymeric inhibitors (eg., F,G,H) tend to have values of ΔH_{ads} near 1200 cal/mole, whereas small molecular weight inhibitors (eg., A,L,N) often have values for ΔH_{ads} closer to 2000 cal/mole.

4 SUMMARY

Effective inhibition can be obtained with low molecular weight inhibitors through endothermic adsorption, driven by favorable entropy conditions associated with the release of associated water molecules. This mechanism, however, is only effective for unseeded conditions. Improved chelating ability, arising from a greater degree of inhibitor ionization, improves adsorption but also increases the endothermicity of adsorption. Thus low molecular weight inhibitors tend to perform best in regions of relatively high pH (>7).

Increasing the hydrogen-bonding capability of a scale inhibitor should increase adsorption to barium sulfate by increasing ΔS_{ads}. The incorporation of heteroatoms in the scale inhibitor structure is thus one way of potentially improving performance at acidic pH.

In situations where a large number of nucleating sites are present, near irreversible adsorption onto active growth sites becomes the primary mechanism for inhibiting scale. Polymeric inhibitors which can adsorb fairly irreversibly, due to the large number of points of attachment, are the ideal candidates for inhibition under seeded growth conditions.

Polymeric inhibitors are likely to show performance maxima as a function of pH due to charge shielding effects which can be controlled by optimization of the number and spacing of chelating functional groups on the backbone of the polymer. Under seeded conditions, polymeric inhibitors with optimized charge density will yield the best performance.

5 REFERENCES

1. G.R. Chesnut, G.D. Chappell, D.H. Emmons, "The Development of Scale Inhibitors and Inhibitor Evaluation Techniques for Carbon Dioxide EOR Floods," Paper SPE 16260, presented at the SPE International Symposium on Oilfield Chemistry, San Antonio, Texas, 1987.
2. E.R. McArtney, A.E. Alexander, J. Colloid Sci., 1958, 13, 383.
3. J.E. Ramsey, L.M. Cenegy, "A Laboratory Evaluation of Barium Sulfate Scale Inhibitors at Low pH for Use in Carbon Dioxide EOR Floods," Paper SPE 14407, presented at the SPE Annual Conference, Las Vegas, Nevada, 1985.
4. D.W. Griffiths, S.D. Roberts, S.-T. Liu, "Inhibition of Calcium Sulfate Dihydrate Crystal Growth by Phosphonic Acids - Influence of Inhibitor Structure and Solution pH," Paper SPE 7862 presented at the SPE International Symposium on Oilfield and Geothermal Chemisty, Houston, Texas, 1979.
5. Naono, H., Bull. Chem. Soc. Jpn., 1967, 40, 1104.
6. A.E. Nielsen, 'Kinetics of Precipitation', Pergamon Press, Oxford, London, 1964.
7. P.A. Williams, R. Harrop, G.O. Phillips, J. Chem. Soc. Faraday Trans. I., 1982, 78, 1733.
8. M.C. van der Leeden, G.M. van Rosmalen, "Inhibition of Barium Sulfate Deposition by Polycarboxylates of Various Molecular Structure," Paper SPE 17914, July 1988.
9. W.H. Leung, G.H. Nancollas, J. Inorg. Nucl. Chem., 1978, 40, 1871.

Derivation of Scale Inhibitors Adsorption Isotherms for Oil Reservoir Squeeze Treatments

K.S. Sorbie, R.M.S. Wat, A.C. Todd, and T. McClosky

DEPARTMENT OF PETROLEUM ENGINEERING, HERIOT-WATT UNIVERSITY, EDINBURGH EH14 4AS, UK

1 INTRODUCTION

The formation of oilfield scales is of major concern in the North Sea. The occurrence and formation of oilfield scales is mainly due to the mixing of incompatible brines after injection water breakthrough. The formation brine, usually has high content of barium and strontium ions, and the injected sea water, has a high concentration of sulphate ions. The release of carbon dioxide during reservoir depletion and the drawdown of production wells also contribute to the formation of carbonate scale. The effect of scale formation and deposition near the wellbore region and within the production tubing can cause severe reduction in oil production, inefficient process plant operation and safety valve malfunction.

The prevention of sulphate scales formation is an ongoing process after injected water has broken through. This is particularly true for some of the mature fields in the North Sea where water injection has been used to maintain oil production for a long period of time. Even in the case of relatively new developed fields where large scale heterogeneity exists with layers of high permeability contrast, this can occur early in their production history. Unlike carbonate scales which can be treated by acidization, the sulphate scales of barium and strontium are highly insoluble and difficult to remove. They are best treated by preventive measures. The most common practice in the North Sea is to carried out a scale inhibitor squeeze treatment [1-7] in which a slug of inhibitor chemical is injected into the producer wells followed by an overflush with sea water and a brief shut-in period. The success of such a squeeze treatment will depend on the inhibitor return once the oil production recommences following the shut-in. Ideally, the effect of the squeeze treatment would be to maintain a long return of inhibitor at a low but effective concentration which would enable scale inhibition. The threshold concentration for the scale inhibitor to remain effective is usually between 5-10 ppm and a retreatment squeeze will be needed once the inhibitor concentration in the producer well drops to or below this threshold level.

Whilst a squeeze treatment is designed to provide a gradual release of inhibitor which has been retained in the formation, the precise retention mechanism that operates initially may differ. The type of treatment most commonly used in the oil industry is the so called adsorption/desorption squeeze [2,3,5,6,7]. In this case, the main mechanism depends on the 'adsorption and desorption' characteristics of the inhibitor with the reservoir rock. The retention of inhibitor within the formation is considered to be through a physical adsorption mechanism and the extent of the process is governed by an adsorption isotherm. A different type of squeeze treatment which has also been widely used in the oil industry is the 'precipitation' squeeze [6,7,8,9]. In this case, apart from the normal adsorption/ desorption process, the composition of the inhibitor slug is adjusted so as to induce in-situ precipitation. The return of inhibitor is due to a combined mechanism of desorption and dissolution and it has been claimed that this process will yield a longer squeeze life. In this paper, we shall present the results from the adsorption/desorption squeeze type experiments and results from the precipitation squeeze will be addressed in a future publication [10].

For the adsorption/desorption process, the extent of the solid/fluid interaction is governed by an adsorption isotherm. Furthermore, we belive that it is the shape of the isotherm at the low concentration region which has the maximum influence on the squeeze life, i.e. the time over which the inhibitor return remains above the threshold value. In order to design a squeeze treatment successfully, field engineers have to know the shape of the adsorption isotherm and in particular, the steeply rising section of the curve between, say 2 - 1000 ppm. Traditional bulk adsorption experiments such as beaker tests and core tests are often used for the evaluation of the adsorption isotherm. These methods are generally adequate when only the overall shape of the isotherm is required. This is because the number of data point available is often limited due to the long period of time for each evaluation and becomes less accurate at low concentration range when good material balance is essential. In this paper, we demonstrate a novel approach in which an approximate method is used for deriving the adsorption isotherm from the effluent data of laboratory core flood experiments. The result is then refined by mathematical modelling to obtain a more accurate representation of the isotherm.

2 EXPERIMENTAL DETAILS

In the course of this work we have carried out an extensive series of inhibitor core flood experiments. The inhibitor used was the phosphonate Diethylenetriaminepenta (methylene-phosphonic acid). Both the acid (I1) and sodium salt (I2) forms of this inhibitor were used in the core flood experiments. In Table 1, the molecular structures and properties of these two variants, I1 and I2, are given. The inhibitor chemical was used as received from the manufacturer. Solutions of specific concentration were made up with artificial seawater brine which composition is given in Table 2. In most of the floods, small quantity of LiCl was added to the inhibitor slugs. Li^+ ion has been used extensively in our core flood experiments and is found to be an excellent inert tracer. This is particular true for core materials with relatively low clay content as in our studies. Of central importance to the analysis of inhibitor floods is that an assay can be performed that is accurate down to inhibitor levels of ~1ppm. In this work, the ultraviolet (UV) photochemical oxidation method was used for the analysis of inhibitor. This technique, often referred to as the Hach method [7], involves a photochemical oxidation of phosphonate to orthophosphate followed by conventional colorimetric determination of the liberated orthophosphate by ascorbic acid. For both inhibitors I1 and I2 the Hach method proved to be sensitive and reproducible with a detection limit of 0.25ppm ± 0.02ppm for ten successive analyses. For the Li^+ tracer, the concentration in the effluent is accurately detected using atomic adsorption spectrophotometry (AAS).

(1) **Inhibitor 1** **(I1)** - Diethylenetriaminepenta (methylenephosphonic acid)

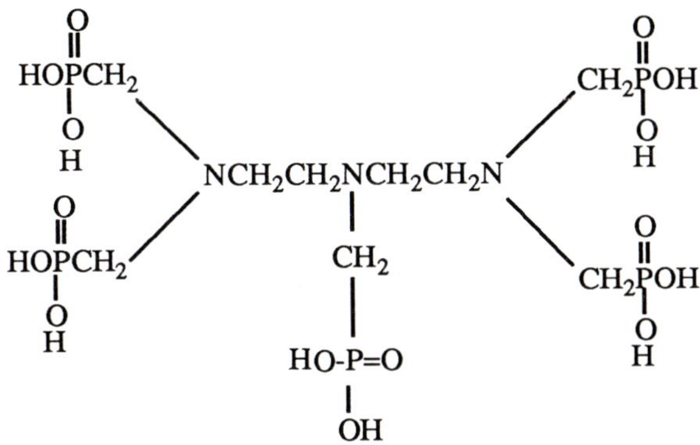

(2) **Inhibitor 2** **(I2)** - (Heptasodium salt of Inhibitor 1)

	Inhibitor 1	Inhibitor 2
Molecular Weight :	573	727
Active acid content :	50. % (by H^+/OH^- titration)	25.% (by H^+/OH^- titration)
pH at 25°C :	< 2	6-8

Table 1 : Inhibitors used in this work

Ion Composition (mg/l)		Compound Composition (g/l)	
Na^+	10,890	NaCl	24.0738
Ca^{2+}	428	$CaCl_2.6H_2O$	2.3395
Mg^{2+}	1,368	$MgCl_2.6H_2O$	11.4362
K^+	460	KCl	0.8771
SO_4^{2-}	2,960	Na_2SO_4	4.376
Cl^-	19,766		

Table 2 : Composition of synthetic brine used in these studies

The core materials used for the inhibitor adsorption experiments are outcrop Clashach and Bentheim sandstone cores. Both types of sandstone are relatively clean, homogeneous and have low clay content. All the core samples used have a diameter of 2.54cm and between 10 to 45 cm in length. The permeability of the core samples varies between 150 to 2500 m Darcy. The details of these core samples are summarised in Table 3. A total of eight inhibitor floods have been carried out at room temperature (20°C ± 1°C) and the details of these are given in Table 3. Although the effect of temperature on the adsorption process has not been included in this study, it is considered that, in the absence of precipitation, the general form of the adsorption isotherm and the dynamics of the tailing is likely to be the same at both high and low temperature.

The injection sequence of each flood is broadly the same. The core was first conditioned by saturation with the artificial brine.A core characterisation stage was then carried out in which the core permeability was evaluated and the dispersion profile was obtained using a Li^+ tracer flood. Apart from providing some indications on the physical state of the core and the epoxy resin coating, the information of the Li^+ dispersion profile was needed for computer simulation. For the actual inhibitor adsorption experiment, an inhibitor/tracer slug was injected into the core until the effluent concentration reached the inlet value for a sustained period of time. Normally, it took between 2 to 3 pv injection of inhibitor, depending on the concentration, to reach the inlet values. During the experiments, however, between 4.5 pv and 8.0 pv of inhibitor was injected in each flood to ensure that full inlet concentration was reached. The core was then shut in for a period of time (normally overnight) before carrying out the long postflush with artificial brine. The desorption stage stopped only after the inhibitor concentration at the effluent could not be detected accurately (~1ppm). In some of the floods additional shut-in periods were included during the postflush period. At the end of the desorption stage the core permeability was again determined before the flood was finally terminated.

3 THEORETICAL MODEL OF INHIBITOR TRANSPORT IN POROUS MEDIUM

For fluid transport in porous medium, the process is often represented by the generalised convection-diffusion (C-D) equation. Very often, however, due to the non equilibrium effect such as adsorption and chemical reaction, a modified equation has to be used [12,13,14]. In this work, in order to better understand the inhibitor adsorption/desorption process, a mathematical model for simulating both equilibrium and non-equilibrium squeeze treatment has been developed. This model [15] is based on the modified C-D equation and is quite similar to that of Hong and Shuler [11] except that we have generalised the numerical model to include:

- (i) the use of either Langmuir or Freudlich adsorption isotherms.
- (ii) adsorption isotherm derived from experimental results and input as a table of numbers.
- (iii) a radial grid system to represent near-wellbore formation.

With the additional term representing the adsorption process, the non-equilibrium transport equation becomes:

$$\frac{\partial C_m}{\partial t}=D\frac{\partial^2 C_m}{\partial x^2}-v\frac{\partial C_m}{\partial x}-\frac{\rho}{\phi}\frac{\partial \Gamma}{\partial t} \tag{1}$$

$$\frac{\partial \Gamma}{\partial t}=r_2\left(\Gamma_{eq}(C_m)-\Gamma\right) \tag{2}$$

where C_m(g/g) is the mobile inhibitor concentration, Γ (g inhibitor/g of rock) is the adsorbed level of inhibitor and $\Gamma_{eq}(C_m)$ is the final equilibrium adsorption level at C_m. The other parameters associated with the C-D equation are D (cm^2/s), the dispersion coefficient, v (cm/s), the superficial velocity, Q (cm^3/s), the injection rate, A (cm^2), the cross sectional area of the core sample, ρ (g/cm^3), the rock density and ϕ, the porosity of the core. In order to solve the equations (1) and (2) the relationship between C_m and Γ, i.e. the isotherm, must be input and then possibly adjusted to match the inhibitor effluent profiles of the experiments. The same isotherm can then be used for designing inhibitor squeeze treatment in the fields with a similar rock/fluid system. Alternatively, using the information from the effluent profile, we can derive the adsorption isotherm of the system and apply it directly to the field. Consider in the case of the long postflush period where equilibrium exists within the core. The concentration gradient is low and the effect of dispersion is negligible. Equation (1) and (2) can be combined and converted to a dimensionless form as:

$$\left\{1+\left(\frac{\rho}{\phi}\frac{\Gamma_{max}}{C_o}\right)\left(\frac{\partial \Gamma'}{\partial C}\right)\right\}\frac{\partial C}{\partial T}=-\frac{\partial C}{\partial X} \tag{3}$$

$$R(C)\frac{\partial C}{\partial T}=-\frac{\partial C}{\partial X} \tag{4}$$

where

$$\Gamma'=\frac{\Gamma}{\Gamma_{max}} \qquad (a)$$

$$C=\frac{C_m}{C_o} \qquad (b)$$

$$X=\frac{x}{L} \qquad (c)$$

$$T=\frac{tv}{L} \text{ (time in pore volumes)} \qquad (d) \quad \text{and}$$

$$R(C)=\left\{1+\left(\frac{\rho\Gamma_{max}}{\phi C_o}\right)\left(\frac{\partial \Gamma'}{\partial C}\right)\right\} \qquad (e) \tag{5}$$

Here the parameter Γ_{max} (g/g) is the maximum adsorption level for a given input concentration C_o (g/g), L (cm) is the length of the core and R(C) is the retardation factor. The parameter R(C) is often used to quantify the extent of retardation experienced by an adsorbed solute (e.g.inhibitor) when compared with a non-adsorbed species (tracer). More importantly, equation (4) now becomes a 'wave' equation of the first order with the propagation velocity equal to 1/R(C) for concentration C. Thus, in order for the low concentration values to have a very low velocity (i.e. that they arrive at the outlet after a long time so that long tailing of inhibitor results) then R(C) must be larger for lower C values. The effect of the shape of the adsorption isotherm on the value of R(C) can be demonstrated when equation (5e) is rearranged to the form of:

$$\left(\frac{\partial \Gamma'}{\partial C}\right) = \frac{\phi C_o}{\rho \Gamma_{max}}(R(C)-1) \tag{6}$$

Equation (6) is an expression for the slope of the adsorption isotherm with respect to the dimensionless parameters C and Γ'. It is clear therefore, from equations (5e) and (6), that in order to have a long tailing in the inhibitor return, the retardation factor R(C) must be large and the slope of the adsorption isotherm must be steep at the low concentration values. It is this steep rise of the isotherm at low C that governs the dynamics of the long tail. From the effluent profiles of both inhibitor and tracer and the above equations, the adsorption isotherm can be 'reconstructed' using the following procedure:

(i) calculate Γ_{max} at C_o by determining the area between the frontal breakthrough profiles of both tracer and inhibitor.

(ii) evaluate R(C) at different concentrations directly from the tail region of the inhibitor effluent profile, i.e. measure the delay of a particular concentration value with respect to the C = 0.5 value in the tracer desorption profile.

(iii) integrate equation (6) in order to obtain a first estimate for the isotherm Γ' vs C and use this isotherm in the simulation model to generate a predicted profile which is compared with the original experimental data.

(iv) adjust Γ_{max} and refine the isotherm to improve the match.

4 EXPERIMENTAL RESULTS

A total of eight inhibitor core floods have been carried out and the details of each flood are given in Table 3. The first five floods involve the phosphonate inhibitor I1 and the remaining three use I2. Figures 1 to 4 show some example normalised effluent profiles of inhibitor, tracer and pH for I1 and I2. Since I1 is the acid form of the phosphonate, the pH closely follows the inhibitor concentration as shown in Figure 1 for flood F5 (see Table 3). The long tail of the inhibitor effluent in the high concentration (19908 ppm) flood F5 is shown in Figure 2 where the effect of a shut in (at 56pv) on both the inhibitor concentration and pH is shown. Figures 3 and 4 show the overall inhibitor and tracer effluents over the first 15pv and 10pv for the lower concentration I2 floods F6 and F8 respectively. Because of the lower inhibitor concentration, the retardation of the I2 front relative to the tracer profile is very clear in both of these figures. It can also be seen that the inhibitor concentration shows an immediate decrease after the first shut-in in Figure 3 indicating some non-equilibrium behaviour. This was observed in the lower concentration phosphonate profiles (F4, F6 and F8) but was not discernible in the higher phosphonate concentration floods (F5 and F7).

Flood	Core & Flood Details	Inhibitor Type & C_o (ppm)	Material Balance of Inhibitor (I) and Tracer (T) In (pv)	Out (pv)	Core Permeability Before (mD)	After (mD)
F1	Clashach core L = 7.62 cm ϕ = 0.14 pv = 5.41 cm^3 Q = 2 cm^3/min	I1 2510 ppm	(I)7.1	-	157	~160
F2	Clashach core L = 7.62 cm ϕ = 0.22 pv = 8.45 cm^3 Q = 2 cm^3/min	I1 2325 ppm	(I)6.7	-	1505	1536
F3	Clashach core L = 7.62 cm ϕ = 0.14 pv = 5.25 cm^3 Q = 2 cm^3/min	I1 2503 ppm	(I)8.0	-	390	377
F4	Bentheim core L = 21.6 cm ϕ = 0.197 pv = 21.56 cm^3 Q = 2 cm^3/min M_{core} = 199.14 g	I1 2517 ppm	(I)4.52	4.34(1)	840	-
F5	Bentheim core L = 18.06 cm ϕ = 0.219 pv = 18.64 cm^3 Q = 2.15 cm^3/min M_{core} = 169.75 g	I1 19908 ppm	(I)7.50 (T)3.46	7.46 3.46(2)	2540	2540
F6	Bentheim core L = 20.08 cm ϕ = 0.22 pv = 21.06 cm^3 Q = 2.16 cm^3/min M_{core} = 192.33 g	I2 2510 ppm + 50 ppm(3) tracer	(I)6.85 (T)6.85	6.76 6.78	2410	2680
F7	Bentheim core L = 19.36 cm ϕ = 0.22 pv = 20.26 cm^3 Q = 2.15 cm^3/min M_{core} = 186.79 g	I2 26150 ppm + 50 ppm tracer	(I)7.29 (T)7.29	7.27 7.30	1810	-
F8	Bentheim core L = 19.36 cm ϕ = 0.22 pv = 20.26 cm^3 Q = 2.15 cm^3/min M_{core} = 186.79 g	I2 2479 ppm + 50 ppm tracer	(I)4.64 (T)4.64	4.62 4.63	1810	1940

NOTES
1. Material produced after 2nd shut in (at 46pv) not included in this figure.
2. Tracer flood performed separately (not in the I1 slug) in this case.
3. In floods F6 - F8, the Li^+ tracer was included in the inhibitor slug

Table 3 : Summary of phosphonate inhibitor floods

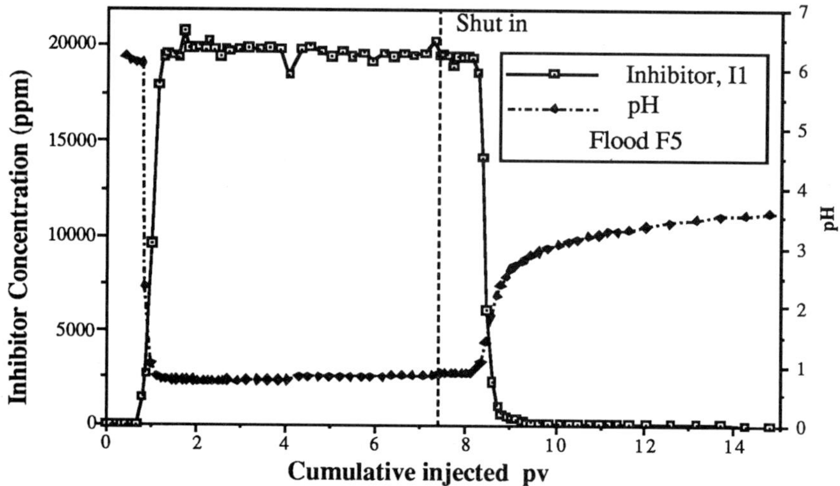

Fig. 1 : Inhibitor and pH effluent profiles for the initial stages of F5

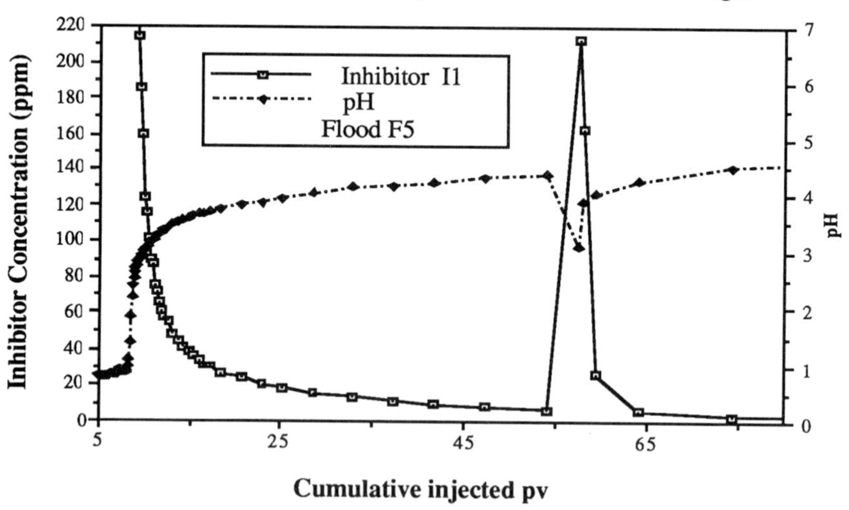

Fig. 2 : Inhibitor (I1) and pH effluent profiles in the tail region of F5

A composite plot showing the tail regions of the first three phosphonate floods in the short cores (F1, F2 and F3) is presented in Figure 5 and for the remaining floods in the longer cores in Figure 6 where, in all cases, the pore volume injected is taken *from the start of the postflush period.* This is necessary in order to be able to compare the results from these various floods and Figures 5 and 6 are plotted using log-log axes as is normal in presenting inhibitor return curves.

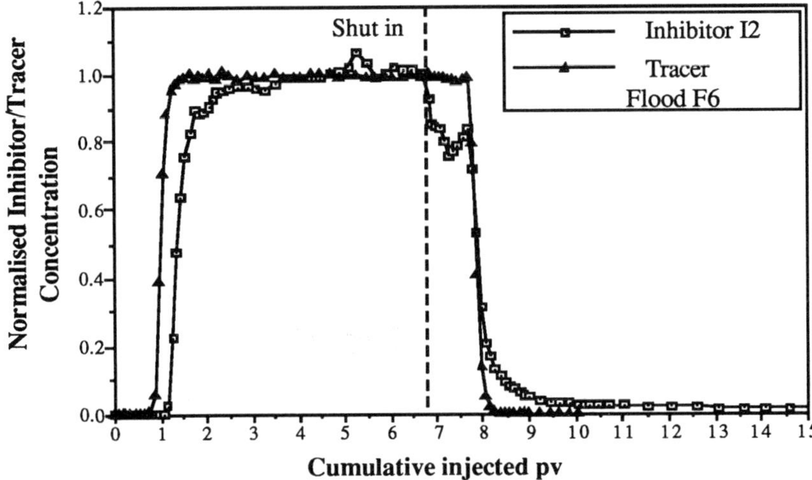

Fig. 3 : Normalised effluent profiles of Inhibitor (I2) and tracer in F6

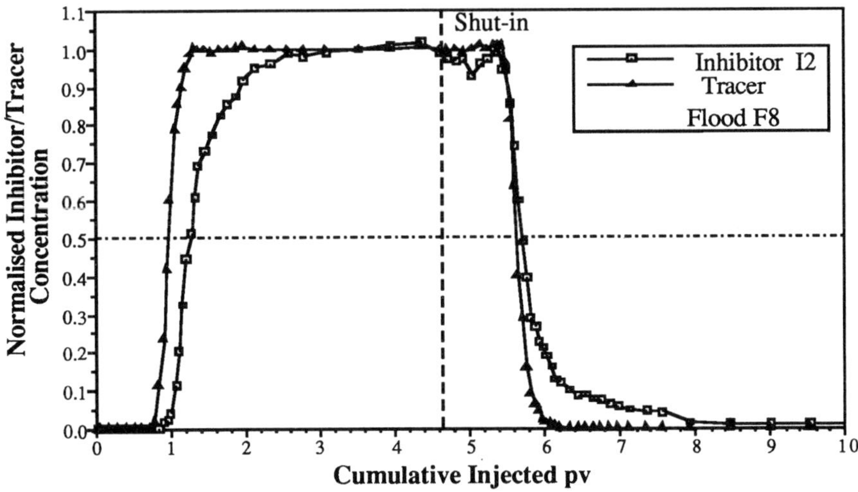

Fig. 4 : Normalised effluent profiles of Inhibitor (I2) and tracer in F8

Some important observations can be made on Figures 4 and 5 which may be interpreted more fully after the application of the isotherm derivation below. Note particularly that:

(i) in all cases in Figures 5 and 6, it can be seen that the return curve is approximately linear in the lower concentration region with respect to the log C vs log PV axes, in accordance with empirical observation [16] and simple analytical models [11].

(ii) the secondary shut-ins in the tailing region appear as peaks in these figures and seem to have a fairly minor effect on overall form of the decline curve.

(iii) in the later time part of the tail, the return curves in Figure 6 are at quite similar inhibitor concentration levels irrespective of whether the initial inhibitor concentration was high (F5 and F7) or low (F4, F6 and F8).

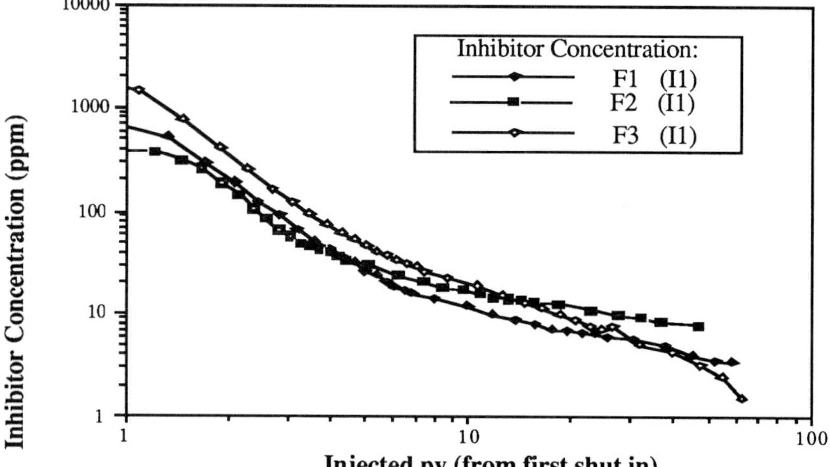

Fig. 5 : I1 Effluent profiles in the tail region in Floods F1, F2 & F3

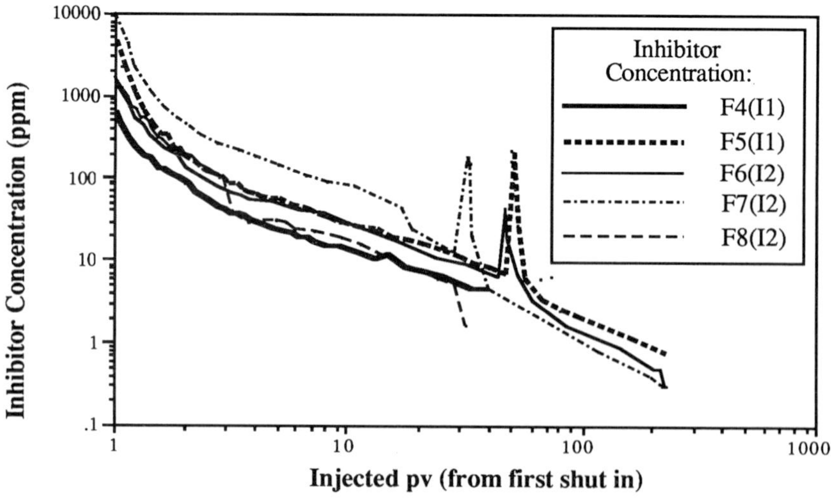

Fig. 6 : Effluent profiles in the tail region of I1, I2 in Floods F4-F8

Finally, a very important feature of the phosphonate floods in the longer cores is that, to a good approximation, nearly all of the injected inhibitor was produced over the course of the flood (see Table 3). Thus, there is little evidence here of irreversibly retained inhibitor and hence - at least in these core floods - it is not necessary to propose some other permanent "loss" mechanism. The phosphonate floods are considered in more detail in the analysis of the inhibitor isotherm in Section 5 below.

5 DERIVATION OF THE ADSORPTION ISOTHERMS

In Section 3, we have describe a procedure by which the adsorption isotherm can be 'reconstructed' from a laboratory core flood experiment. In this section, we are going to demonstrate the application of this procedure by using flood F5 as an example although good results have been obtained for all the phosphonate floods. The level of inhibitor, Γ_{max}, adsorbed in F5 is calculated by examining the amount of injected and produced inhibitor before the shut-in period. For example, if V_1 and V_2 (cm^3) of inhibitor at the input concentration have been injected and produced respectively before the shut-in, then the amount adsorbed on the core can be represented as:

$$\Gamma(C_o) = \frac{(V_1 - V_2 - PV)C_o}{M_{core}} \qquad (7)$$

where $\Gamma(C_o)$ is in mg inhibitor per g of rock, PV is the core pore volume in cm^3, C_o is the input concentration and M_{core} is the mass of the core in gm. Using the above equation, the value of $\Gamma(C_o)$ for flood F5 is found to be 0.24 mg/g.

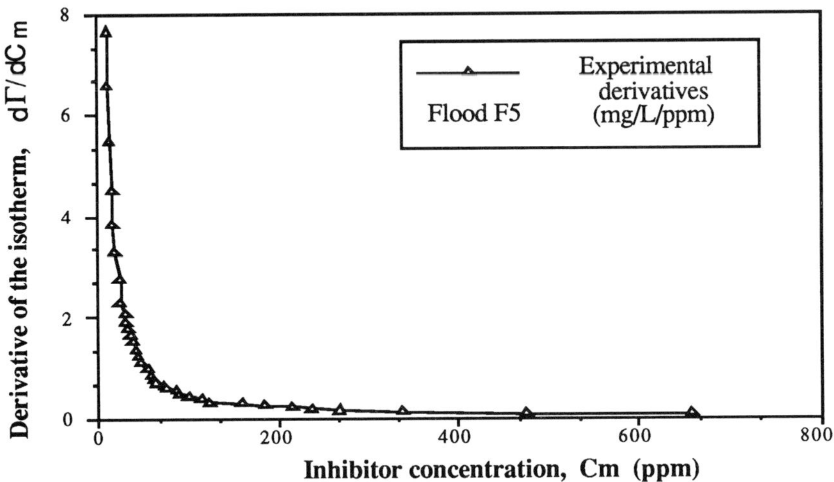

Fig 7 : The slope of isotherm vs. inhibitor conc^n. using F5 data

In Figure 7, the curve of $(\partial\Gamma'/\partial C)$ is obtained by measuring the retardation of various concentration values in the effluent tail of flood F5 as expressed by equation (6). Note that a reliable value of this derivative is only shown up to a dimensionless concentration of $C \sim 0.035$ (ie ~ 700 ppm for F5) although in practice we can obtain an approximate slope up to around 2000 ppm. Since we must have the value of the isotherm up to $\Gamma' = 1$ (ie up to Γ_{max} in dimensional terms) there is a considerable concentration region in this experiment, between ~ 1000 ppm and 19908 ppm, where we have very little information on the inhibitor adsorption isotherm from the effluent profile. However, this is not a serious problem in terms of matching the long tail although it does make it more difficult to match the frontal profile of the inhibitor flood as we explain below.

Integrating equation (6) directly leads to a first approximation to the adsorption isotherm which is bound to be inaccurate in the ways described above but which, nevertheless, contains the most important information regarding the long tailing in the inhibitor effluent; that is, on *the steepness and shape of the isotherm at low inhibitor concentration*. Using this approximate isotherm as a starting point, we have found that, by adjusting the following two simple physical parameters, we can obtain a good estimate of the inhibitor adsorption isotherm:

(i) the actual level of the maximum adsorption, Γ_{max}, for which we have a first estimate by calculating the differences between the frontal profiles of tracer and inhibitor

(ii) the "plateau" concentration, C_p, at which the isotherm reaches the Γ_{max} (or $\Gamma' = 1$)

However, the key issue is that the **shape** of the isotherm between $C = 0$ and $C = C_p$ must be preserved. Applying this procedure, we derive the isotherm shown in Figures 8(a) and 8(b) for the inhibitor adsorption in flood F5. In order to obtain this isotherm it was necessary to rescale the directly estimated isotherm (integration of equation (6)) as described above where:

(i) the maximum adsorption level had to be reduced to a value of $\Gamma_{max} =$ 0.12 mg/g.

(ii) the isotherm reached this maximum level at a "plateau" concentration level of $C_p = 0.055$ (i.e. $C_m = 1100$ ppm) as shown in Figure 8.

The form of the derived isotherm in Figure 8 is very similar to those presented by Meyers et al [7] in that it rises very steeply in the low concentration region and it reaches its maximum adsorption level at a relatively low value of inhibitor concentration. The isotherms presented by Meyers et al [7] were, in fact, for static adsorption onto crushed core material and the isotherms presented in this work are the dynamic curves which are more relevant for describing inhibitor squeeze processes in porous media. Only one result is presented here, although we have derived isotherms for all of the floods in the longer cores. As we would expect [7], there is some core to core variation in the actual adsorption level but the central features of the resulting inhibitor isotherms are very similar for all cases.

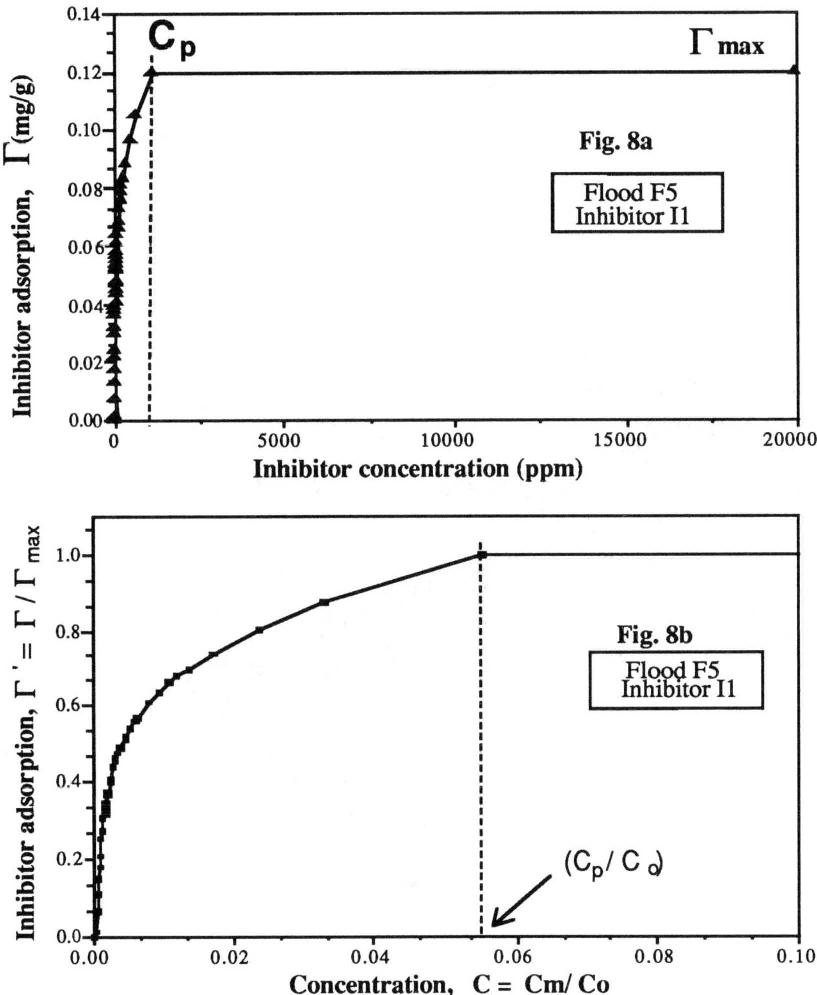

Fig. 8 : Dynamic adsorption isotherm derived from Flood F5 data

When the isotherm in Figure 8 was used in the numerical model, the F5 inhibitor effluent is matched quite well as is shown in Figures 9 and 10. The position and shape of the frontal breakthrough profile in Figure 9 is controlled by **all** of the isotherm and the actual level of inhibitor adsorption, Γ_{max}. However, the dynamics of the long tail region is governed almost entirely by the sharply rising initial region of the inhibitor adsorption isotherm and the match between the experimental and calculated profiles in Figure 10 is excellent. The particularly accurate modelling of the long tail region is as we expect since there are many points on the isotherm in the steeply rising low concentration regime (see Figure 8) which were, of course, derived from the experimental data in this region. Thus, we may view the very good match to the tailing behaviour as a confirmation of the method being proposed in this paper and of the characteristics of the resulting inhibitor adsorption isotherm that has been derived.

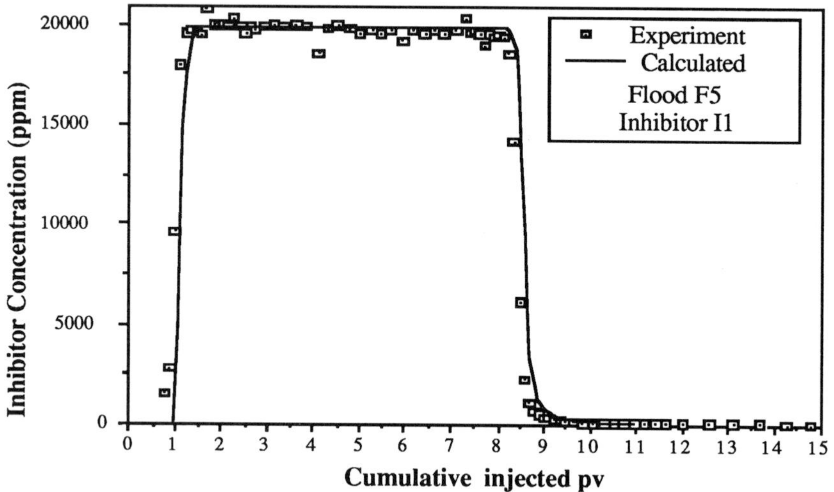

Fig. 9 : Calculated and experimental I1 profiles using the derived adsorption isotherm

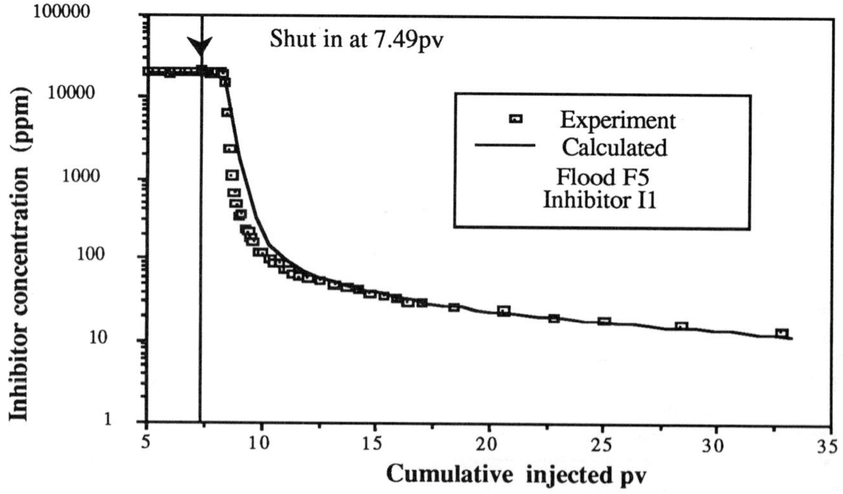

Fig. 10 : Calculated and experimental tail profiles using the derived adsorption isotherm

6 CONCLUSIONS

We have demonstrated in the previous sections a procedure by which the adsorption isotherm of a particular rock/fluid system can be determined from the effluents of inhibitor core floods. In the course of this work, we have carried out a series of phosphonate inhibitor core flood experiments and the main findings and conclusions arising from our study can be summarised as follows:

(i) the isotherm of the adsorption process can be 'reconstructed' by an approximate procedure described in this paper. Further refinement using computer modelling may be needed to improve the representation of the isotherm at the high concentration region. The procedure has been applied successfully to several of our experimental floods and an example using the results from flood F5 is presented here.

(ii) the most critical region of the inhibitor adsorption isotherm which governs the detailed dynamics of the long tail is the steeply rising part at very low, near threshold ($C_m \sim 5$ to 10 ppm) concentration; this is due to the fact that the velocity of this concentration level, v_c, is given by $v/\{1 + (\rho/\phi).(\partial\Gamma/\partial c)\}$ and hence the steeper the slope of the isotherm, the slower will be the return of that concentration level. The result of this initially steep isotherm is to make it **appear** as if the inhibitor is desorbing slowly although, in fact, it is simply that the phase velocity of concentration C (v_c) decreases as C itself decreases.

(iii) in field application, the most beneficial parameter to adjust may be the amount of brine overflush. This has the effect of maximising the part of formation which are treated with inhibitor as it is propagated out to a longer distance from the treated well. In the case where the isotherm shape at low concentration is the key factor, the effect of the actual inhibitor concentration and the slug size will become less significant.

(iv) although some non-equilibrium behaviour was observed after the shut ins in all of these floods, the dynamics of the long tailing is principally determined by the adsorption isotherm.

(v) inhibitors I1 and I2 show the same qualitative characteristics in their effluent tailing behaviour in this type of core flood since the rock/fluid interaction is through an adsorption/desorption mechanism in both cases.

NOMENCLATURE

A	cross-sectional area of the core	(cm^2)
C	dimensionless concentration of inhibitor	(C_m/C_o)
C_m	mobile phase inhibitor concentration	(g/g)
C_o	injected inhibitor concentration	(g/cm^3)
D	dispersion coefficient of inhibitor or tracer in core	(cm^2/s)
I1, I2	inhibitors 1 and 2 - see Table 1	
L	length of the core	(cm)
M_{core}	mass of core	(g)
PV	one core pore volume, PV = A.L. ϕ	(cm^3)
Q	injection rate	(cm^3/s)
r_2	inhibitor desorption rate in non-equilibrium model	(s^{-1})
R(C)	retardation factor - defined in equation (6)	
t, T	time (s); dimensionless time in pv (tv/L)	

v	fluid superficial velocity, $(Q/A\phi)$	
V_1, V_2	fluid volumes - see text	(cm^3)
x, X	distance (cm); dimensionless distance, (x/L)	
ρ	density of rock	(g/cm^3)
ϕ	porosity	
Γ, Γ_{max}	inhibitor adsorption level; maximum adsorption level	(g/g)
Γ'	dimensionless inhibitor adsorption level, (Γ/Γ_{max})	
$\Gamma_{eq}(C_m)$	equilibrium adsorption level for mobile phase concentration of C_m	(g/g)

ACKNOWLEDGEMENTS

The authors would like to thank: the member organisations funding the Heriot-Watt University Oilfield Scale Research Project which includes B.P., Shell UK Exploration and Production, Marathon, Chevron UK Ltd, Texaco, Statoil, Ciba-Geigy Industrial Chemicals and M.T.D. Ltd/SERC (Grant no. GR/E63322); Mrs Karen Morton for typing the paper.

REFERENCES

1. KERVER, J K and HEILHECKER, J K, J. Can. Pet. Tech., 15-23, January-March 1969.
2. SPRIGGS, D M and HOVER, G W, J. Pet. Tech., 812-816, July 1972.
3. VETTER, O J, J. Pet. Tech., 339-353, March 1973.
4. KING, G E and WARDEN, S L, SPE18485, presented at the SPE International Symposium on Oilfield Chemistry, Houston, TX, 8-10 February 1989.
5. PRZYBYLINSKI, J L, SPE18486, presented at the SPE International Symposium on Oilfield Chemistry, Houston, TX, 8-10 February 1989.
6. SHULER, P J and JENKINS, W H, presented at the SPE Offshore Europe Conference and Exhibition, Aberdeen, Scotland, 4-8 September 1989.
7. MEYERS, K O, SKILLMAN, H L AND HERRING, G D, SPE12472, presented at the SPE Formation Damage Control Symposium, Bakersfield, CA, 13-14 February 1984.
8. CARLBERG, B L, Oil and Gas J., 152-154, December 1983.
9. BERKSHIRE, D C, LAWSON, J B and RICHARDSON, E A, US patent 4,357,248, 1982.
10. SORBIE, K S, WAT, R M S, TODD, A C and McCLOSKY, T, to be published.
11. HONG, S A and SHULER, P J, SPE16263, presented at the SPE International Symposium on Oilfield Chemistry, San Antonio, TX, 4-6 February 1987.
12. RAMIREZ, W F, SHULER, P J and FRIEDMANN, F, Soc. Pet. Eng. J., 430-438, December 1980.
13. VAN GENEUCHTEN, M T, Research Report 119, US Department of Agriculture Salinity Laboratory, Riverside, CA, February 1981.
14. SORBIE, K S, JOHNSON, P V A, HUBBARD, S and TEMPLE, J, In Situ, 13 (3), 121-163, 1989.
15. SORBIE, K S, "SQUEEZE User Manual", Department of Petroleum Engineering, Heriot-Watt University, February 1990.
16. DURHAM, D K, SPE11708, presented at the SPE California Regional Meeting, Ventura, CA, 22-25 March 1983.

Tailoring Aromatic Hydrocarbons for Asphaltene Removal

Charles W. Benson, Richard A. Simcox, and Ingrid C. Huldal

ROEMEX LIMITED, CROMBIE ROAD, TORRIE, ABERDEEN, UK

1 SUMMARY

The deposition of ASPHALTENES, a naturally occurring polycyclic hydrocarbon present in many hydrocarbon deposits, has been the plague of the oil industry for many years.

In the absence of suitable chemical compounds to inhibit this problem, a study of chemical species which could provide swift and effective removal of deposited ASPHALTENE has been made. Chemicals, especially aromatics, that demonstrate good solvation characteristics at ambient, as well as at elevated temperatures are identified. Emphasis was placed on handling and safety aspects and studies were focused on aromatics with high flash points.

The effects of varying different parameters including temperature, the number of rings in the molecule and hydrocarbon side-chains, are illustrated.

2 INTRODUCTION

The presence and associated problems of asphalts in crude oil have been reported as far back as 1945 and before. Indeed a tremendous amount of data has been written on the subject. However, little is really understood about the way in which asphaltenes are held in solution, the method of deposition and routes to prevention.

A number of models are presented, the most commonly agreed being that asphaltenes are, broadly, n-heptane insoluble[1] and aromatic soluble, and are believed to be held in solution within crude oil by the presence of resins and aromatics. The resins act as peptizing agents, protecting the asphaltene particles and preventing their flocculation in the presence of aliphatic hydrocarbons. As the physical characteristics of crude oil alter during production near the bubble point of the crude, there is a shift in the aliphatic: aromatic balance and the peptizing layer around the asphaltene particles is stripped away. In the presence of the aliphatic hydrocarbons, the resultant naked asphaltene particles precipitate and accumulate on the surfaces of the production equipment.

This phenomena has been cause for much concern in the oil industry for as long as data on its presence has been reported. The resultant problems associated with asphaltene deposition are most commonly: restricted access of wells, inhibition of production and reduced plant capacity.

Without suitable chemistry to prevent the deposition of asphaltinic material, an efficient route to remove the deposits is required. Until recently only mechanical stripping and the application of simple solvents such as xylene, toluene and basic alkyl benzenes, have been available. Reports indicate only limited success with treatments using these solvents owing to incomplete removal of the deposits.

This paper details work carried out to identify chemicals, particularly aromatics, that demonstrate good asphaltene solvation characteristics and provide swift and relatively safe removal of asphaltene deposits.

3 PROCEDURES

Samples of asphaltene deposits were received from a number of wells from different North Sea Fields. At the time, severe problems associated with asphaltene deposition were encountered by British Petroleum operating the CLYDE field. BP kindly made a 10 kilogram sample of CLYDE asphaltene available and this was used as the standard source for laboratory work.

A standard procedure was adopted which would enable consistent results to be achieved being relatively easy to carry out and interpret and which would best reflect the application of a solvent in the field.

Sample Preparation

Large aggregates of asphaltene were removed from the main sample, leaving asphaltene of a standard particle size for the work.

Method

1. 1 gram aliquots of the asphaltene were placed in pre-weighed, acetone dried, 30 ml screw-top glass bottles.

2. 5 mls of the candidate solvent was added to the bottle and incubated at the test temperature for the requisite time. The samples at 18° C were constantly shaken during the test. Samples at elevated temperatures were hand-shaken every 15 minutes.

3. On completion of the test, the free solvent was carefully decanted off and the residual asphaltene flushed with acetone and left to dry over-night.

4. Finally, the bottles were re-weighed and the percentages of asphaltene dissolved were calculated and recorded.

NB

1. All reagents used during this work, excluding the commercial solvents and those produced by refineries on our behalf, were of A R GRADE.

2. A standard test period of 4 hours was adopted for the work (with intervals of one hour). Additional tests were carried out over an extended period (8 hours) to observe the performance of poorer solvent types in more detail.

Also tests were performed to see the effect of increasing the asphaltene:solvent ratio on the rate and extent of dissolution.

4 PARAMETERS INVESTIGATED

Asphaltene is made up of complex, polar, polycyclical structures. Asphaltenes have a molecular weight in the order of 10^3 but tend to exist in agglomerated form having higher molecular weight distributions of 10^3 - 10^5[3]. The associated resins have a molecular weight in the order of 500[4]. Asphalt is used as a general term given to a mixture of both asphaltenes and resins[5].

Accepting this and bearing in mind solvation mechanisms, it was clear that aromatic solvents would figure prominently in the development of suitable products for asphaltene dissolution. However, it was also considered that simple, single-ringed structures such as benzene, xylene, toluene etc, could only partially provide the key to dissolution of such complex structures.

The aim of the project was to identify the relevant parameters, so that a product (or products) could be formulated to be wholly efficient in the dissolution of the complex structures described. Parameters including the content of paraffinic, bi-cyclic and tri-cyclic structures were investigated.

In addition, the effects of contact time and temperature were considered to be very significant, particularly bearing in mind the variety of conditions encountered during field application. Depending on production conditions, asphaltene can be deposited in the reservoir, around the perforations, in the production tubing, or in the surface equipment. Clearly it was important to correlate the attributes of the substance to be applied with the physical characteristics of the treatment.

5 RESULTS

Table 1. Initial studies were confined to a number of readily available solvents to provide a basis from which to work; these included the following:

SOLVENT	COMPOSITION (%V/V)	FLASH POINT (°C)
BENZENE	99% mono-cyclic	-11
TOLUENE	99% methylbenzene	5
XYLENE	99% di-methylbenzene	27
XRS-1A	>99% alkylbenzenes	40
	<1% paraffinic	
XRS-2A	96% alkylated mono-cyclic	60
	4% alkylated bi-cyclic	
XRS-3A	>99% aromatic	65
	<1% paraffinic	
XRS-4A	50% alkylated mono-cyclic	70
	30% alkylated bi-cyclic	
	<1% alkylated tri-cyclic	
	19% paraffinic	

The commercially available alkylbenzenes were selected on the basis of their flash point, aromatic content and proven record as effective solvents for wax deposits and demulsifier resins.

Graphs: 1a and 1b show their relative performance at 18° C.

Graphs: 2a and 2b show their relative performance at 40° C.

In each case the rate of dissolution and ultimate asphaltene uptake was enhanced to some extent by temperature. However, none of the solvents gave 100% dissolution after 4 hours contact time even at the higher temperature. It transpired that in repeated studies the solvents based solely on mono-cyclic aromatics never gave 100% dissolution, irrespective of contact time or asphaltene:solvent ratio.

As the testing programme continued, results showed there to be little difference in performance between these solvents at ambient. At elevated temperatures they all gave better results, particularly XRS-2A, which has no paraffinic content.

1. From the results of the first series it was decided to investigate the effects of combining XRS-2A and XRS-4A with xylene and benzene.

Graphs: 3a shows XRS-2A 50% with xylene
and 50% with benzene

Graphs: 3b shows XRS-4A 50% with xylene
and 50% with benzene

each test being carried out at 18° C.

In each case, the combination of xylene with XRS-2A or XRS-4A produced a synergistic effect, whereas, benzene with XRS-2A or XRS-4A produced an inhibitive effect.

Further work carried out with different ratios and other solvents (including XRS-1A and XRS-3A) gave the same effects - positive synergism in the case of xylene and negative in the case of benzene.

2. From the results of the first and second test series, together with a re-evaluation of the programme, it

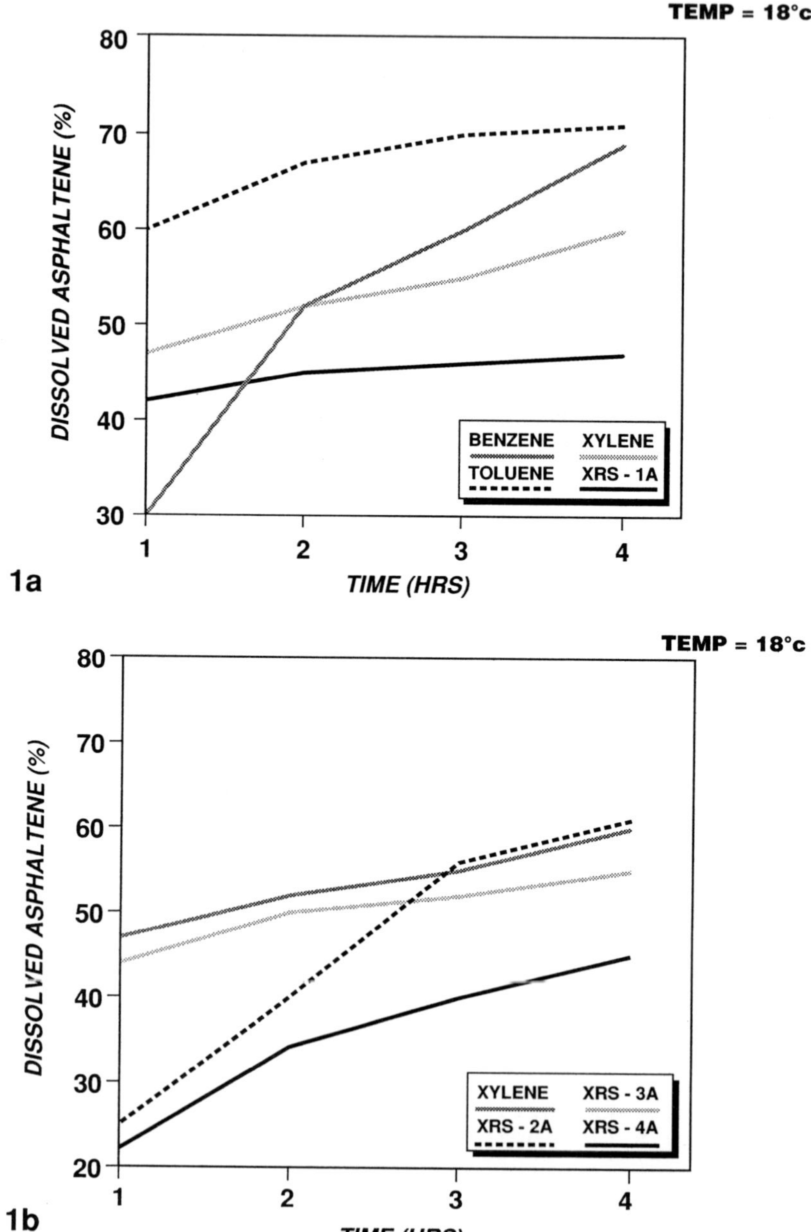
TEMP = 18°c
80
70
60
50
40
30
DISSOLVED ASPHALTENE (%)
BENZENE
XYLENE
TOLUENE
XRS - 1A
1
2
3
4
TIME (HRS)
1a
TEMP = 18°c
80
70
60
50
40
30
20
DISSOLVED ASPHALTENE (%)
XYLENE
XRS - 3A
XRS - 2A
XRS - 4A
1
2
3
4
TIME (HRS)
1b

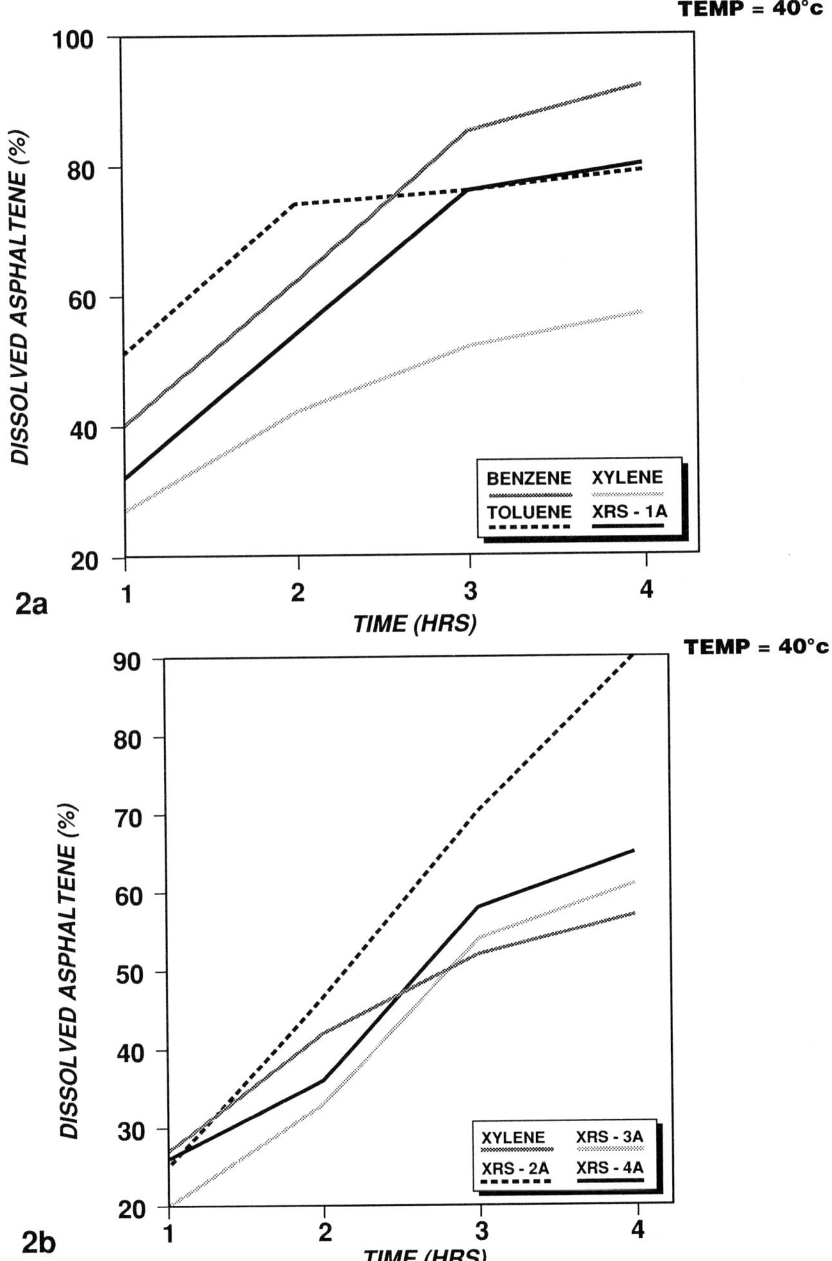
TEMP = 40°c
100
80
60
40
20
DISSOLVED ASPHALTENE (%)
BENZENE
XYLENE
TOLUENE
XRS - 1A
1
2
3
4
2a
TIME (HRS)
TEMP = 40°c
90
80
70
60
50
40
30
20
DISSOLVED ASPHALTENE (%)
XYLENE
XRS - 3A
XRS - 2A
XRS - 4A
1
2
3
4
2b
TIME (HRS)

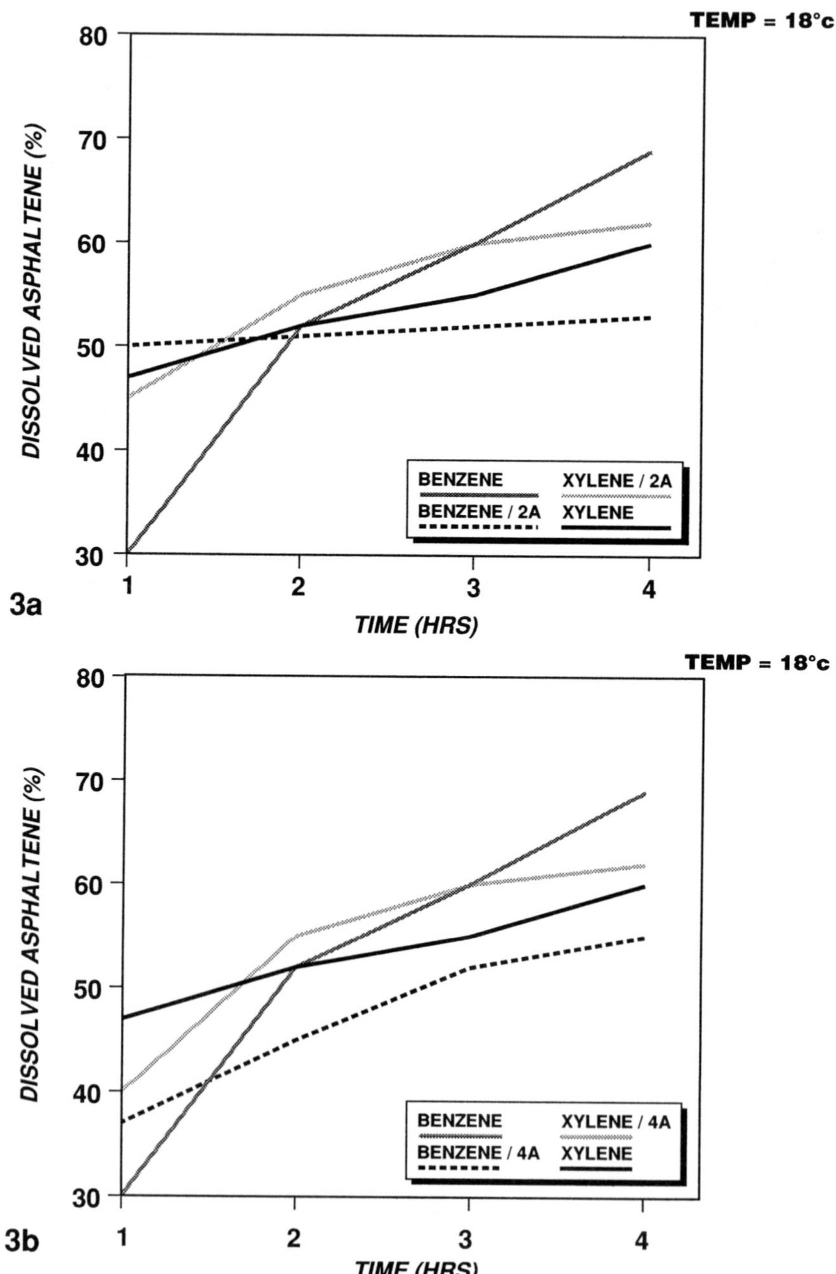
TEMP = 18°c
80
70
60
50
40
30
DISSOLVED ASPHALTENE (%)
BENZENE
XYLENE / 2A
BENZENE / 2A
XYLENE
1
2
3
4
3a
TIME (HRS)
TEMP = 18°c
80
70
60
50
40
30
DISSOLVED ASPHALTENE (%)
BENZENE
XYLENE / 4A
BENZENE / 4A
XYLENE
3b
1
2
3
4
TIME (HRS)

was decided, especially on the basis of results seen with XRS-4A, to investigate the effects of incorporating additional 2 and 3 ringed structures, as well as altering the paraffinic content.

Kerosene was used as a source of paraffin, naphthalene for bi-cyclic aromatics, and anthracene for tri-cyclic aromatics.

Table 2

SOLVENT	COMPOSITION (%V/V) MONO-	BI-	TRI-	PARAFFIN
XRS-4A	50	30	<1	19
XRS-4B	45	27	<1	29
XRS-4C	40	24	<1	39
XRS-4D	35	21	<1	49

Graph: 4a shows the effect of increased paraffin content at 18° C.

Graph: 4b shows the effect of increased paraffin content at 40° C.

3. It could be seen that increasing the paraffinic content had a detrimental effect on performance both at 18° C and 40° C. As such, the bi-cyclic content of XRS-4A was then increased.

Table 3

SOLVENT	COMPOSITION (%V/V) MONO-	BI-	TRI-	PARAFFIN
XRS-4A	50	30	<1	19
XRS-4E	45	37	<1	17
XRS-4F	42.5	40	<1	16.5
XRS-4G	37.5	47.5	<1	14.25
XRS-4H	32.5	54.5	<1	6.5

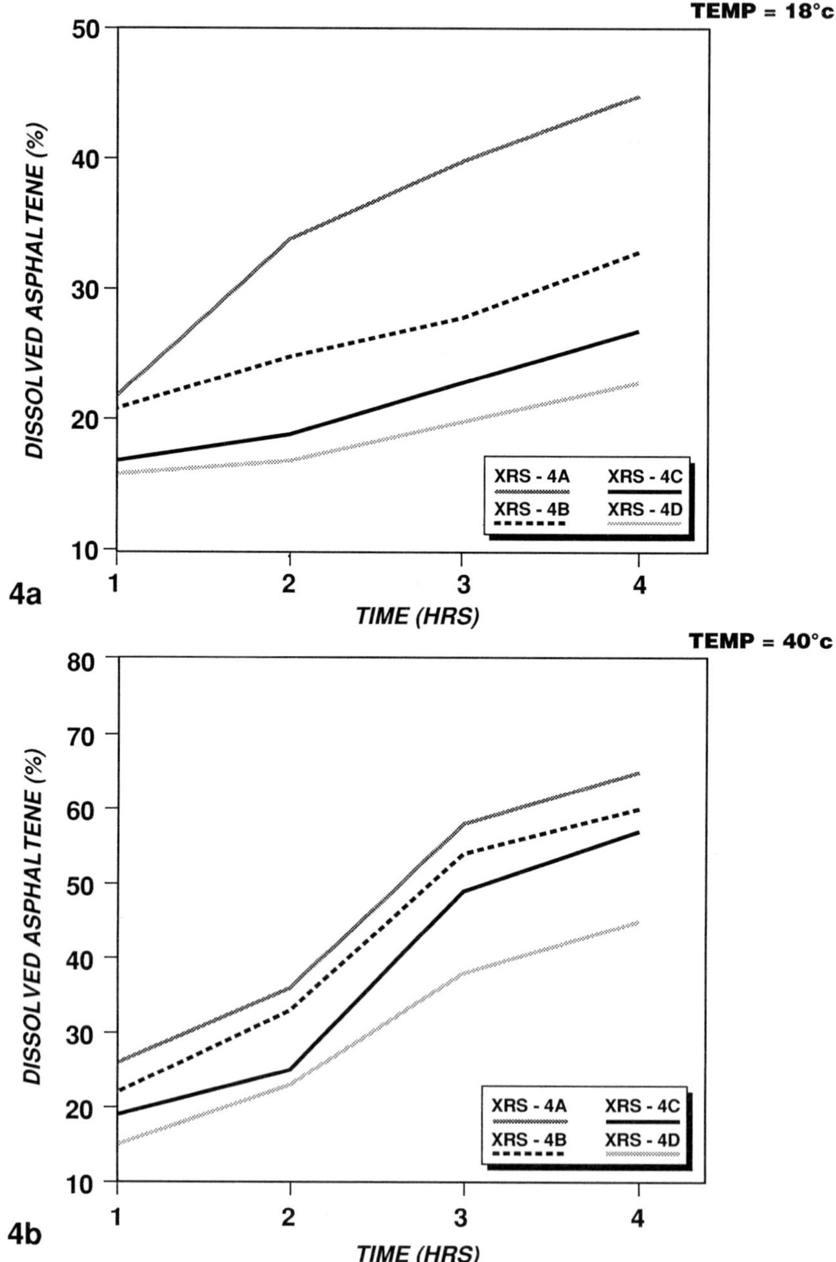
TEMP = 18°c
50
40
30
20
10
DISSOLVED ASPHALTENE (%)
XRS - 4A
XRS - 4B
XRS - 4C
XRS - 4D
1
2
3
4
4a
TIME (HRS)
TEMP = 40°c
80
70
60
50
40
30
20
10
DISSOLVED ASPHALTENE (%)
XRS - 4A
XRS - 4B
XRS - 4C
XRS - 4D
1
2
3
4
4b
TIME (HRS)

The saturation level of naphthalene in XRS-4A was attained after the addition of 15% naphthalene at 18° C, and 35% at 40° C. As will be shown later, it was possible, by selection of suitable feedstocks, to produce a fraction which contained a total of 65% bi-cyclics which is stable to less than -20° C.

Graph: 5a shows the effect of increasing bi-cyclic content at 18° C.

Graph: 5b shows the effect of increasing bi-cyclic content at 40° C.

It could be seen clearly that raising the content of bi-cyclic groups increased both the rate and degree of dissolution - the latter to 100%.

At this juncture the introduction of tri-cyclic groups was investigated; anthracene was used as the standard source. With XRS-4A, it was not possible to incorporate even 1% anthracene; this was also the case for the other more complex solvents. It was, however, possible to make up a 1% solution of anthracene in benzene. Series of tests were carried out and in each case the performance of benzene was impaired by the presence of the 3-ringed structure.

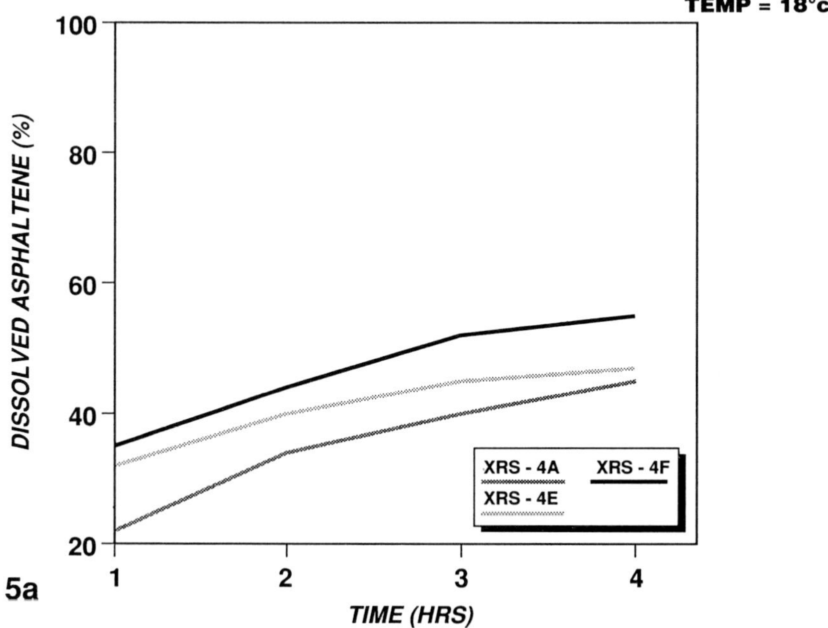

TEMP = 40°c
100
80
60
40
20
DISSOLVED ASPHALTENE (%)
1
2
3
4
TIME (HRS)
XRS - 4A
XRS - 4G
XRS - 4E
XRS - 4H
TEMP = 18°c
100
80
60
40
20
DISSOLVED ASPHALTENE (%)
1
2
3
4
TIME (HRS)
XRS - 5X
XRS - 5A

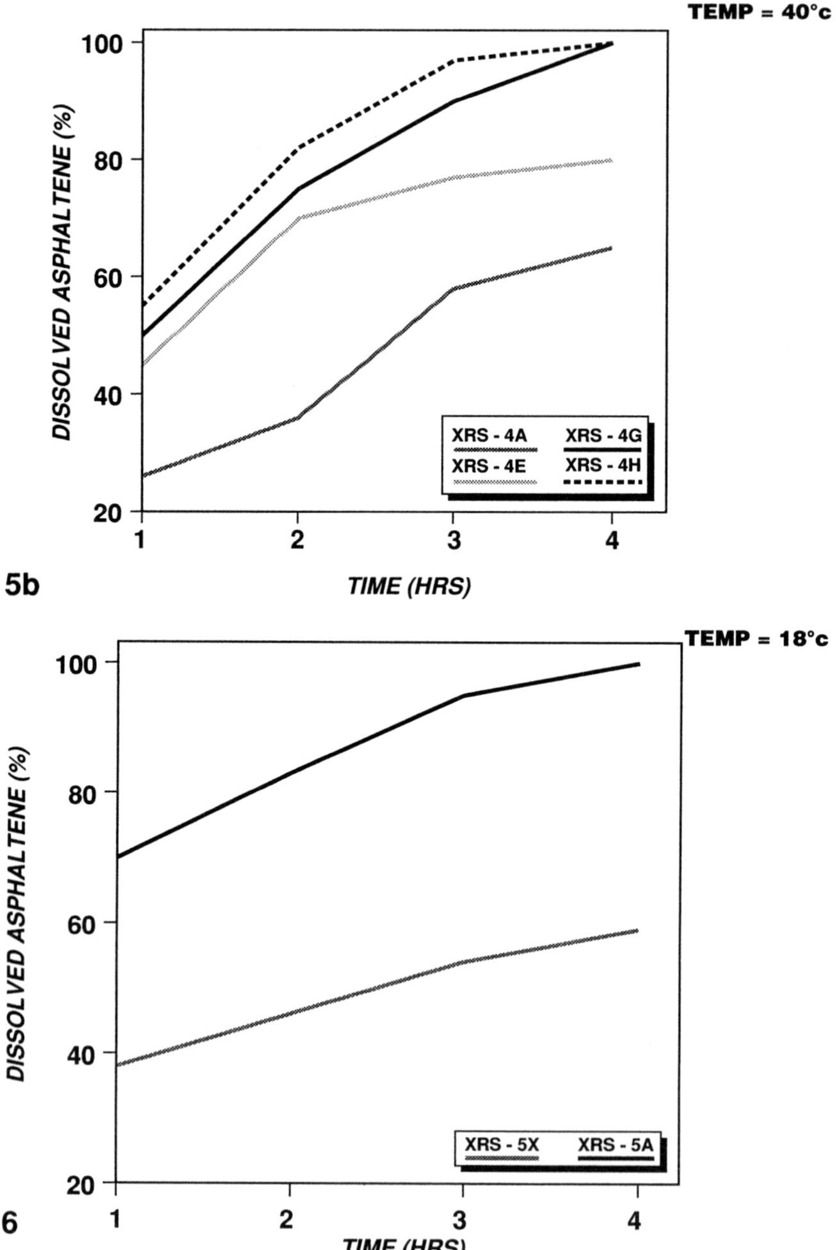

5b

6

4. Having established that paraffinic content has a detrimental effect on asphaltene dissolution and bi-cyclic content a beneficial effect, a number of refineries were approached to supply specific aromatic solvents with very low or zero paraffinic contents, produced from selected feedstocks.

A series of tests was then performed on this new developmental solvent blend XRS-5X; unfortunately the results were not as good as anticipated. Further analyical work was carried out which identified tertiary components. It was discovered that if these substances were removed through additional refining, the resulting solvent blend, namely XRS-5A, demonstrated far superior solvency characteristics. These differences are not detectable on basic HPLC analysis.

Graph: 6 shows the relative performance of XRS-5X and XRS-5A.

Having achieved our original goal, a series of evaluation tests were run on the XRS-5A incorporating kerosene and/or naphthalene.

<u>Table 4</u>

	COMPOSITION (%V/V)			
SOLVENT	MONO-	BI-	TRI-	PARAFFIN
XRS-5A	33	66	1	NIL
XR5-5B	29.7	59.5	<1	10
XR5-5C	26.4	53	<1	20
XR5-5D	29.7	69.5	<1	NIL

Graph: 7a shows the effect of addition of paraffin and/or naphthalene on XRS-5A @ 18° C.

Graph: 7b shows the effect of addition of paraffin and/or naphthalene on XRS-5A @ 40° C.

Graph: 7c shows the effect of addition of paraffin and/or naphthalene on XRS-5A @ 70° C.

Two traits are evident from these results, namely that increasing the level of bi-cyclics enhances performance, and increasing the level of paraffinics reduces it.

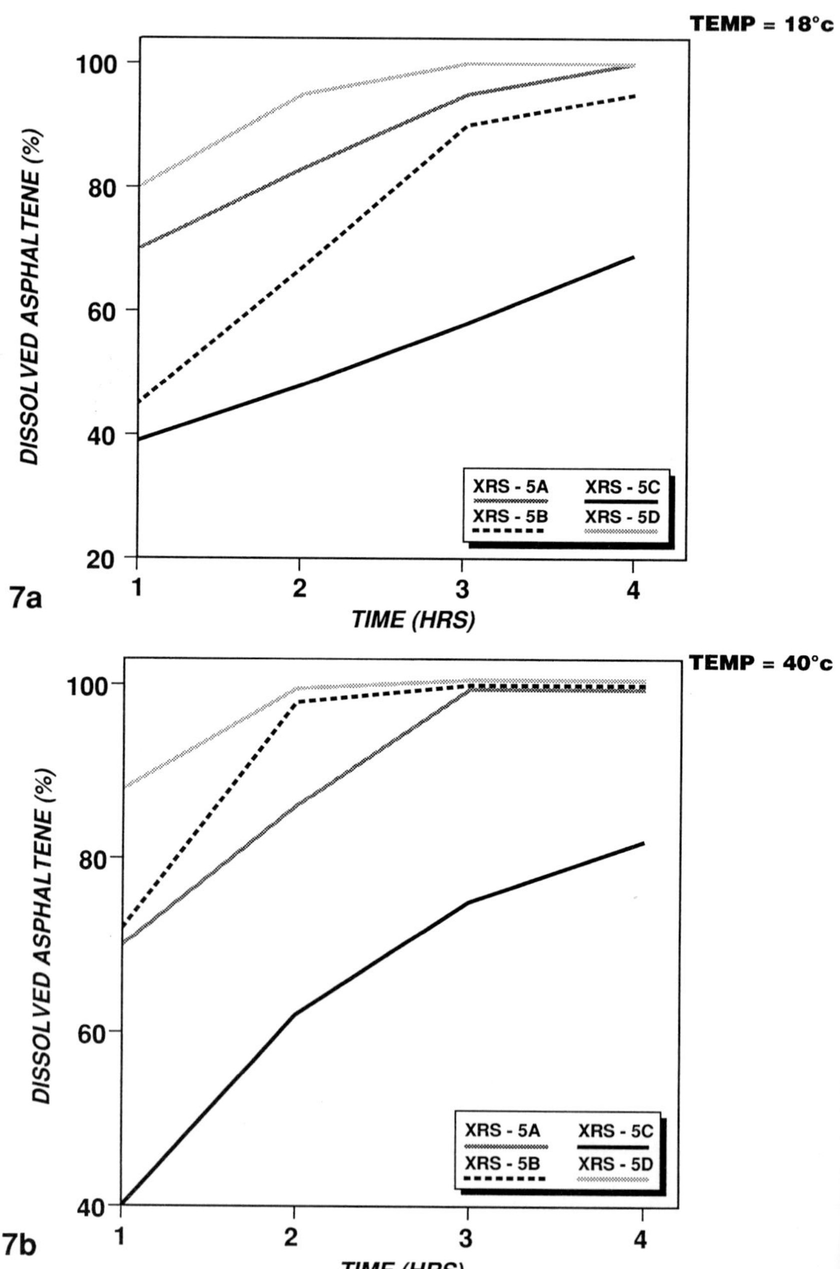

TEMP = 18°c
100
80
60
40
20
DISSOLVED ASPHALTENE (%)
XRS - 5A
XRS - 5B
XRS - 5C
XRS - 5D
1
2
3
4
TIME (HRS)
7a
TEMP = 40°c
100
80
60
40
DISSOLVED ASPHALTENE (%)
XRS - 5A
XRS - 5B
XRS - 5C
XRS - 5D
1
2
3
4
TIME (HRS)
7b

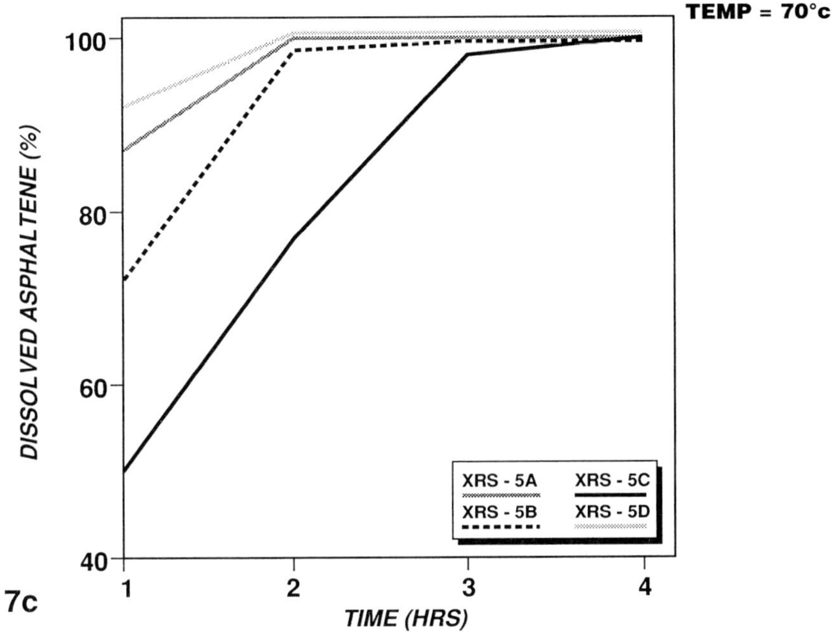

5. A series of tests were then set up to observe the effect of increasing the asphaltene:solvent ratio to 2 g per 5 ml. At this stage it was deemed only worthwhile looking at the 2 best products ie XRS-5A and XRS-5D, alongside 2 other previously screened solvents - xylene and XRS-4A were chosen. XRS-5A and XRS-5D performed equally well, and all 4 selected products showed a similar rate and degree of uptake of asphaltene as with the standard asphaltene:solvent ratio of 1 g per 5 ml.

Graph: 8a shows the results of test using the higher asphaltene:solvent ratio at 18° C.

Graph: 8b shows the results of test using the higher asphaltene:solvent ratio at 40° C.

6. To obtain more insight into the performance of various products not demonstrating total asphaltene uptake after 4 hours, extended tests were carried out on several appropriate solvents. Improvements of between 5% only for XRS-2A and over 30% in the case of XRS-4A were achieved after 8 hours. XRS-5C gave over 90% removal indicating that total dissolution would be achieved after approximately 11 hours total exposure time.

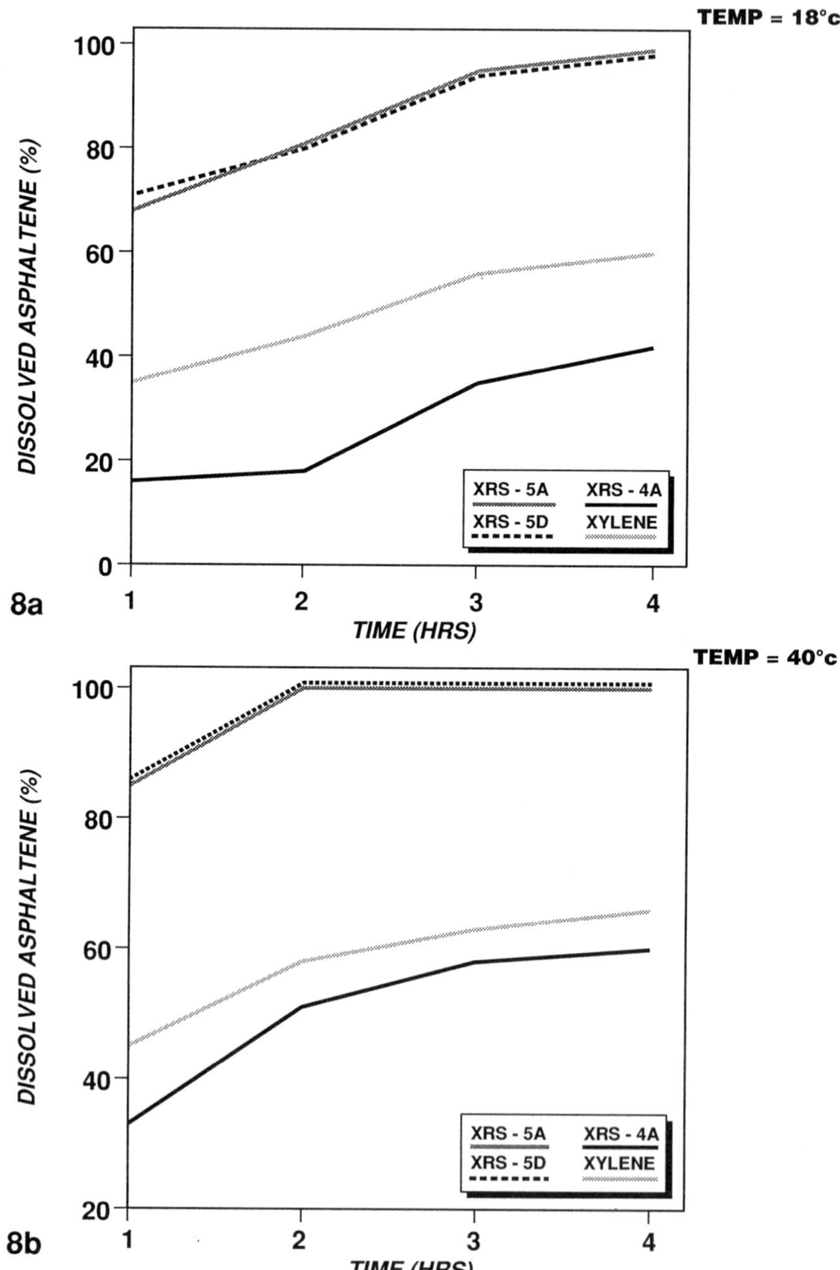
TEMP = 18°c
100
80
60
40
20
0
DISSOLVED ASPHALTENE (%)
XRS - 5A
XRS - 4A
XRS - 5D
XYLENE
8a
1
2
3
4
TIME (HRS)
TEMP = 40°c
100
80
60
40
20
DISSOLVED ASPHALTENE (%)
XRS - 5A
XRS - 4A
XRS - 5D
XYLENE
8b
1
2
3
4
TIME (HRS)

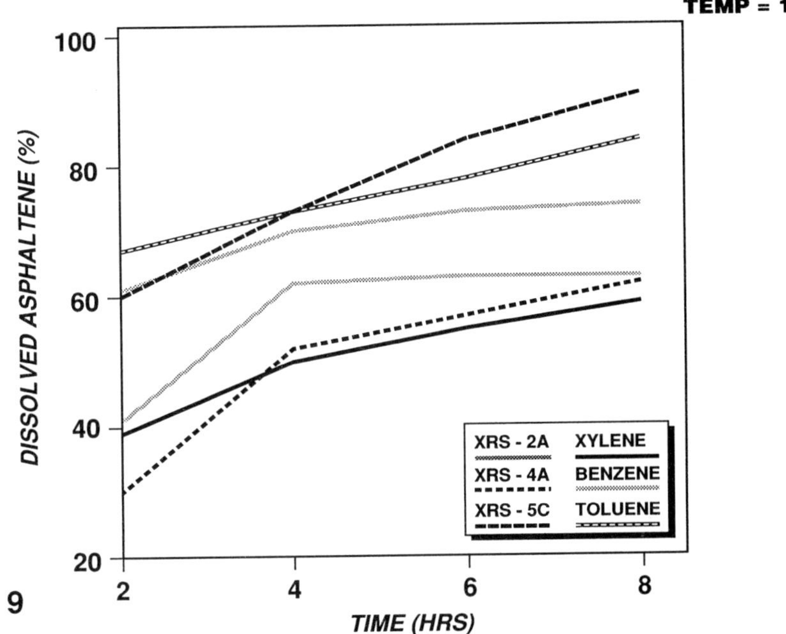

Graph: 9 shows the percentage uptake of asphaltene over 8 hours by xylene, XRS-4A, XRS-2A, benzene, toluene and XRS-5C at 18° C.

6 CONCLUSIONS

1. Presence of paraffinics detrimental.

2. Presence of mono-cyclics beneficial.

3. Presence of bi-cyclics beneficial.

4. Presence of tri-cyclics non-contributory.

5. Dissolution rate is $\pm$ constant with varying asphaltene:solvent ratio.

6. Rate of asphaltene uptake generally increases with temperature.

Note: Presence of polycyclics (>4 rings) believed to be detrimental.

7 APPLICATION - CASE HISTORY

From the work described in this paper was borne the product:

ROEMEX ASPHALTENE DISSOLVER RX-33/1

Its use in the UK and Norwegian sectors of the North Sea has been extensive over the past 2 years. During that time, a variety of problems have been broached, 2 examples of which are given here.

Asphaltene Deposition in Tubing Near to Elastomeric Seals

Compatibility tests had been carried out on the effect of RX-33/1, together with other solvents such as xylene, vs elastomeric seals. Whilst in general xylene tends to be much more aggressive than RX-33/1, nitrile seals are particularly susceptible to attack by both xylene and RX-33/1. As such, the treatment was designed so that contact between the RX-33/1 and the down-hole seals would be avoided. To this end, a viscous pill was bull-headed down the tubing first to act as a protective interphase between the RX-33/1 and elastomeric seals.

The deposit itself was spread over an interval of some 2,000 m. The mass of asphaltene deposited was estimated based on depth of deposit, ID of tubing and period of interval. The volume of RX-33/1 recommended was approximately twice that of the asphaltene to be removed, this being enough to cover one third of the interval of deposit. Thus, the pill had to be "spotted" by pumping it to the top section of the deposit; the pill was then pushed down to the middle section and finally to the bottom of the deposit, allowing a one hour soak time after each movement. With the treatment complete, the well was brought on to production to displace the spent RX-33/1 pill and was produced via the test separator.

Deposition Beneath Tubing Mule Shoe and Across the Perforations

Again, the treatment was designed to be approximately twice the estimated quantity of asphaltene present. Since the asphaltene deposit was located across the perforations , it was considered too much of a risk to use the "bull-head" technique which might force debris into the formation. Thus, spotting via coiled tubing was selected in preference.

As the void within the perforations/liner interval where the asphaltene was present was so small, the treatment had to be carried out in 3 separate stages, in this case with 3 x 1 m^3 pills.

The deposit was so abundant that the 1 1/2" Coiled Tubing (CT) could not pass through. Thus, once a restriction was encountered during the introduction of the CT, the RX-33/1 was slowly pumped via the CT to dissolve away the restriction allowing positioning of the CT to 10 m below the bottom of the deposit.

As the uptake of asphaltene by RX-33/1 is so rapid at bottom hole temperature, it was decided to spot the first treatment pill (1 x 1 m^3) for 10 minutes, the second for 45 minutes and the third for 60 minutes. The CT was pulled out of the interval each time whilst spotting the pill, and the well was produced between pills to displace the spent RX-33/1 out of the hole. The RX-33/1 pill was, again, produced via the test separator but flared off rather than exported via the pipeline.

Finally, a 3 1/2" drift run was carried out and no presence of asphaltene was detected whatsoever.

A rubber dart (spacer) was placed behind the RX-33/1 in the CT and displaced with sea water, ensuring efficient use of the RX-33/1 and no intermixing with sea water during displacement.

REFERENCES

1. Institute of Petroleum, London. "Standard Methods for Analysis and Testing of Petroleum and Related Products". IP143/84 (1988).

2. A. Hirschberg et al SPEJ. "Influence of Temperature and Pressure on Asphaltene Flocculation". (June 1984) 283-93.

3. D. H. Katz and K. E. Ben, Ind and Eng Chem. "Nature of Asphaltic Substances" (1945) 37,195.

4. J. G. Speight. "Solvent Effects in the Molecular Weights of Petroleum Asphaltenes".

5. J. A. Koots and J. G. Speight. "Relation of Petroleum Resins to Asphaltenes".

The Use of Oilfield Chemicals on the Norwegian Shelf

Peter A. Read

DRILLING AND WELL TECHNOLOGY DEPARTMENT, DEN NORSKE STATS OLJESELSKAP, STATOIL, STAVANGER, NORWAY

1 INTRODUCTION

The Norwegian sector of the European continental shelf currently produces around 110 million oil equivalent tonnes of hydrocarbon - mainly oil. Of the 15 oil fields currently in production, one has been on production for 20 years (Ekofisk), 2 for between 10 and 15 years (Frigg and Statfjord) and six for more than 5 years. In the coming 5 years a further 7 fields will be added, making a total of 22 fields in production by 1996.

The increasing age of some of Norway's largest producing fields, plus the rapid rate at which new fields with new problems, are being developed places increasing operator dependance on chemical applications. Fields developed during the 1960's utilised technology now over a quarter of a century old. Fields being developed today can benefit from the experience gained in operating these earlier installations. Though often complex, the analysis of problems requiring a chemical solution coupled with improved chemical products can lead to considerable cost savings in production operations.

The foregoing contribute to making the North Sea one of oil field chemistry's fastest growing markets. The Norwegian sector consumed production chemicals to a value of over 100 mill. Norwegian kroner in 1990, and this expenditure is expected to grow at an estimated 20% per year for the foreseeable future.

Society's increased environmental concerns have caused both the oil producing industry, as well as the chemical industry, to review our activities. Some of our previous freedoms have been restricted by legislation. We may expect more, rather than fewer, restrictions in time. However, by self-initiated action, and the application of good science, we may avert the introduction of unnecessarily restrictive legislation.

Most operating oil companies in Norway have a policy of minimising their dependance on chemicals in production.

It is a sad fact that many chemicals consumed in the oil field are the result of sub-optimal process solutions. Greater attention to process performance at the design stage would go a long way to minimising unnecessary chemical consumption. Improved management of waste releases, such as the re-injection of produced water, would further reduce our releases to the environment. The aim should be to have zero environmental releases. It is against this background that oil field chemists meet the challenges of the last decade of the 20th century. This paper intends to review past chemical usage and show how our chemical consumption patterns are subtly changing.

The 1990's will be a decade of increasing reliance on production chemicals and a decreasing inclination to do so!

2 CHEMICAL CONSUMPTION

By far the greatest chemical consumption is connected to water handling, both water injection and produced water. Taking specific field cases, in 1989, Statfjord field consumed about 3000 Tonnes of production chemicals in total. Process related consumption accounted for, including water injected, 2,500 Tonnes, or 83%. The same story is found in Gullfaks where the fields total consumption in 1989 was 600 tonnes, 500 tonnes being process related. A percentage of again 83%. Both fields are producing water, some of this being re-cycled injection water. Fig. 1.

The largest consumption on Statfjord of individual chemical is that of biocide. By tonnage biocide accounted for 28.5% of Statfjords total consumption of process/-production chemicals. Gullfaks consumed somewhat less biocide at 20.3% of total, and we will discuss some of the reasons for this later. On Gullfaks the largest individual consumption of any chemical was for oxygen scavenger, accounting for 26.2% of consumption. The oxygen scavenger consumption on Statfjord was only 19% of total.

OIL, WATER PRODUCTION
& WATER INJECTION
FOR A TYPICAL NORTH SEA FIELD
WATER INJECTION
WATER PRODUCTION
OIL PRODUCTION
m3 / d
(x1000)
90
80
70
60
50
40
30
20
10
0
0
5
10
15
20
YEARS PRODUCTION

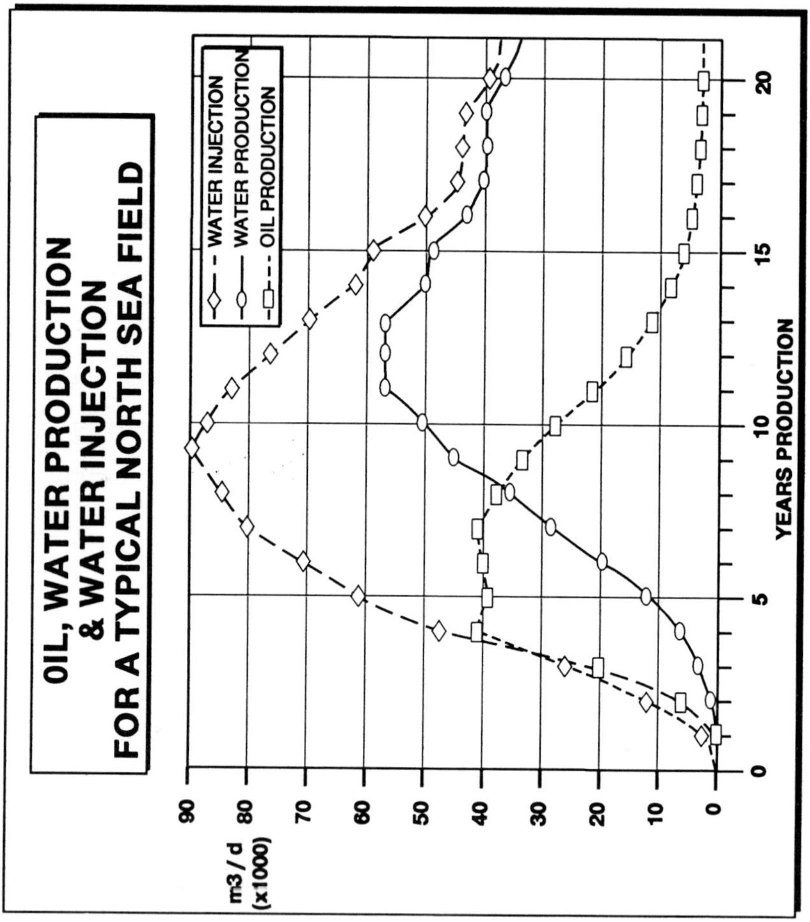

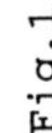

Fig.1

It must be added that for a period in 1989, the vacuum towers at Gullfaks were not operating at peak efficiency resulting in a need to apply more oxygen scavenger. Again we will return to this theme later. These 2 chemicals alone account for about half of all consumption offshore. It should be emphasised that these chemicals are primarily intended to follow the injection water, and as such have little environmental impact. Apart from some polyelectrolyte as filter aid, and foam breaker a biocide for the vacuum towers, the remaining chemical consumption is applied to the produced water side.

Scale inhibitors are largely applied in 2 ways:

(a) Directly to the produced water system
(b) Squeezed into producing wells and applied downhole.

At Statfjord 322 tonnes of scale inhibitor were applied to the process stream. In addition, 244 tonnes were applied in 9 well treatments. All these materials are phosphonates. However, at Gullfaks, 111 tonnes of process scale inhibitor was consumed, whilst one well squeeze treatment consumed 23 tonnes of a phosphino-carboxylic acid.

Both Statfjord and Gullfaks require flocculants to achieve oil in water release specifications. Due to the installation of hydrocyclones at Statfjord, the use of both emulsion breakers and flocculants has been reduced. The same strategy is planned for Gullfaks. To illustrate the potential for chemical savings, Statfjord consumed 148 tonnes in total of demulsifier and flocculant in 1989 to polish 7,702,000 Sm^3 of water.

Gullfaks required 108.5 tonnes to polish 2,704,000 Sm^3 water. Gullfaks has a more persistent emulsion problem and no hydrocyclones. The advantages of hydrocyclones on oils whose emulsions can be treated in this manner can thereby lead to considerable reduction in chemical additives. It must be pointed out that these chemicals in part are released to the marine environment although some follow the oil. Correct design of the separator/water polishing system can lead to large reductions in chemical consumption. The relative distributions are shown in Fig.2.

It is a sad fact that many of the production and process chemicals are used in order to assist installed equipment meet design specifications.

DISTRIBUTION OF CHEMICAL ADDITIVES
STATFJORD
33%
29%
19%
11%
2%
2%
4%
GULLFAKS
20%
27%
9%
8%
7%
1%
12%
14%
O_2 SCAVENGER
DEMULSIFIER
FLOCCULANT
CORR. INH.
BIOCIDE
SCALE INH.
ANTI FOAM
GAS TREAT.

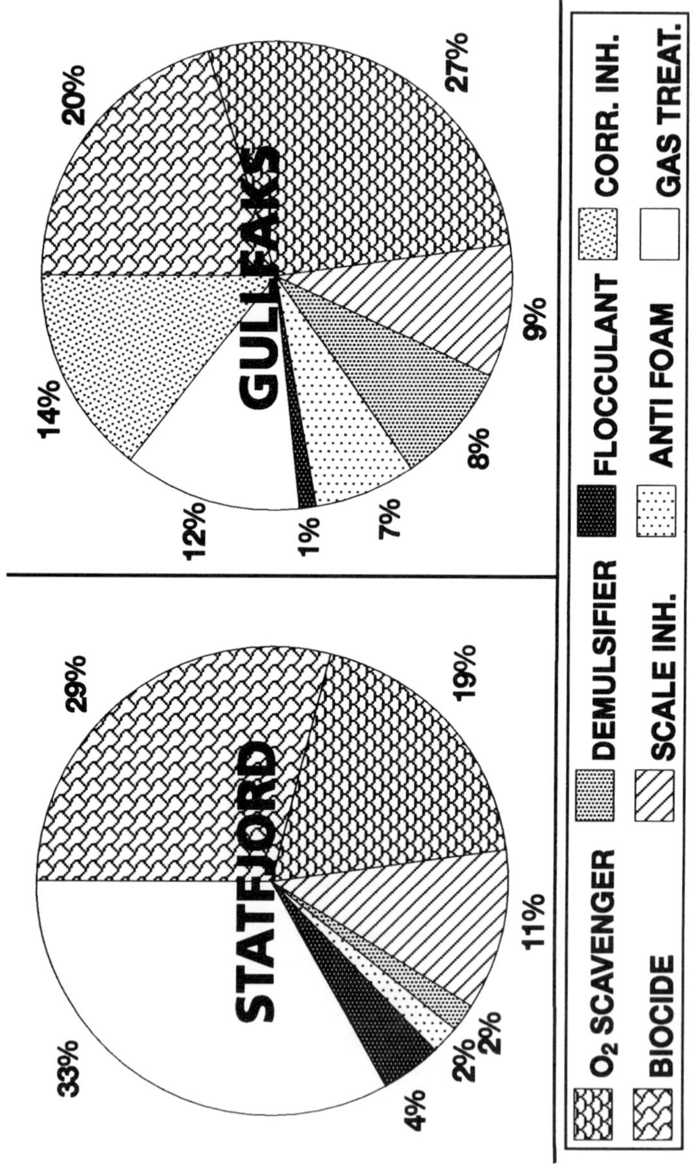

Fig.2

For example, failure of vacuum pumps on a deaerator tower lead to increased consumption of oxygen scavenger. Indeed, some equipment is designed with the intention of applying chemical additives to achieve specifications. Floatation oil/water separators are perhaps a classic example of this. Without the correct chemical additives, these separators just will not achieve adequate release levels.

This means that as environmental release specifications are tightened, process equipment becomes more dependant on applied chemistry. To simultaneously restrict the release of the chemicals needed to achieve the new specification would seem to demand the impossible. This is, in effect, what the industrial chemist is being asked to do.

It is perhaps opportune to review this cyclic dilemma. Maybe the problem should be returned to the process engineer with design specifications on produced water handling equipment which does not permit reliance on chemistry to achieve functional integrity.

3 CHEMICAL RELEASES

Statfjord released to the marine environment about 480 tonnes of production chemicals in 1989. This was mainly scale inhibitor (320 tonnes) together with a small quantity of biocide (10 tonnes) used to combat H_2S generation by hypothermophilic bacteria in the production train.

The release profiles for Gullfaks were similar with a total of 90 tonnes released of which 50 tonnes were scale inhibitor and 10 tonnes were polyelectrolyte. The remaining chemicals were largely surface active compounds such as emulsion breakers and flocculant.

4 IMPROVEMENTS IN INJECTION WATER TREATMENT

A vital use of production chemicals is in connection with the injection water system. Experience dictates that most of the chemical additives are used to minimise corrosion of installed tubulars.

Remarkably, the addition of certain chemicals may actually stimulate the very phenomenon they are intended to prevent. In designing the chemical additive package for

Gullfaks, a particularly insidious form of bacterial corrosion was identified as being of special threat. Experience from Statfjord showed that inappropriately treated injection water could encourage the growth of very thick biofilms, below which corrosion could be very aggressive. The biofilms were particularly persistent, and their removal a difficult undertaking.

Some of the more popular additive chemicals were identified as potential nutrients for biofilms. Ammonium Bisulphite was eliminated as being a very ready source of nutrient. Statfjord injects 534 Tonnes of this nutrient chemical annually. At Gullfaks, the Ammonium Bisulphite was replaced by Sodium Bisulphite, initially catalysed using cobalt but now an iron catalyst is used.

The method of deaeration was identified as a potential stimulus to bacterial activity. The introduction of soluble carbon compounds from the fuel gas of conventional gas deaerators was seen as undesirable. Therefore, a vacuum deaerator was chosen for Gullfaks. Though problems have been experienced with the vacuum pumps, this problem has only given rise to slightly increased consumption of oxygen scavenger. A second tower problem was reduced efficiency due to foaming. Application of a defoamer resulted in improved efficiency, but increased bacterial activity. It transpired that the solvent used to apply the antifoam was being used as a carbon source by anaerobic bacteria, and on changing this solvent bacterial activity reverted to previously low levels.

A vital component of the Gullfaks water injection facility is its filtration system. The filters chosen were sand filters for the initial platforms. These require polyelectrolyte during bloom periods. The filtration specification demands removal down to 7 microns in order to remove the major content of dead planktonic material which could otherwise be utilised by bacteria.

The sand filters have caused problems by leaking filter media to the downstream system. Though not affecting water quality, some problems have been experienced in removing this sand.

A water treatment system is currently being installed on Gullfaks B to cope with increased water injection demand. The filter chosen for this system was an acid washable cartridge filter. This is in part due to experience with our dual media filters, but they also have weight and space savings. The chemical advantages are that polyelectrolyte is replaced by sulphuric acid. Spent acid will, on release, revert to sulphate and the excess hydrogen will be taken up by the seas natural buffer capacity.

The philosophy of screening and choosing chemicals which are not utilised by sulphate reducing and other anaerobic bacteria has given dividends. Chemical dosages have been reduced, especially of biocide. This is due to the natural limits to bacterial growth in the injection water. Corrosion rates are low, and considerably less than comparable systems using 'conventional' chemical treatments. The operational and maintenance costs are lower, and we believe that the chances of severe reservoir souring are reduced. Backflow of an injection well showed no signs of sour production or of unusually high SRB or GAB populations.

5 IMPROVED WELL TREATMENTS - SCALE SQUEEZES

Increased efficiency of production and reductions in both chemical consumption and releases to the environment can be effected by improving the application techniques of the chemicals we use. A routine operation is that of sqeezing wells with scale inhibitor. An adsorption squeeze treatment is extremely wasteful of treatment chemicals. After placement, immediate returns of well fluids show huge concentrations of inhibitor - often comparable to initially applied concentrations.

Improved squeeze lifetimes have been achieved by applying phosphino-carboxylic acid in a precipitation squeeze mode. This has been effected by providing calcium ion at pH's where some precipitation occurs at reservoir temperatures. The effect has been to reduce by about 20% the amount of chemical required to be applied in the initial squeeze. Secondly, squeeze lifetime has been considerably improved with re-squeeze times of a year, or more, possible in typical low barium reservoirs.

There has been a trend away from use of the traditional phosphonate scale inhibitors in favour of the phosphino polycarboxylic acids. The growth in phosphino polycarboxylic acid consumption is shown in Fig. 3. for the last 5 years. In 1989 some 559 tonnes of phosphino polycarboxylic acid was sold in total, 20% of which (59 tonnes) was used in the Norwegian Sector. In the coming years, these inhibitors are expected to gradually replace the phosphonates in well treatments, as we learn to apply

CONSUMPTION OF PHOSPHINO ACID SCALE INHIBITOR
WORLDWIDE & IN NORWEGIAN NORTH SEA
TONNES NORWAY
100
90
80
70
60
50
40
30
20
10
0
TONNES WORLDWIDE
1000
900
800
700
600
500
400
300
200
100
0
1985
1986
1987
1988
1989
1990
1991
NORWAY
WORLDWIDE

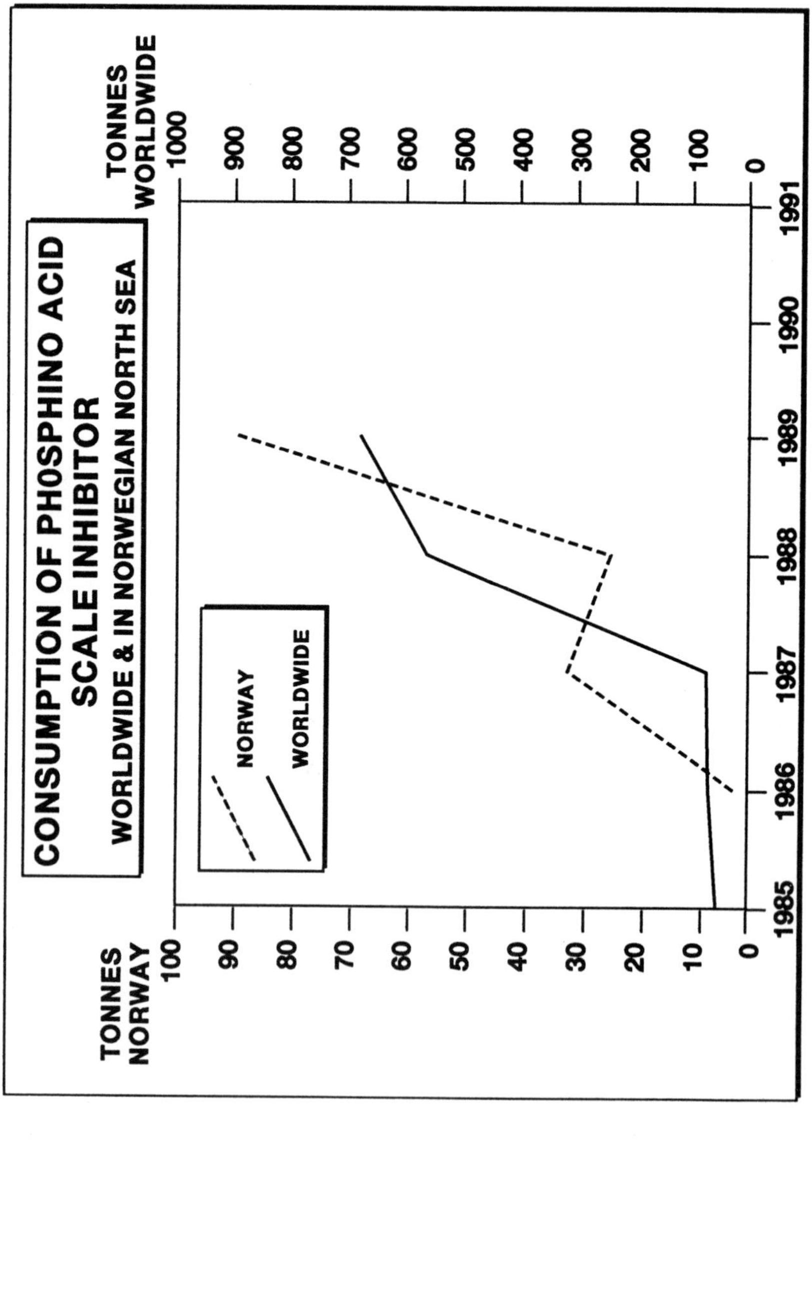

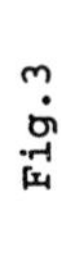

Fig.3

them to more field situations. The benefits of simplifying the chemical package are shown in comparing the treatments on Gullfaks wells with those on Statfjord, Fig.4.

6 SUBSEA WELLS - SPECIAL PROBLEMS

Several Norwegian fields have in operation, or plan to have, subsea satellites. These developments may be many miles from their parent platforms. The furthest distance planned to date is 15 kms. Service to wells of this nature places particular constraints on chemical treatments. Lack of freedom to access the well makes scale squeeze treatments very difficult to schedule not least due to difficulties in monitoring.

The application of corrosion, hydrate and scale inhibitors becomes an acute problem when limited by both the number and size of access lines. Merely the energy required to pump the chemicals produces a problem which is by no means trivial.

To avoid the more common problems, operators have reviewed using membrane based sulphate removal plants to eliminate the need for scale inhibitors, and reduce problems from sulphate reducing bacteria in subsea developments.

To date, one sulphate removal plant is in operation in the North Sea and one further plant is in the advanced planning stages. Several other operators are evaluating sulphate removal for fields with particularly severe scale problems. Their use in satellite developments will clear the way for the simplification and reduction of chemical treatments to these remote wells. Further simplification of the diversity of required chemicals will be achieved by multi-functional chemicals - products combining several activities in one material - and is an area currently under investigation.

7 WELL MAINTENANCE/COMPLETION

As particularly difficult well situations become apparent so chemical consumption in connection with well workovers and completion will tend to increase. A conventionally completed well in Statfjord field will consume about 2 1/2 Tonnes of chemical additives during the completion phase. Higher reservoir pressures and more demanding operating conditions leads to an increase in consumption of

DISTRIBUTION OF CHEMICAL ADDITIVES
USED DURING SCALE SQUEEZE TREATMENT
STATFJORD
38%
11%
13%
18%
4%
7%
9%
GULLFAKS
33%
23%
43%
OX. SCAVENGER
ANTIFOAM
SURFACTANT
BIOCIDE
DEMULSIFIER
OIL SOL.
VISCOSIFIER

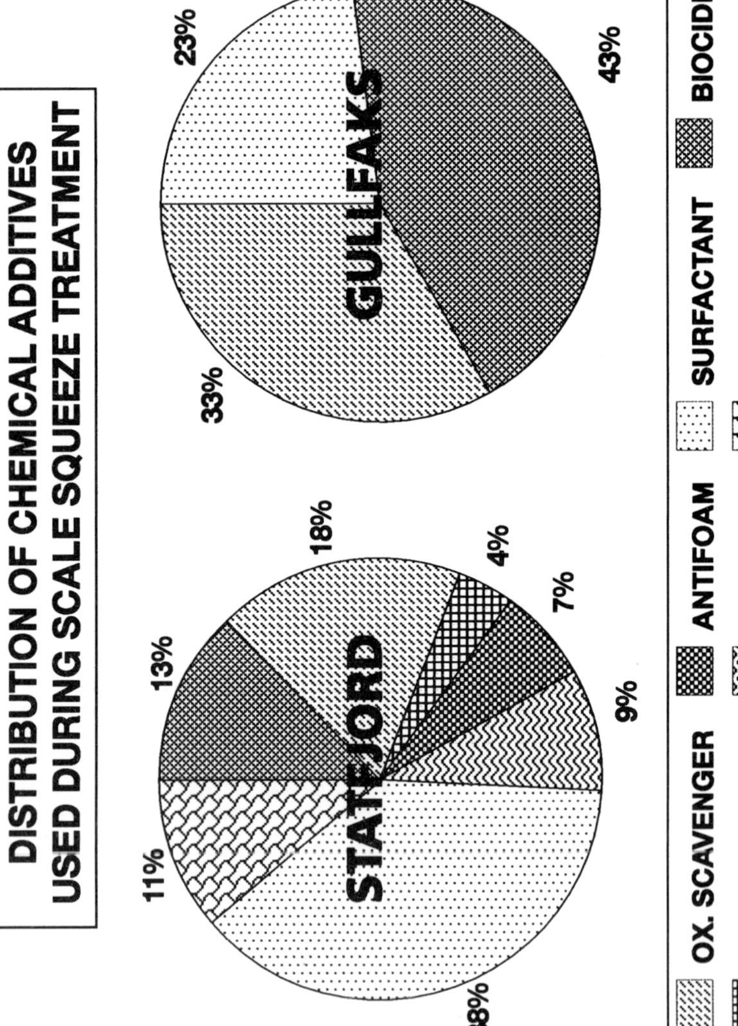

Fig.4

chemicals in Gullfaks field. On average a conventional well in Gullfaks will consume about 4 Tonnes of chemicals.

Huge increases in consumption of chemicals however occurs when special well conditions demand the installation of gravel packs to control sand production. A gravel packed Gullfaks well on average consumes about 30 Tonnes of chemicals. These comprise largely oil soluble resins, viscosifiers, surfactants and dense brine components. Indeed, the large increase in chemical consumption for gravel packed wells, and the attendant risks of environmental releases, encourages activity to find alternatives to gravel packing as a means of controlling sand. Currently, methods to control sand production by using unconventional oil field chemical systems are under evaluation coupled with selective placement techniques. This will, if successful, contribute significantly to reduce total chemical consumption as well as releases to the environment. The chemical systems under study are resins, polymers and silicate based chemicals which will have a high added value when combined with selective placement techniques such as coiled tubing.

8 REINJECTION OF PRODUCED WATER

Many Tonnes of chemicals are consumed in de-oiling produced water. Of the chemicals released to the environment, by far the largest amounts are linked to the need to polish produced water. The current release levels of oil in produced water of 40 ppm are under review by the Paris Convention. In addition, as fields in the Norwegian sector come off plateau production, lack of water handling capacity figures centrally as one of the underlying reasons. A field such as Statfjord releases a large proportion of its scale inhibitor, flocculant and demulsifier to the environment via the produced water system.

In addition to suspended oil at about 30-40 ppm, variable amounts of volatile fatty acids are also present in produced water. Values vary, but amounts from several hundred up to 1000 mg/l are not unknown. Hydrocyclones have relieved pressure on the conventional water handling systems on many platforms. Pioneered on the UK sector, their use has greatly contributed to reduced chemical dependance. Being an efficient separator operating on physical concepts permits rational handling of increasing produced water quantities. However this increase in produced water necessarily means that less injected water enters the reservoir to replace voidage. Therefore, we have seen a need to increase injection water capacity to keep pace with the increases in produced water quantity.

This cycle is doomed to spiral upwards as fields come off plateau production levels. In recognition of this, a strategy of reinjecting produced water is being considered. Some trials have already been conducted on the UK Sector. However, studies are currently under way to evaluate the advantages of a bolder plan to reinject a large proportion of that water produced in several of our fields. Fig.5.

Immediate advantages are that environmental releases will be considerably reduced. It may be expected that unconstrained by the 40 ppm limit on suspended oil that many of the chemicals used in operating the conventional flotation cells may be reduced. They will, however, not be completely eliminated, as some overboard dumping of water will always be needed. The constraints imposed by environmental toxicity considerations need not be a restriction on our choice of products if a reinjection strategy is adopted. The use of small, simpler process equipment may also release platform capacity for other novel equipment - such as sulphate removal facilities or membrane water de-oiling systems.

The re-injection of produced water is, however, a radical departure from established North Sea practice, and it is not without difficulty. The combating of scale, increased corrosion, risk of reservoir souring, effect of injecting hot water and maintenance of long term injectivity are issues now being studied. To effect changes on a field wide scale will necessarily involve time and investment. None the less, there is a recognition that much can be achieved to satisfy environmental demands, which also can have a positive economic advantage to operators on the Norwegian Shelf.

The patterns of chemical usage are changing. For some, the news may not, at first sight, be all good. However, it must be recognised that the uncritical use of chemicals serves nobody's interests, neither supplier nor consumer. The environmental damage colours all, if not directly then at least by implication.

FRACTION OF PRODUCED WATER
RELATIVE TO INJECTION WATER DEMAND
FOR A TYPICAL NORTH SEA FIELD
%
PRODUCED
WATER
100
50
0
0
5
10
15
20
PRODUCTION YEARS
PHASE ONE
PHASE TWO

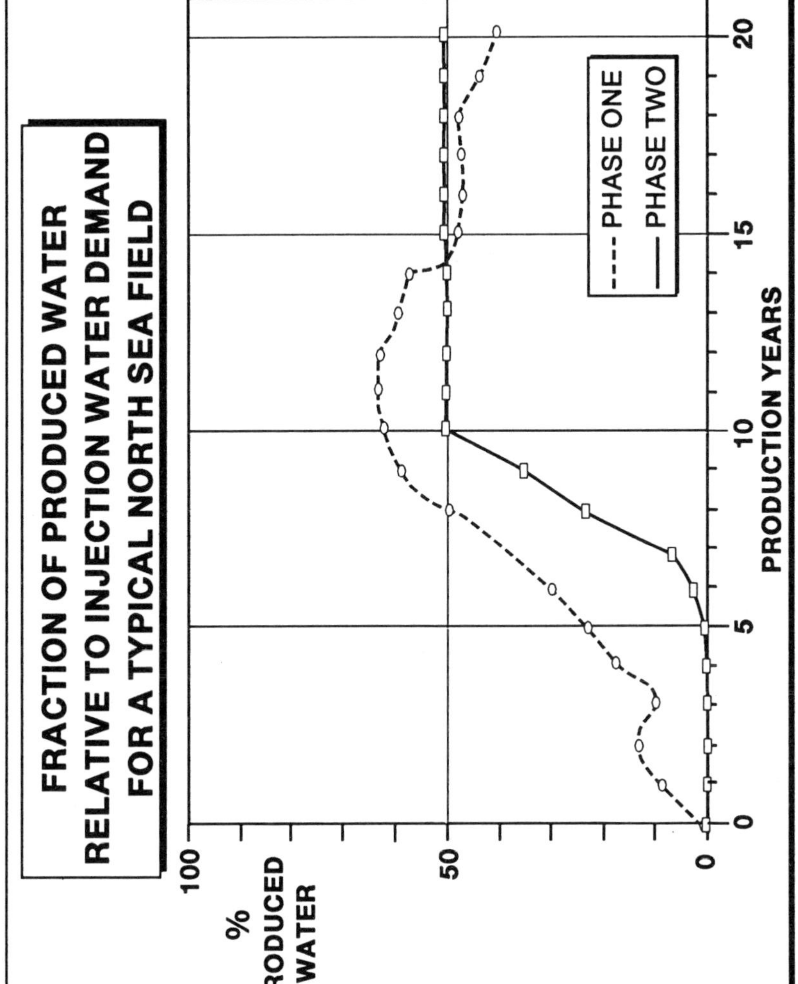

Fig.5

9 CONCLUSIONS

In line with national policy, and as a Company, our aim is to operate our facilities with zero releases to the environment. Some progress has been made through an increased awareness of the cause-effect loops and their impact on chemical application. The philosophy of "if a little is good then a lot must be fantastic" has severe pitfalls. But by a slight shift in approach, much may be achieved. There are considerable challenges to the oil industry and those who sell us chemicals and service. There will be a continued, and growing, need to rely on chemical additives and products in a manner which reduces waste by closing material cycles. A healthier image in matters of environment and occupational health may well permit us to regain greater freedom in solving our problems. Nobody is served by tougher legislation. However, a responsible approach in the coming decade may allow us to repossess the professional freedoms to address the 21st Century's challenges to increase further our recoverable oil fraction from economically important fields such as Statfjord and Gullfaks. To achieve those aims will result in a reliance on chemicals.

The hope is that, by then, Society better understands our chemical dependance and does not treat the oil industry like an addict - hooked and unwilling to kick the habit.

REFERENCES

1. T.A. Lawless, 'The Evaluation of Chemicals for North Sea Oilfield Application', A.E.A. Petroleum Services, Winfrith.

2. Sunde, Thorstenson and Torsvik, 'Growth of Bacteria on Water Injection Additives', S.P.E. 5th Annual Technical Conference and Exhibition, New Orleans 1990.

3. Tjomsland, 'Use of Chemicals in Statoil', Statoil report, 1990.

4. Sunde, Read, Thorstenson and Torsvik, 'Selection of Chemicals to minimise Bacterial Growth in Gullfaks Water Injection System', Oilfield Chemistry Symposium, Geilo 1991.

5. Charles M. Hudgins jr., 'Chemical Usage in North Sea Oil and Gas Production and Exploration Operations', Oilfield Chemistry Symposium, Geilo.

6. Georgie and Read, 'Reinjection of Produced Water' Oilfield Chemistry Symposium, Geilo.

ACKNOWLEDGEMENTS

I wish to thank Ciba Geigy, Dyno Oilfield Chemicals, Norol Hoechst, Nalfloc and Exxon Chemicals for furnishing data, and Statoil for their kind permission to publish this paper.

Corrosion Inhibitors in Oil and Gas Production: Critical Notes on Application

J. van Dijk

SERVO DELDEN BV, PO BOX 1, NL-7490 AA DELDEN, THE NETHERLANDS

INTRODUCTION.

The use of corrosion inhibitors as a tool in corrosion control in oil and gas production systems has a long tradition. It is a respected method of corrosion prevention and requires, its own specific control and monitoring techniques.

A question, quite often raised, especially when the chemical is working well, is if it is indeed necessary to use a chemical at all. This reaction is similar to that found amongst buyers of insurance policies, they pay for the policy, but as long as nothing happens, they develop the feeling that they are paying for nothing.

Never-the-less, there are numerous case histories where the value of using corrosion inhibitors has been proven beyond doubt.

As a corrosion inhibitor manufacturer, supplier and service company we naturally receive a lot of opinions and questions concerning corrosion inhibitor application. In some quarters the chemical is still considered as a sort of magic potion, whose composition is a well guarded secret by the supplier. In extreme cases, it is hard for some users to accept that a chemical used in such very low concentrations can indeed slow down severe corrosion reactions.
What is in fact "magic" in that "potion" is not the type of chemical being used, one can find numerous chemicals which could be used successfully as corrosion inhibitors, in the extensive literature available. However in practical applications, these chemicals have to be carefully selected and precisely formulated to be not only effective of course, but also able to

control all kinds of possible side effects. This is the formulation technique expertise and know-how of the supplier and it is this knowledge which is naturally well guarded.

In most reviews of the use of corrosion inhibitors only the success stories are presented and seldom are the difficulties or misunderstandings even mentioned. We think it is time to break with this tradition.

In this paper we discuss several subjects related to the application of corrosion inhibitors, and about which controversial opinions exist.

It is our submission that by an open exchange of views on these subjects, a more effective team can be formed to combat corrosion.

This paper is not intended to be a close scientific study of these topics but we hope that by highlighting areas of controversy it will contribute to a better understanding of the application of corrosion inhibitors.

The subjects under discussion are:

1. **The use of three phase inhibitors.**
2. **The partitioning coefficient of a corrosion inhibitor over the oil and water phases.**
3. **Measuring residual corrosion inhibitor concentrations as a corrosion monitoring method.**
4. **The relationship between the user and the service company.**

1. THE USE OF THREE PHASE INHIBITORS.

In gas producing systems there are usually three phases, gas, liquid hydrocarbons (indicated as condensate) and water. If the production fluid is corrosive and a corrosion inhibitor is to be used, one of the main questions which has to be answered is the transport of the corrosion inhibitor.

Corrosion inhibitors are liquids with quite low vapour pressures and in relation to transport within gas systems they should be considered as liquids. As a consequence they can only reach and wet the entire metal surface if they are adequately transported by the production fluid. For this reason, it is important to check the flow pattern in a multiphase fluid system, using one of the various calculation and determination methods.

If a segregated, stratified or wave flow pattern is predicted, it is obvious that the liquids can only reach the upper part of the pipe via condensation. But, if as already stated, corrosion inhibitors have low vapour pressures, they can only be present in minute concentrations in the condensed phase on the upper part of the pipe. As a consequence of this, corrosion in this part of the pipe will not be inhibited.
In order to overcome these transport difficulties, different application methods have been devised. One method is to create an inhibitor film by periodically batching the system with a corrosion inhibitor. This is a rather complicated method and is not popular with operations personnel.

A concept launched in the seventies to perhaps overcome this transport problem was the idea of a "three phase inhibitor". From an operational point of view it sounded an attractive alternative, being able to remove or reduce the need for batching. Such a product would need of course to have a composition containing chemicals with high vapour pressure. This would result, in theory, in protection also being afforded to the metal surface in contact with the gas phase. (Of course we are still talking about wet corrosive gas).

In our opinion this entire concept is a misleading one!
From the three phase inhibitor any light components will indeed be evaporated and transported by the gas. However, the heavy components, the real film-formers, will stay behind as a liquid and will still have the same difficulties in being transported to the metal surface in the upper part of the pipe, as do traditional inhibitor products.
This inhibition process in the gas phase by the light components should be recognised as a simple raising of the pH in the condensed water phase rather than as classic inhibition via film-forming. Therefore this type of inhibitor is better indicated as a neutraliser which is a more accurate name than the commercially attractive sounding term of "three phase inhibitor".

Moreover, if the inhibitor is indeed only acting as a neutraliser then of course the film-forming components are not needed in the formulation!

If a neutraliser is to be used, we have to face other problems, e.g. determination of the concentration necessary to give effective protection. In order to do this, we have to deal with a rather complex equilibrium system and different physical constants of the product - the vapour pressure, the partitioning across the gas and liquid phases (Henry's law), and the neutralising properties etc. These constants are not always available nor easy to measure.

A schematic outline of the reactions needed to reach an equilibrium condition is given:

$$\underset{\text{(liquid)}}{\textbf{Inh.}} \xrightleftharpoons[]{\text{vapour pressure}} \underset{\text{(gas)}}{\textbf{Inh.}} \xrightleftharpoons[]{\text{Henry's law}} \underset{\text{(adsorbed)}}{\textbf{Inh.}} \xrightleftharpoons[]{\text{neutralising}} \underset{\text{(reaction product)}}{\textbf{Inh.}}$$

As well as the aforementioned physical constants there are several variables which influence the amount of neutralising liquid necessary to reach the desired equilibrium. These variables include the volume of condensed water, the system temperature and the concentration of the components to be neutralised, e.g. CO_2 and H_2S derived acids. Calculations are difficult to make and in practice a trial and error procedure is followed, e.g. by monitoring the pH.
As far as we know this approach is only worthwhile in systems with very low amounts of condensed liquids, particularly condensed water.

There is another group of products frequently suggested for gas systems with flow patterns not adequate for continuous injection, the so-called Vapour Phase Inhibitors "VPI's". However, this type of product is used in closed systems where the atmosphere remains unchanged. The product must have a high vapour pressure, unlike traditional film-forming inhibitors. The VPI's act by first evaporating, then saturating the gas phase or environment and finally adsorbing onto the metal surface. They are not used in systems with free liquids, but only in environments with a rather high humidity where a water film could adsorb and cause corrosion. Suitable applications are the protection of metal artefacts in musea cabinets, and the protection of packaged metal and electronic spare parts and components whilst in transport or storage.
This type of product has no practical application in a gas producing system as the gas stream would have to be saturated continuously with the inhibitor and therefore huge volumes of these products would be necessary. Besides this the products are ten to twenty times more expensive than the traditional film-forming inhibitors.

2. THE PARTITIONING COEFFICIENT OF A CORROSION INHIBITOR OVER THE OIL AND WATER PHASES.

In an oil-water system where a corrosion inhibitor is used, the question of the solubility of the corrosion inhibitor in the oil and/or the water phase, is frequently asked. Without debating in which phase the inhibitor would be most effective, the question of distribution or partitioning across the oil-water phase boundary, is often posed. This question is asked both in relation to corrosion inhibition efficiency, and environmental regulations, e.g., residual inhibitor concentrations in effluent waters.
Usually the question is formulated as follows:
"What is the partitioning coefficient of the corrosion inhibitor?".
This question cannot be answered! The reasoning behind this strong statement is as follows:
The distribution law states that at equilibrium at constant temperature, the ratio [f] of the concentrations of the solute in the oil [C_{oil}] and in the water [C_{water}] phase is a constant, irrespective of the total quantity in which the solute is present:

$$f = \frac{C_{oil}}{C_{water}} = \text{constant}$$

The restrictions for this law are;
* it must be a real solution
* it must be a dilute solution
* the solute must have the same molecular form in both phases or if an alteration occurs, a correction factor has to be applied

Given corrosion inhibitor formulations it is immediately apparent that these restrictions cannot be fulfilled as inhibitors are mainly mixtures of ionic and non-ionic surfactants in solvents and the raw materials are not well defined, single molecules. Adjusting the ratio and/or changing the nature of the components, does give the manufacturer many possibilities to effect the solubility or dispersibility properties of the product and thus the actual partitioning across both phases.
Surfactants don't give real solutions, they can form micelles and they can emulsify oil in water and water in oil.

In figure 1 a scheme of a possible equilibrium of the distribution of a surfactant in a oil-water system is given.

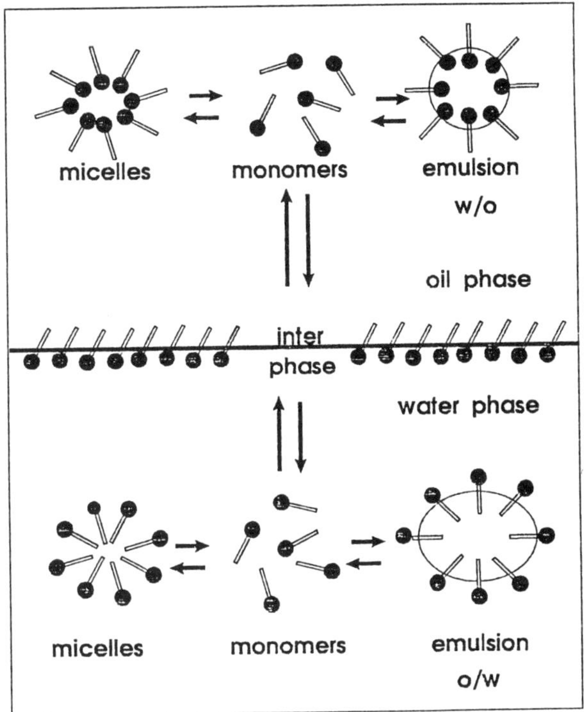

figure 1.

For a complex system such as this, it is impossible to apply or to determine a partitioning coefficient. Based on our experience, only an indication can be obtained regarding the preference of the inhibitor for one or other of the two phases in a specific system, and no accurate figure calculated.

We agree that from the point of view of the user and the authorities involved in environmental control, this is an unsatisfactory situation. However, all parties have to accept this as a reality - and try to find other ways to express partitioning of corrosion inhibitors over the oil and water phase boundaries.

3. MEASURING RESIDUAL CORROSION INHIBITOR CONCENTRATIONS AS A CORROSION MONITORING METHOD.

Unsupported residual corrosion inhibitor concentrations are frequently misused as a tool in corrosion monitoring. Their presence and concentration levels are used as an indication of the corrosion protection and as a check on the operation of the injection system.

We wish to emphasis that this type of analysis is in fact not a corrosion monitoring method in itself. The data obtained say nothing about the actual corrosion condition of the material. To be meaningful, this data must be supported by other corrosion data from weight-loss coupons, electrical or electrochemical corrosion measurements or data from inspections by X-rays and/or wall thickness measurements etc. Only after the relationship between these sets of data is established and the user is sufficiently experienced in interpreting the significance of residuals, can this method be used to its best advantage.

Further, a complication arising when using this method is that it is based on the determination of a specific component in the inhibitor. This means that if the inhibitor is a composite and the different components are not distributed over the phases in a multiphase system in the same ratio as they are in the formulation, only a certain part of the inhibitor will be measured. This can be very confusing if a material balance is of interest. For the same reason, the interpretation of the residuals in waste water systems can be misleading.

Another problem which quite often occurs is that in production streams, natural components are present which may interfere with the analysis method. These interferences can only be detected in uninhibited samples (blanks) of the fluid, and unfortunately these are rarely available. If the residual corrosion inhibitor concentrations are to be used to monitor the operation of the injection system we must be very careful in interpreting the results. An example of an actual mis-interpretation follows.

A corrosion inhibitor was injected "continuously" into the annular space, prior to downhole injection into an oil production stream. A decision was later taken to monitor the injection system by measuring corrosion residuals in the production stream. At that time it was not fully realised that the downhole injection was in fact "discontinuous".
This "discontinuity" is caused by the fact that the downhole injection valve is activated only intermittently, resulting in a subsequent intermittent injection

into the production fluid. The frequency of valve opening is therefore dependent upon the fluctuating pressure drop across the valve. The actual frequency of the valve opening can vary from twice daily down to once every two days or even less frequently.
Hence you cannot expect to measure a steady residual inhibitor concentration. The concentration found in fact may vary between zero and the maximum concentration of the inhibitor in the annulus.

In our opinion the significance of residual inhibitor concentrations alone is often over emphasised. One possible reason for this is that other methods of corrosion monitoring in a multiphase system can be even more complicated or unreliable. As stated earlier the method has its value but only if used with care.

4. THE RELATIONSHIP BETWEEN THE USER AND THE SERVICE COMPANY.

Prior to any corrosion inhibitor being accepted it is of course necessary to have an extensive exchange of information and carry out a number of investigations. Preferably this should be done during the design phase. We would like to stress the point that it is very important to be as open as possible about the expected process conditions and the inhibitor selection methods. This gives the service company the opportunity to select the appropriate product. Occasionally we are not informed of the selection criteria beforehand, perhaps because the user suspects that it will then be very easy for suppliers to formulate an inhibitor which will pass those tests but may not work well in practice. This attitude may of course lead to inadequate proposals based on incomplete information. One may deduce further that there is little or no confidence on the part of the user in the reliability of their own selection criteria.

Occasionally the technical selection of the inhibitor is based on the offer with the lowest injection rate, below traditionally accepted values.
In our opinion we consider this a dangerous approach. Reliance upon an untested low dosage proposal, means operating on the very borderline of the effectivity of the corrosion inhibitor.
We believe this is an irresponsible approach to a matter as important as corrosion protection.
It should be recognised that it is very difficult to recommend the correct injection rate of a corrosion inhibitor, even on those occasions when a short field trial is allowed. Optimisation can only be achieved if reliable corrosion

monitoring methods are in use. Unfortunately these monitoring methods are still complex and under-used in multiphase systems.
Therefore in our opinion, the technical selection should be based on the realistic dosages currently being used in similar systems.

Every corrosion protection system demands continuous monitoring. Corrosion inhibitors are no exception.
When the service company is not directly involved in the monitoring from the outset, unexpected results may be brought to the attention of the service company too late to prevent severe damage to the system and this may lead to friction between the user and the service company.

A better approach would be that prior to the start of the corrosion inhibition programme, the user and supplier agree to an exchange of corrosion monitoring data and field results at regular intervals, for example once every one or two months.
The input of the supplier should not simply stop at delivery of the corrosion inhibitor. Service should preferably be a part of the contract. In this way it is possible to form a team which has a greater chance of achieving the desired goals, namely, an acceptable level of corrosion and uninterrupted production.

SUMMARY.

The erroneous ideas concerning three phase inhibitors and partitioning coefficients, discussed in this paper, are fortunately not widely held in the corrosion control community. However, these misconceptions do still exist.

The use of residual corrosion inhibitor concentrations as a corrosion monitoring method can be misleading when used in isolation. There does not exist a corrosion monitoring method which is sufficient when used on its own. It is more acceptable to use a combination of various methods including residual inhibitor concentration determinations. New, more reliable monitoring methods, especially in multiphase systems would be most welcome to the industry.

A close relationship between the user and the service company is to our mutual benefit, bearing in mind that the control of the corrosion problem itself, remains the responsibility of the operating company.

Membranes and Their Use in the Treatment of Oily Water

J. Koning and D.M. Koenhen

X-FLOW BV, ALMELO, THE NETHERLANDS

1 INTRODUCTION

Natural gas and oil contain a certain amount of water. This water is separated at the production well and comes free as "production water". The search for gas and oil leads to drilling wastes and waste water from the drilling location.
These waste streams contain oil, that has to be removed before discharge into surface or coastal waters.

The conventional techniques are not able to remove oil from these waste streams with constant efficiency, due to their sensitivity to variations of the quality and the quantity of the waste streams. Furthermore the present discharge limits are about the best these technologies can achieve.
The offshore industry is faced with stricter effluent discharge requirements in future.
One way to remove oil from waste streams is cross flow microfiltration by hydrophilic membranes with separation on 0.2 micrometer (1). The separation efficiency is comparable with that of ultrafiltration, but the capacity is higher.

Here some pilot and full scale examples for treatment of production water and of waste water with drilling mud components are discussed.

2 OIL IN PRODUCTION WATER AND IN WASTE WATER FROM DRILLING OPERATIONS

Production water from natural gas and oil fields

The water coming from natural gas originates from condensation. The oil consists of hydrocarbons, that are condensed together with the water to form a mixture, that contains partly free floating oil, partly dissolved oil compounds as light aromatics and partly emulsified oil. The free floating oil can be easily removed by traditional methods like flotation, dispersed gas flotation and coalescence.

Production water from oil fields was present as an emulsion of water droplets in oil. After chemical addition it is separated as a water stream with mainly free floating oil and some emulsified oil and dissolved oil. The hydrocarbons have a higher molecular weight than those in water from gas fields.

Emulsified oil droplets vary in size from 1 to 50 micrometer depending on temperature, gas composition and condensation parameters. Also the fraction of emulsified oil varies.
The emulsion droplets can be formed in a mechanical way and by chemical stabilization. The stability of the oil emulsion depends on the presence of oil compounds that have emulsifying properties by themselves and of chemicals that have been added to the gas or oil production system like corrosion inhibitors (2).

The oil concentration in produced water from gas fields varies from several grams per liter to a few percent. After conventional treatment the concentration can be as low as the present discharge limit for the North Sea of 40 ppm, but values of 2000 ppm are measured as well.

Production water from oil fields is less emulsified and therefore conventional treatment often leads to satisfactory results provided the residence time in the installation is sufficiently long. This however might lead to installation sizes, that are not acceptable in offshore industry. Several oil production wells produce water, that is well emulsified. The effluent of the flotation unit may contain 1000 ppm of oil.

<u>Waste water from drilling operations</u>

Waste water from drilling operations consists of rai water, spills, rust, cuttings, drilling mud residues formation compounds, salt and some oil. Treatment o location is often not appropriate and the transport fro the drilling location to a central treatment plant i costy. With hydrophilic microfiltration these waste streams should be concentrated such, that a net saving c transport and treatment costs is realized, whilst th removed water meets present and future discharge limits.

3 THE HYDROPHILIC MICROFILTRATION MEMBRANE

The varying fraction and consistence of the emulsion is the main cause for the variations in efficiency of the conventional treatment methods. Hydrophilic microfiltration membranes act like a barrier, through which only water and dissolved material can pass and that retains the oil droplets emulsified or not.

The hydrophilic microfiltration membrane is an open structure with pores of 0.1 - 0.2 micrometer in diameter. The membrane consists of capillaries of 1 m length with an internal diameter of 1.5 mm and external diameter of 2.5 mm. The wall of the capillary is the actual membrane. Figure 1 shows a Scanning Electro Microscope photo.
The treatment of oily water streams is based upon cross flow membrane filtration. The oily water enters a capillary on one end and flows through the bore of the membrane. Only a part of the water is pressed through the capillary wall. The oil is retained and remains emulsified in the not filtered water, that leaves the capillary at the other end. To filter all of the water out, the water is circulated several times through the capillary, thereby concentrating the retained oil to a level that makes it suitable for disposal or further treatment.

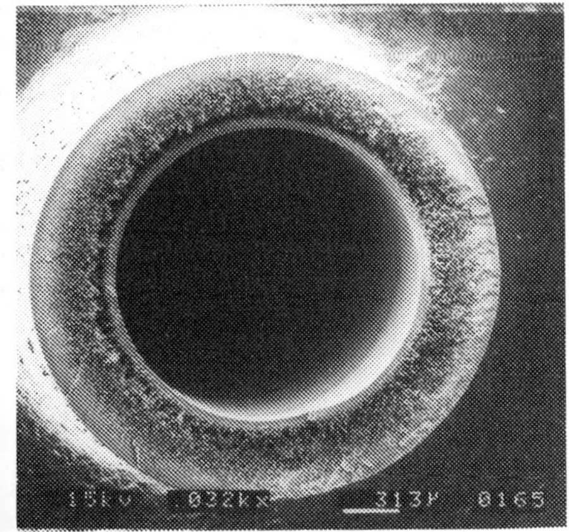

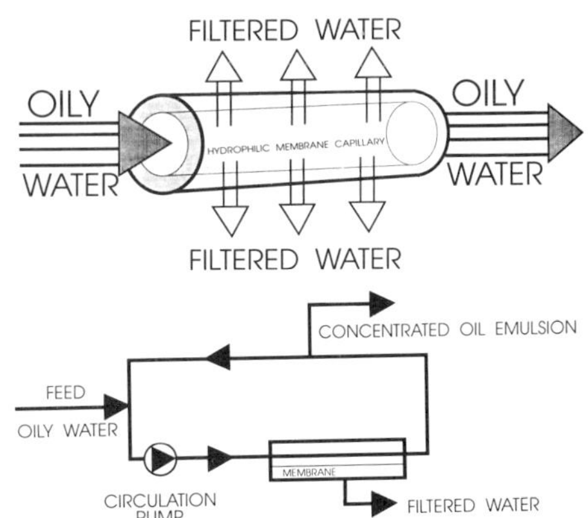

Figure 1 : Microfiltration capillary

Figure 2 : Cross Flow Microfiltration

The microfiltration membrane in the treatment installation presented here is made of a blend of a hydrophobic and a hydrophilic polymer. The hydrophobic polymer gives the membrane its structure and strength, while the hydrophilic polymer is responsible for the membrane properties. The membrane is permanently hydrophilic through the whole of its matrix; a property, that is not lost when the membrane is dried. It can withstand temperatures upto 125 ^{o}C and is resistant to the compounds usually found in oily waste streams (1).

The hydrophilic nature of the membrane is the main feature, that prevents the adsorption of hydrophobic compounds for instance oil onto the membrane surface. In fact it acts like a barrier to membrane fouling, which is regarded as one of the drawbacks for using membrane filtration in the treatment of waste streams(3).

Two other advantages contribute to the feasibility of microfiltration. Firstly the hydrophilic surface enables the operator to clean the membrane with common detergents in case of accidental fouling. Secondly the walls of the pores in the membrane are covered preferentially by water molecules resulting in a narrower passage for hydrophobic material and as such leading to a finer filtration, than expected according to the pore size of 0.2 micrometer. Figure 3 presents this phenomenon schematically.

4 MICROFILTRATION MODULE AND INSTALLATION

A bundle of membrane capillaries form a microfiltration module. It is provided with connections for the incoming waste stream and the outgoing more concentrated stream and for the collection and discharge of filtered water.

A microfiltration installation contains several modules, a pump to circulate the water through the modules a pump to inject the production water into the circulating stream and the necessary measuring and control devices.

Normally these installations run fully automatically.

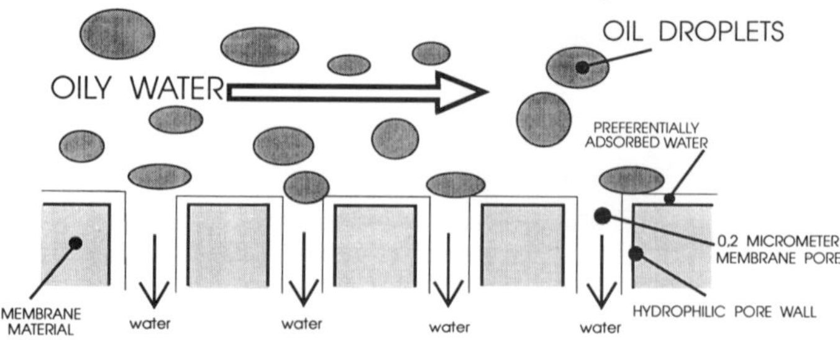

Figure 3 : The water adsorption on the hydrophilic pore walls influences the passage of oil droplets.

Figure 4 : microfiltration module.

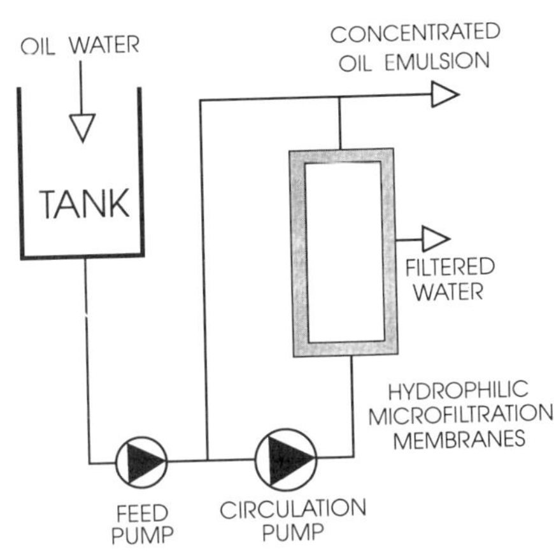

Figure 5 : Flow scheme microfiltration

5 TREATMENT OF PRODUCTION WATER

Production water has been treated on bench, pilot and full scale. The experience shows, that no general rule of the thumb for the capacity can be given. However the membranes perform well, if the microfiltration process is well controlled and operated.
The results of the bench and pilot scale runs are given in table 1, that clearly demonstrates the good filtration performance with oil concentrations in the filtered production water far below present and expected future discharge limits. The oil can be concentrated upto a few percent resulting in less than one percent of the original stream to be finally disposed off.

Table 1 : Bench and pilot scale microfiltration of production water from natural gas and oil fields

		Natural gas	Oil
Installation			
. filtration area	m^2	0.1	5
. capacity	l/h	10-20	500-1000
Pretreatment		Flotation Coalescence filter	Unknown
Membrane capacity	$l/(m^2.h)$	110-200	100-220
Temperature	^oC	15-20	20-55
Influent oil	ppm	500	50-450
Effluent oil	ppm	1-5	<10
Concentrate oil	%	>1	-

The full scale installation has treated production water from natural gas as well as from oil and has treated well conservation and cleaning fluids from several oil wells that were taken in operation again after a temporary close down.
The installation has operated offshore on different locations. It is mounted in a 20 foot closed and locked container. The operation is fully automatic and controlled by satellite telephone connection onshore.
A process flow diagram shows the simplicity of the unit, that has been integrated with the already existing conventional treatment by flotation. The concentrated oily water stream from the microfiltration unit flows back to the flotation tank, where it is mixed and treated together with the incoming production water. The oil is skimmed off. This leads to an almost 100 % oil waste for incineration or cotransport with oil in the pipeline to shore.
The application of hydrophilic microfiltration has been tested parallel to hydrophobic ultrafiltration. As can be expected in general, the capacity of ultrafiltration with 30-65 l/(m2.h) was lower than of microfiltration. The oil content of the filtered water was comparable. The parallel operation showed, that the hydrophobic membranes were sensitive to fouling and required weekly cleaning.

The hydrophilic membranes are running since May 1990. They perform well. A few times the operation of the installation has been upset by accidental discharge of oil well chemicals into the production water line. After cleaning the installation performed as before.

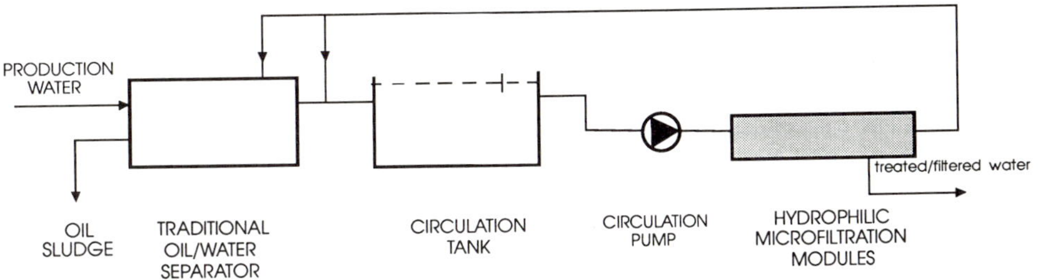

Figure 6 : Flow diagram of the full scale installation for treatment of production water (4).

Table 2 : Efficiency of full scale production water treatment on offshore gas and oil fields (4).

		Gas field	Oil field
Filtration area	m^2	45	45
Capacity	m^3/h	7.2	3.5-9
Pretreatment		Flocculation Plate separator	Flotation Plate separator
Membrane capacity	$m^3/(m^2.h)$	160	90-200
Temperature	^oC	20-30	10-25
Influent oil	ppm	1000	2000
Effluent oil	ppm	<10	<10
Concentrate oil	%	5-10	10-20

6 WASTE WATER FROM DRILLING OPERATIONS

Waste water from drilling operations varies continuously. It may consist of just rain water being slightly polluted, but more often it contains all the components related to drilling. Therefore the bench scale tests have been executed with very contaminated waste water.
The results as presented in table 3 were encouraging enough to justify further technology development with a full scale installation offshore.
The offshore installation has been built in a container and is operated manually. This requires skilled personnel. The unit consists of two completely separated installations with each one X-Flow hydrophilic microfiltration module of 10 m2 filtration area.

It can be expected, that by the filtrating activities the high concentration of particles leads to deposits on the membrane surface, thereby forming a layer, that reduces the membrane capacity. This process can occur in spite of the high cross flow velocity of the circulating waste water.
For this reason the installation is provided with back flush. Back flush is initiated by increasing the pressure of the filtered water to above the filtration pressure. The filtered water is pressed in a reversed direction through the membrane. This back flowing water lifts the layer of retained material, that is redispersed in the circulating waste water.
Back flush requires only a few seconds. The back flush frequency depends on the capacity decline. It may vary from once each few minutes to each hour. For this reason the back flush provisions are automatic.

Figure 7 shows the simplified flow scheme.

Table 3 : Bench scale microfiltration of waste water from drilling operations (5). Membrane area 0.5 m^2. Pretreatment with 100 micrometer bag filter.

		Waste water	Filtered water	Concentrate
Membrane capacity	$l/(m^2.h)$	-	40-120	-
Installation cap.	l/h	-	20- 60	-
Temperature	^oC	16-22	16- 22	-
Suspended solids	%	0.02	nil	14-18
Oil	mg/l	200	0.1- 1	-
	%	-	-	16-42

Offshore the full scale installation is combined with a three-phase decanter, that separates the waste water in a solid phase, a mud water phase and an oil in water phase. The first phase has to be disposed of as solid waste. The solids have been removed down to 5 micrometer. The two other phases are presented to the microfiltration installation as a mixture (5).

The hydrophilic microfiltration membranes concentrate the oil and solids such, that transport and final disposal costs are reduced considerably. The investment of the microfiltration installation pays itself back in a short period.

The installation is offshore from March 1991. The performance will be presented a publication to follow.

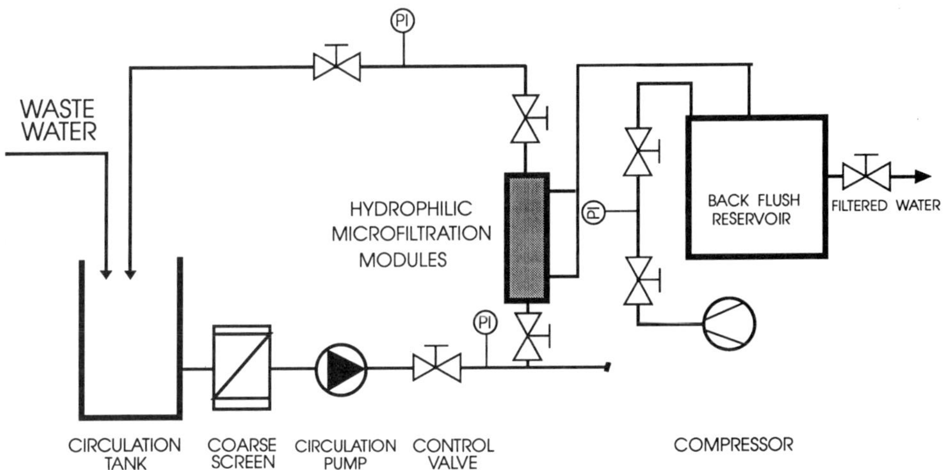

Figure 7: Simplified flowdiagram of the full scale installation for waste water from drilling operations offshore.

7 CONCLUSION

The experience from operation under practical conditions shows, that microfiltration with hydrophilic membranes is a young but proven technology for offshore and onshore oil and gas industry.

The microfiltration technology can be combined with the existing conventional installations.

Hydrophilic microfiltration is capable of treating production water, that contains oil emulsified by the original as well as added components. The effluent meets all discharge requirements. The removed and concentrated oil can be led back to the conventional treatment unit for total recovery.

The same membrane characteristics are responsible for the good performance in treating waste streams from drilling operations. Here too the effluent oil content is far below discharge limits. The retained oil and solids can be concentrated to high levels, which reduces the amount of waste to be transported and disposed off considerably.

REFERENCES

1. USA Patent 4,789,847 and European Patent 0261734
2. J. van Dijk,Corrosion Inhibitors for Oil Production, this congress, 1991
3. E.H.W.Roesink, PhD Thesis, University of Twente NL, 1989
4. E. Dall, 3S Kopenhagen DK, personal communication, February 1991
5. R. van Elsen and M. Smid, SCS Heerhugowaard NL, personal communication, March 1991

European Patents in the Oil Industry

E. Waeckerlin

EUROPEAN PATENT OFFICE, ERHARDSTRASSE 27, W-8000 MUNICH 2, GERMANY

1 INTRODUCTION

A patent system seeks to contribute to technological development by providing a legal context for the transfer of new technologies. Patents have several functions: (1) They provide a just reward for the inventor; (2) they ensure an adequate return on investment in research and development; (3) they offer to the expert a wealth of technical information; and (4) they help to avoid double investments in research projects.

It goes without saying that a modern and efficient patent system is necessary to provide inventors and industry with these important advantages and services. The procedure set up within the framework of the European Patent Convention (EPC)[1] is, indeed, such a system, and perhaps the most successful one that there is. The scope of my contribution is the function of the European Patent Office (EPO) with a special reference to the Oil Industry.

The principal function of the EPO is to grant patents on the basis of the EPC which came into force in 1977. The Office also has certain responsibilities under the Patent Cooperation Treaty (PCT), acting in particular as an International Searching and Preliminary Examining Authority. The EPO also performs national searches for several States and the public. Finally the dissemination of patent information is fostered by various activities including the publication of the European Patent Bulletin and the Official Journal, as well as the running of the International Patent Documentation Center (INPADOC) in Vienna.

2 THE EUROPEAN PATENT GRANT PROCEDURE

The European patent is granted following an examination designed to check whether the invention complies with the patentability requirements laid down in the EPC.

Procedural Stages

In the first phase, the EPO carries out a formalities examination, followed by a search to determine the relevant prior art. The results of the search are set out in a search report. The application is published, usually together with the search report, 18 months after the date of filing (or date of priority).

In the second phase each application is examined by an Examining Division consisting normally of three examiners who are experts in the field concerned. They verify whether the formal and substantive requirements for patentability have been met. There are three main criteria for patentability, namely novelty, inventive step and industrial application. If the Examining Division considers that the invention is patentable, the patent is granted; if it concludes that the requirements are not fulfilled, the application is refused.

Grant of the patent is followed by a nine-month period during which any party may file notice of opposition in a reasoned statement. This leads to opposition proceedings as a third phase, resulting in either revocation of the patent, rejection of the opposition or maintainance of the patent in an amended form. There is a right of appeal against decisions taken by the Receiving Section, Examining Divisions and Opposition Divisions of the EPO. The Boards of Appeal are independent and composed of technically and legally qualified members.

Processing Time

The processing time depends largely on the workload of the EPO and the response of the applicant. The average time taken from the date of filing of an application until the drafting of the search report is about nine months. The time-lag between the filing and the final decision made by the Examining Division is 44 months on average. In a typical case the length of the European grant procedure is something between 3 and 5 years[2].

Applicants may wish, depending on the circumstances of the case, to obtain accelerated processing of their application. With this in mind, the EPO in 1989 introduced a seven-point programme by which the processing time for applications can be considerably shortened on the applicant's request[3]. The number of such requests for accelerated search or examination has been far lower than expected.

Fees and Costs

The structure of the European grant procedure provides for fee-payment to be spread out over the time of the examination procedure[4]. In particular, the formal separation of search and substantive examination means that the applicant may decide in the light of the search report whether to continue with the application.

The European procedure is more expensive than the national procedure for the grant of a patent in each separate EPC Contracting State. This means that the national patent authorities will still be of advantage to applicants who wish to use an invention in only one or two States. In the Oil Industry this is certainly an exceptional case. If the applicant requires protection in more States, the European procedure is generally more economical. A European patent costs less, as a rule, than three separate national patents. The cost advantage grows with the number of states designated in the application. If a patent is filed for all 14 EPC States, a European patent can cut costs by up to 50 %.

3 STATISTICS[5]

From 1978 to the end of June 1990 some 375000 applications have been filed at the EPO, a clear indication that the European patent procedure is widely used by the industry in highly industrialised and economically strong countries. Approximately 33 % of the applications are in Chemistry, 35 % in Mechanics, and the remaining 32 % in Physics and Electricity.

The general trends observed during the past few years are set out in Table 1. It can be seen from the Table that the rate of filings and the number of European patents are steadily increasing. The figures of Table 1 relate to all technical fields.

Table 1 European and Euro-PCT Applications filed, Requests for Examination and European Patents granted, 1985 - 1990 (provisional figures are in brackets)

	Year: 1985	1986	1987	1988	1989	1990
European applications filed	33748	36783	39961	44773	49280	(52378)
Euro-PCT (regional phase)	3244	4675	5299	7539	8485	(10400)
Total European and Euro-PCT	36992	41458	45260	52312	57765	(62778)
Requests for examination	27304	32518	34171	38959	46151	(53100)
European patents granted	15117	18471	17143	19750	22587	(24758)

Of course the figures for a particular technical field are much smaller and can vary depending on the area. Thus the number of applications coming from the the Oil Industry hardly exceeds a hundred per year. In fact the Oil technology is a highly specialised but relatively small area and as such inevitably less active than, for example, petrochemistry. Further it is important to bear in mind that there are many inventions belonging to mechanics, chemical engineering and general chemistry which are of interest for the Oil Industry.

Geographical Distribution

An interesting aspect is the geographical distribution of the applications. In general, 50.7 % of all European applications come from the Contracting States. These applications are mainly from the Federal Republic of Germany (21.4 %), France (8.2 %) and the United Kingdom (6.5 %). Of the applications from non-member States, 26 % come from the USA and 19 % from Japan. This illustrates that patent activity is concentrated in three main economic areas - Europe, the USA and Japan - which account for 96 % of all applications.

On average, the designation rate is 7.3 Contracting States per application. The Federal Republic of Germany, France and the United Kingdom are designated in more than 90 % of all applications, followed by Italy (74 %) and the Netherlands (63 %). The remaining Contracting States are less often designated.

In the field of Oil Technology the geographical distribution is quite different. If the drilling fluids are taken as representative, the following picture emerges. In terms of the countries of origin of European patent applications, the largest number is filed by applicants having their residence in the USA (53 %), followed by the Federal Republic of Germany (22 %), France (7 %), the Netherlands (7 %) and the United Kingdom (5 %). Applications from Japan represent a comparatively small share (2 %)[6].

Analysis of the Patent Grant Procedure

Of every 100 applications an average of 67.6 result in grant of a patent. The unsuccessful applications are either withdrawn by the applicant (24 %), refused by the Receiving Section or the Examining Divisions (5.1 %), or revoked by the Opposition Divisions (2.3 %) (Table 2).

Table 2 Analysis of the European Patent Grant Procedure (percentages are based on applications filed)

Applications filed	100 %
Withdrawals before substantive examination	10 %
Refusals before substantive examination	0.1 %
Withdrawals in substantive examination	15 %
Refusals in substantive examination	5 %
Patents granted in substantive examination	69.9 %
Revocations after opposition proceedings	2.3 %
Total of patents granted	67.6 %

The proportion of European patents opposed is 8.5 % on average. The opposition rate has diminished steadily over the years. Obviously the opponents came to the conclusion that there is no point in opposing if no new facts, evidence or arguments can be presented. In 38 % of the cases the oppositions result in the revocation of the patent; in the remaining 62 % of the cases the patent is maintained, often after substantial amendments (21 %).

On the basis of these figures it can be concluded that, as a whole, the European grant procedure has a highly selective effect.

4 ACTIVE FIELDS OF OIL TECHNOLOGY

Advances are steadily being made in the Oil Industry. It would be pointless to try to produce a comprehensive study of the areas of innovation within the field of Oil Technology. The following are only a few typical examples:

An area which is particularly active is the development of additives for use as components in drilling fluids. It is essential that the formulation of the drilling fluid be such that its principle functions are optimized during the drilling operations. These functions are transport of the cuttings, cooling and lubrication of the drilling equipment, control of subsurface pressures and maintainance of the stability of the borehole. It is also important that the fluid loss be low and that the fluid does not corrode the drilling equipment. Other limitations are the toxicity and the compatibility with the environment.

The composition and properties of drilling muds are therefore very complex. Current fluids may contain a dispersed aqueous phase, together with emulsifiers, surfactants, viscosity-reducing agents, gel-forming components, fillers, fluid loss additives and corrosion inhibitors. All these components have been the subject of extensive research and patent activity. In particular the inclusion of new polymeric additives has contributed much to the solution of the technical problems[7].

Other areas which have attracted interest are spacer compositions and crosslinking gels for the treatment of subterranean formations[8,9].

It is interesting that microbiological methods for producing biopolymers are becoming increasingly important. Such biopolymers, for example polysaccharides prepared by fermentation with microorganisms of the Xanthomonas campestris type[10], can be used in well-treatments and enhanced oil recovery (EOR). Microbiology is also a factor in remediation of oil spills[11].

In general EOR techniques and stimulation of wells are focused by research activities[12]. The same applies to the development of new catalysts, for example modified zeolites; but this technology belongs more to Petrochemistry than to the Oil Industry.

5 CONCLUDING REMARKS

Summing up, it may be said that the European patent procedure has attracted much attention. It is a successful example of European co-operation. The EPC has brought about a modernization and harmonization of patent law. But time flies and nothing is stable. It is necessary to continue to adapt the patent system to the changing needs of the applicants and the general public. In particular further simplifications of the system are envisaged.

Some recent developments will certainly affect the Oil Industry. The patent policy is, of course, a matter which is defined by the applicants and not by the Patent Offices. Nevertheless, by the very nature of their services, the Offices exercise an influence.

A first aspect is the development of case law by the Boards of Appeal. It is of fundamental significance for the day-to-day practice of the EPO. The Office has published comprehensive guidelines for examination which are regularly revised in the light of leading decisions taken by the Boards of Appeal[13].

A second element is the dissemination of patent information. In the past the patent information has been exploited using mainly conventional data carriers such as paper. However, the handling and storage of these large volumes of paper is time-consuming and cumbersome. The trend is increasingly for paper to be replaced by more efficient and economical storage means.

The EPO has therefore recently begun to publish patent information on CD-ROM optical disks. The range of products includes a complete annual collection of the European patent application documents (ESPACE-EP) and the PCT International applications (ESPACE-WORLD). The EPO also offers an additional service on CD-ROM, called ESPACE-FIRST. This series records the first page of the European and PCT-applications with their bibliographic data. Any user can use the special software to carry out searches using a number of bibliographic data elements including the technical fields (IPC classification), the applicant name and the title of the invention. A special CD-ROM search tool that offers bibliographic information on all European patent applications filed since the foundation of the EPO in 1978 is also available (ESPACE-ACCESS). This means that libraries and external firms have now the possibility to build up their own documentation on European patent applications at low costs.

Apart from this, another important event should be mentioned. In 1989 an agreement on the Community patent has been reached by the EC's 12 Member States. It can be expected that, after years of preparation, the Community Patent Convention (CPC) will enter into force for most of the Community in 1992. This step is often regarded as the realization of the second pillar of the European patent system and a major contribution to a common code of protection for industrial property in Europe.

ACKNOWLEDGMENT

Thanks are due to Mr. F. Falls for critical reading of the manuscript and valuable suggestions.

REFERENCES AND FOOTNOTES

1. Convention on the Grant of European Patents (European Patent Convention), published by the European Patent Office, 5th edition, Munich, (1989).
2. Statistical figures are regularly published in the "Annual Report" obtainable from the EPO.
3. Official Journal EPO, 1989, 523.
4. Schedule of Fees, Costs and Prices of the EPO, published by the EPO.
5. EPO Annual Report 1989, Munich, 1990.
6. These figures are derived from a random sample of 175 applications filed from 1985 to 1990 and relating to drilling fluids.
7. Many examples for such patent applications can be found in the Main Group C09K7/00 of the International Patent Classification.
8. EP-A-0 243 067.
9. EP-A-0 338 161.
10. EP-A-0 130 647; EP-A-0 209 277.
11. EP-A-0 402 158
12. EP-A-0 150 112; WO89/01491.
13. Guidelines for Examination in the EPO, obtainable from the EPO.

Surfactants in Oil Production

H.M. Muijs

SHELL RESEARCH BV, KONINKLIJKE/SHELL LABORATORIUM, BADHUISWEG 3, 1031 CM AMSTERDAM, THE NETHERLANDS

1 INTRODUCTION

The world market for oil-field chemicals is strongly influenced by the price of crude oil. The collapse of this price from a booming level of over US $ 30/bbl to slightly under $ 10/bbl for a short period in time and the subsequent rebound to its current strongly fluctuating level have had an impact on many activities in the oil industry, with recent reports showing a renewed increase in drilling activity. In Western Europe, the market for oil field chemicals is dominated by the North Sea activities; of course, on a world wide basis this market represents only a small fraction.[1]

The chemical spectrum associated with oil and gas exploration and production is highly diversified. Reviews of oil producers' needs have been given at two previous symposia organised by the Royal Society of Chemistry.[2,3] The current situation will be covered at this congress. In the field of oil-field chemicals, surfactants serve a wide variety of purposes and so play an important role. According to the study referred to previously,[1] surfactants account for 60% of the overall consumption of speciality chemicals in the oil industry in the US. Together with polymers, surfactants are probably the most significant organic speciality chemicals used by this industry, and are of vital importance in the fields of exploration, drilling and production.

This paper discusses the role of surfactants in various sectors of oil and gas production together with some recent developments and potential other applications.

2 DRILLING CHEMICALS

Drilling a well requires the application of a drilling fluid, generally known as "drilling mud". A significant amount of inorganic commodities (bentonite, barytes) is used for this purpose, in addition to polymers (viscosifiers) and emulsifiers. These materials are used both in water-based (fresh water, seawater, brine, mixed salt solutions) and in oil-based (diesel or hydrotreated hydrocarbons) drilling fluids. For foam or mist drilling only foaming agents are employed. The function of the drilling fluids is to keep the borehole wall stable and thus to ensure optimal economic drilling results. The basic functions of a drilling fluid are listed in Table 1.

The selection and design of drilling fluids is increasingly being dictated by environmental considerations and constraints. This resulted already a couple of years ago in the abandonment of the use of diesel in favour of the so-called low-tox invert oil emulsion muds. The oils in these muds are mostly hydrotreated and have a very low aromatics content. The mud consists of a water/oil emulsion, in a 30/70 ratio with a suitable emulsifier package consisting of oil- and water-soluble emulsifiers and high-temperature stabilisers. Recent developments tend to favour a water/oil ratio of 50/50 as this reduces the amount of oil adhering to the cuttings. The maximum amount of oil allowed on the cuttings in various parts of the North Sea offshore sectors is now set at 10% but steps are being taken to achieve zero discharge in a couple of years.[4,5] The costs involved in the disposal of mud and drilling waste (i.e. used mud and drilled cuttings) are increasing as a result of the tightening environmental legislation. Remarkably this legislation often varies from country to country. This aspect was discussed from the point of view of a service company at a recent

Table 1 Functions of a drilling fluid

- Cooling and lubrication
- Solid suspension
- Cuttings transport
- Fluid loss prevention (filter cake)
- Pressure balance
- Prevention of interaction with the drilled cuttings and the formation
- Corrosion control

conference.[6] The pros and cons of oil-based and water-based drilling fluids are outlined in Table 2.

To meet the required oil-on-cuttings level, a surfactant system could be used to wash (or another solvent could be used to displace) the mud from the cuttings. Experiments carried out recently in this field in Norway and in the UK, together with several other ways to minimise the environmental impact of drilling fluids, will be discussed at this conference. To avoid the use of hydrocarbons altogether, invert oil emulsion mud systems based on synthetic lubricants (esters) have been proposed.[7,8]

A recent development in the area of drilling fluids is a class of modified water-based muds, which can deposit a tiny film of water-insoluble fluid on metal and rock surfaces. This type of mud may be formed from a water-based mud by the addition of some low-tox oil or synthetic oil and an emulsifier package.
The emulsifying agent must be present to prevent coalescence of the oil drops, give a stable emulsion at high (downhole) temperatures and cover the cuttings to prevent disintegration of the shale, as is usually observed in water-based muds. It is clear that such an emulsifier should be hydrophilic. The (preliminary) properties of such a system as compared to those of a water-based system are given in Table 3.

For foam or mist drilling, widely used in e.g. Canada, the study of gas blockage by foam (see below) has resulted in a better understanding and selection of foamers for use under certain conditions.

Table 2 Pros and cons of water-based and oil-based muds

Properties	Water-based	Oil-Based
Lubrication	-	++
Shale hydration	--	++
Anti-corrosion	--	++
Logging	++	--
"Oil" on cuttings	none	10% max
Environmental aspect	++	--

Table 3 Properties of a water-based emulsion and oil-based drilling fluids

Properties	Water-based	Oil-based
Lubrication	++	++
Shale hydration	+	++
Anti-corrosion	+	++
Logging	++	--
"Oil" on cuttings	2-3%	10% max
Environmental aspect	+/?	--
Re-use	?	-

2 PRODUCTION CHEMICALS

The production chemicals employed in the oil industry usually comprise all the chemicals used for hydrocarbon processing and production. These chemicals are associated with crude oil dehydration, foaming in gas-oil separation, paraffin wax deposition, corrosion, scaling, de-oiling of water, bacterial activity, etc. The majority, by far, of the active ingredients (incorporated in a wide variety of proprietary formulations) often designed to cope with a particular problem are speciality surfactants, as given in Table 4.

It is beyond the scope of this paper to discuss in detail all the surfactants used in these areas. Three classes of surfactants offering new(er) opportunities are highlighted in some detail below.

Table 4 Speciality surfactant-type production chemicals

Product class	Description and type
Demulsifiers	Emulsion breakers or dehydration chemicals, propylene oxide/ethylene oxide (PO/EO) products, oxyalkylenes of alkylformaldehyde resins, sulphonates
Defoamers	Modified silicon-based PO/EO adducts
Corrosion inhibitors	(Di)amines, polyamines,imidazolines (as such or modified), phosphates and phosphonates
De-oiling agents	Cationics

Foaming agents

Foam is already being applied in a number of oil-field operations such as air drilling,[9-11] well cleaning,[12] and fracturing and in foamed cementing.[13-15] Because of its high apparent viscosity foam has also been suggested for use in the control and modification of injection profiles and as a drive fluid in oil-displacement processes such as carbon-dioxide flooding[16] and chemical flooding. The most efficient foaming agents in terms of oil displacement are the foamers of medium stability.[17] Highly stable foamers are more suitable for permeability reduction and core plugging. A detailed literature review is given in reference [18].

We have built a special apparatus to assess the gas mobility-reducing properties of surfactants. The pressure drop generated over a sandpack core during steady-state flow is taken as a yardstick of the foamability and the efficacy of the product. The set-up is shown schematically in Figure 1. Effects of temperature, (back)pressure level, flow rate, foam quality, surfactant type, concentration and salinity can be chosen freely and have been reported in the literature.[19] The results obtained with a matrix of alcoholethoxy sulphates, enabling one to select for a given condition the optimum product in this class, is illustrated in Figure 2. It should be emphasised that these tests in a porous medium have been done in the absence of oil. For a foam-drive recovery process (and cases where the foaming agent is used in so-called foam or mist drilling) the effect of oil should also be included since foaming and foam propagation are highly surfactant specific.[20]

In order to gain an idea of the effect of oil on the action of various foamers we have designed a foam titration test. This test, developed in our Laboratory measures the foam stability upon gradual addition of a soil mixture.[21] In our experiments we selected North Sea crude as "soil". Special surfactant blends and formulations with stabilisers can greatly enhance the performance, as is shown in Figure 3.

Gas-hydrate inhibitors

Triggered by oil prices the newer offshore sub-sea developments have become important. According to the

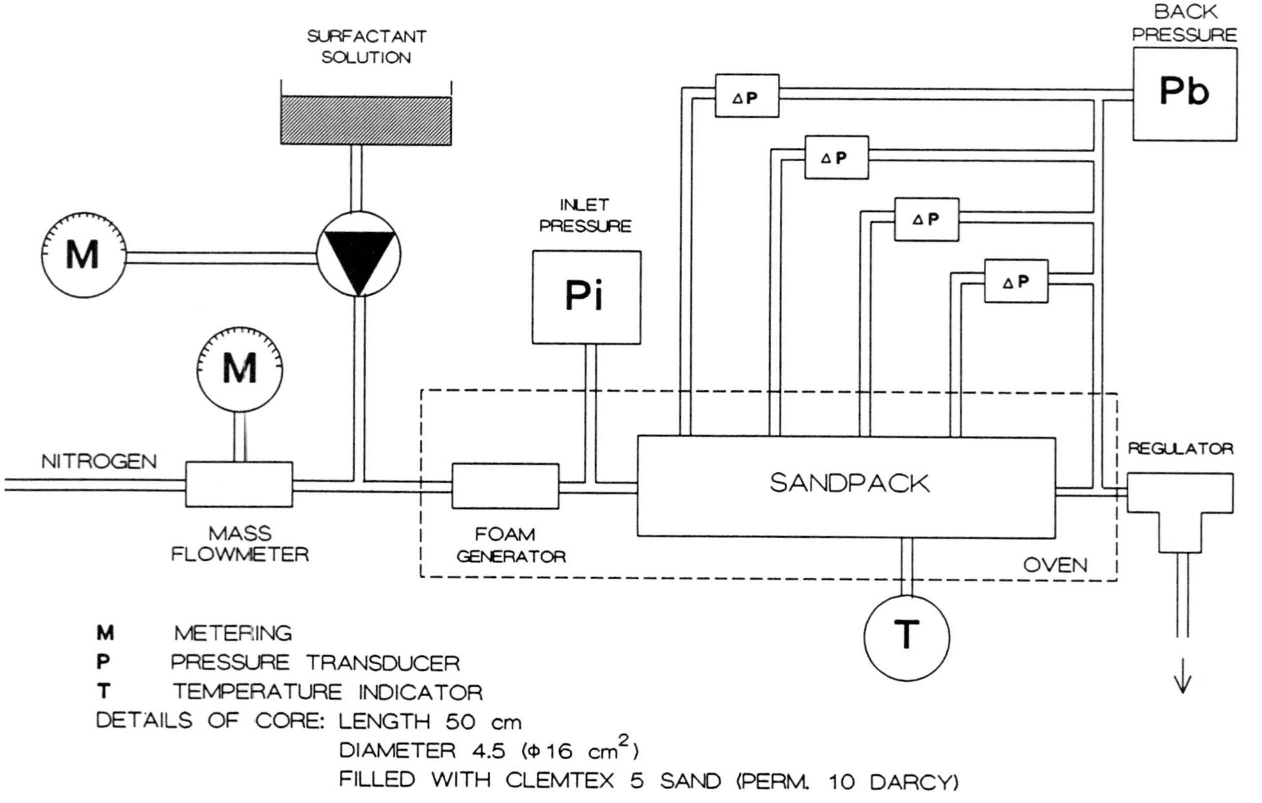

FIGURE 1: SCHEME OF FOAM FLOODING APPARATUS

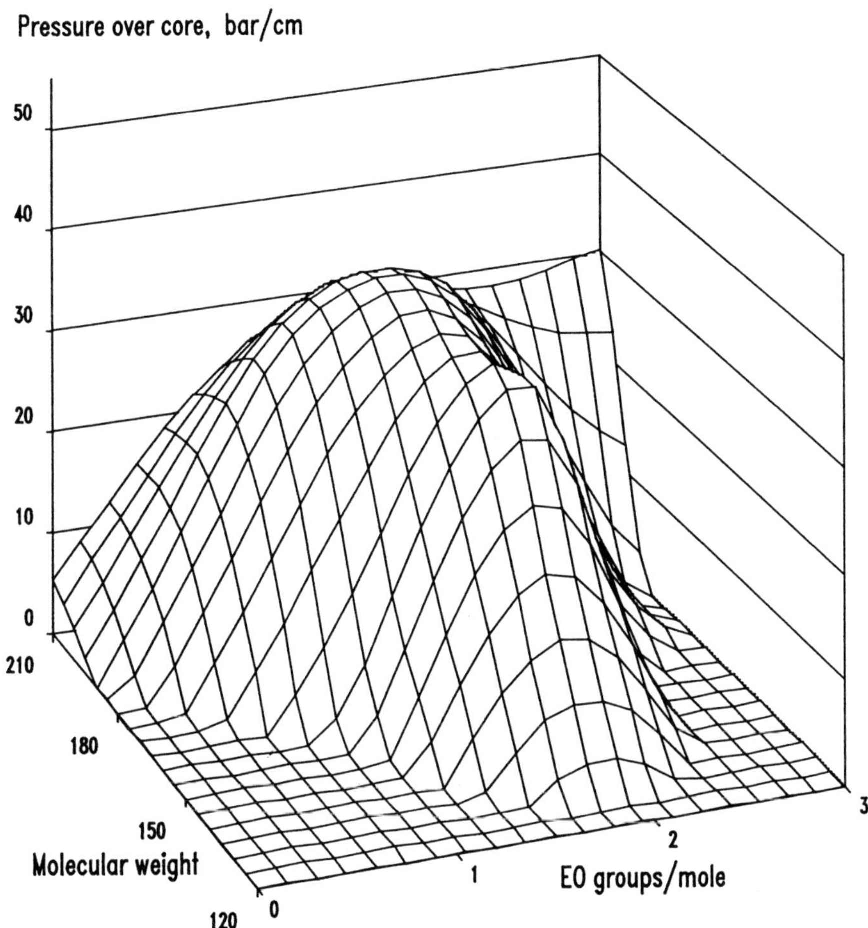

FIGURE 2 : FOAM GENERATION OF ETHOXYSULPHATES IN SPRING WATER

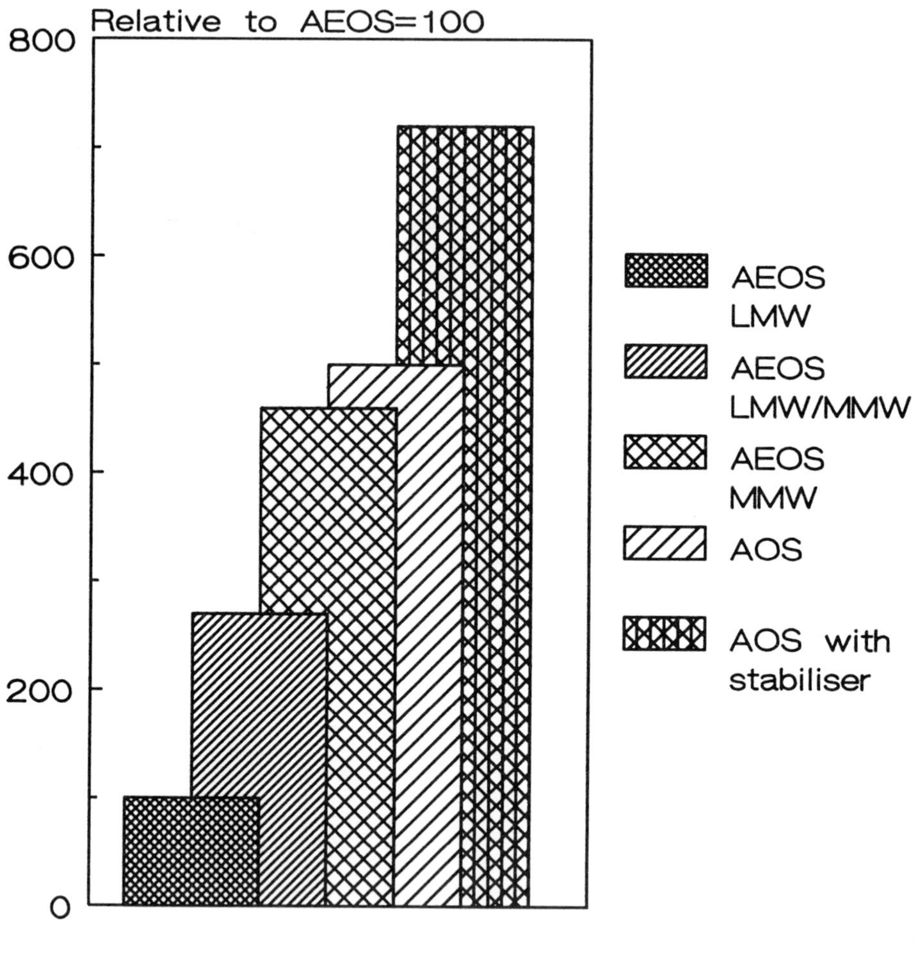
FIG.3:EFFECT OF OIL ON FOAM
(oil titration test)
Relative to AEOS=100
800
600
400
200
0
AEOS LMW
AEOS LMW/MMW
AEOS MMW
AOS
AOS with stabiliser
Product type

literature it seems likely that a substantial part of future developments would make use of this sub-sea technology for satellite field developments. One of the production difficulties would be the low temperatures prevailing in e.g. non-insulated sub-sea pipelines.[22,23] This could lead to hydrate formation and possibly plugging, an aspect discussed in some detail below.

Hydrates are crystalline structures composed of a water lattice and at least one other compound.[24] The lattice is constituted of cages formed by hydrogen bonding of the water molecules. Some of the cages are occupied by molecules of the "guest" compound. Methane, ethane, propane, isopropane, isobutane, but also carbon dioxide and hydrogen sulphide are possible "guests" of a hydrate structure. Two hydrate structures are of interest to the oil industry: Hydrate I, with two types of voids, 12-hedra and 14-hedra, and Hydrate II, also with two types of voids, 12-hedra, as in I, and 16-hedra. The 16-hedra void of Hydrate type II is larger than the 14-hedra void of I, so it can accommodate larger "guests" (propane), while smaller "guests" like methane reside in Hydrates of Type I. (See Figure 4). Currently, gas hydration formation is prevented by using methanol or glycol; drawback is that both compounds have to be used in large quantities (20 up to 50%).

Our aim is to study and develop additives which affect the morphology of the hydrates in such a way that a hydrate plug will not be formed in a pipeline. For this purpose we have built a high pressure cell; a high pressure-loop is under construction.

The pressure cell, constructed of stainless steel with two sapphire windows, can be operated under pressure and vacuum. Fluid from a thermostatic bath can be circulated in the jacket of the cell in the temperature range from -20 °C to +20°C. Precision high-pressure pumps allow easy and accurate dosing of liquids and liquefied gas. Cell temperature and pressure are recorded as a function of time by a data acquisition system. The system progressively lowers the temperature from 20 to 1 °C and then increases it back again. Several such cycles are carried out consecutively. Hydrate nucleation is detected visually (characterisation of the crystals) and from the pressure and temperature recordings. Upon nucleation the temperature rises temporarily (heat of crystal formation) and the pressure drops since gas molecules

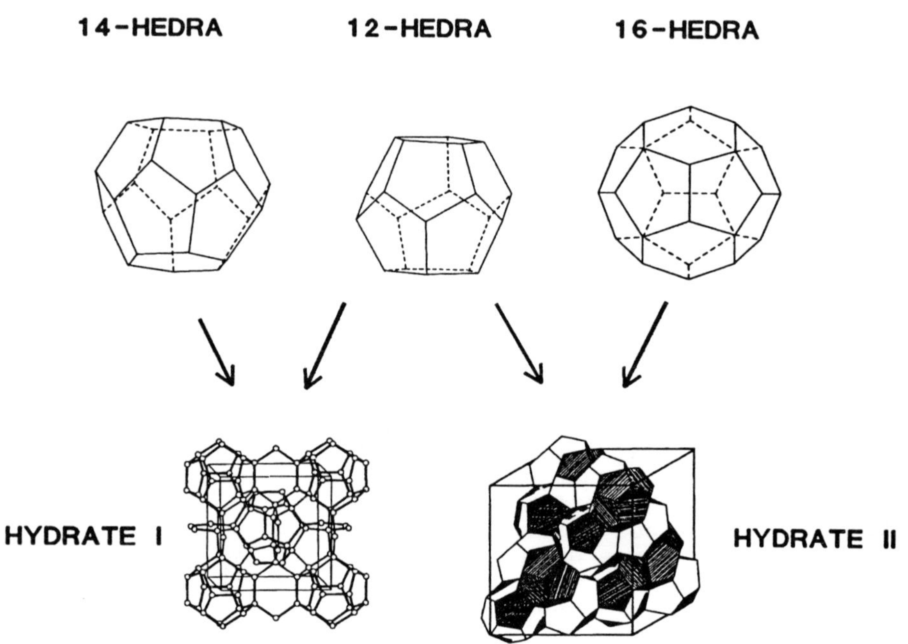

FIGURE 4: HYDRATE STRUCTURES

are trapped in the hydrate cages. As the temperature is increased the melting point of the system is detected. This procedure is illustrated in Figure 5.

Preliminary results obtained with this new approach, using small amounts of amphiphilic substances to modify crystallisation so as to prevent sticking of the hydrates and to turn them into a more "mushy" pumpable consistency are encouraging.

Asphaltene inhibitors

The deposition of asphaltenes is sometimes a problem in production operations (and in refineries).[25] As a remedial action, the use of e.g. aromatic solvents, also covered in a paper at this congress, is often the only solution.

Asphaltenes are defined as the n-heptane-insoluble fraction of crude oil (Method IP 143). Resins are defined as the propane-insoluble fraction of crude oil; they are soluble in n-heptane. Asphaltenes are generally thought to be colloidally dispersed in crudes rather than in solution.[26] These asphaltenes are probably dispersed by the adsorption of the more soluble maltene molecules. Low-molecular-weight paraffinic solvents remove the maltenes and subsequently the asphaltenes flocculate and precipitate. Controversy exists as to whether or not the asphaltene deposition process is irreversible.[27]

We have used the asphaltene flocculation test to assess the influence of inhibitors. The oil with inhibitor was destabilised by addition of n-heptane and the flocculation process was monitored by an ultrasonic particle counter. The first scouting results obtained with various anionic surfactants, given in Figure 6, revealed that the use of such additives could be an attractive route to solve the problem. Further optimisation of the most promising candidates, to improve the dose-response relationship, is still in progress. The results obtained with some development products are illustrated in Figure 7.

3 GENERAL CLEANERS FOR OFFSHORE USE

On account of regulations on safety and environmental protection and for reasons of economy, oil-field

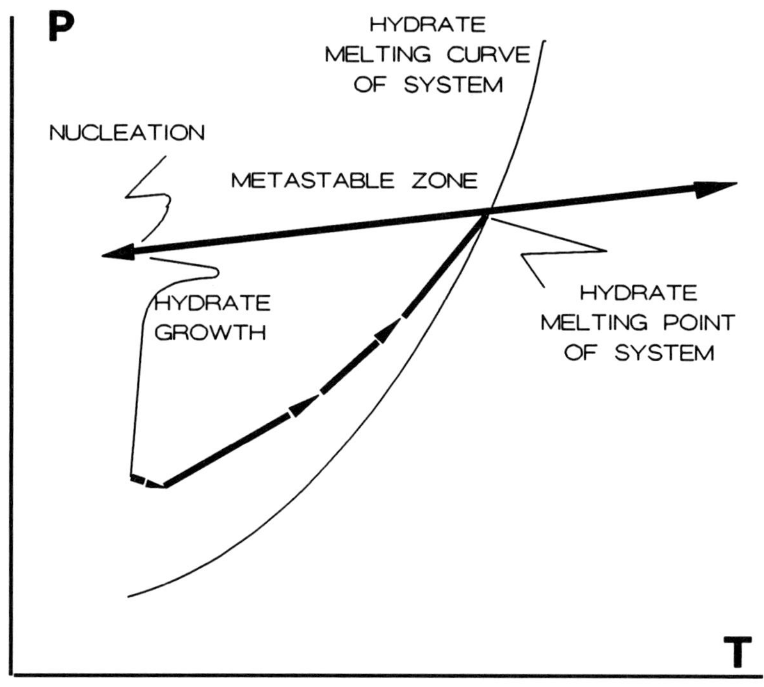

FIGURE 5: HIGH PRESSURE CEL DATA RECORDING PROCEDURE

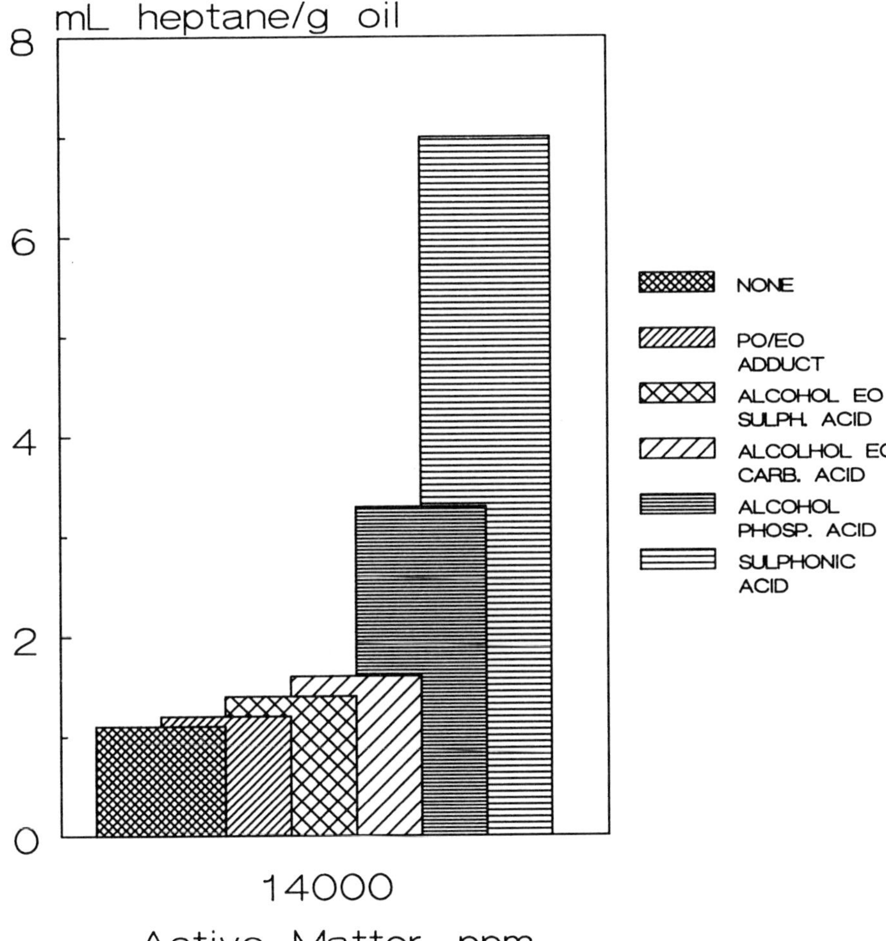
FIG.6:ADDITIVE EFFECT ON ASPHALTENE FLOCCULATION IN HEPTANE TITRATION
mL heptane/g oil
8
6
4
2
0
14000
Active Matter, ppm
NONE
PO/EO ADDUCT
ALCOHOL EO SULPH. ACID
ALCOLHOL EO CARB. ACID
ALCOHOL PHOSP. ACID
SULPHONIC ACID

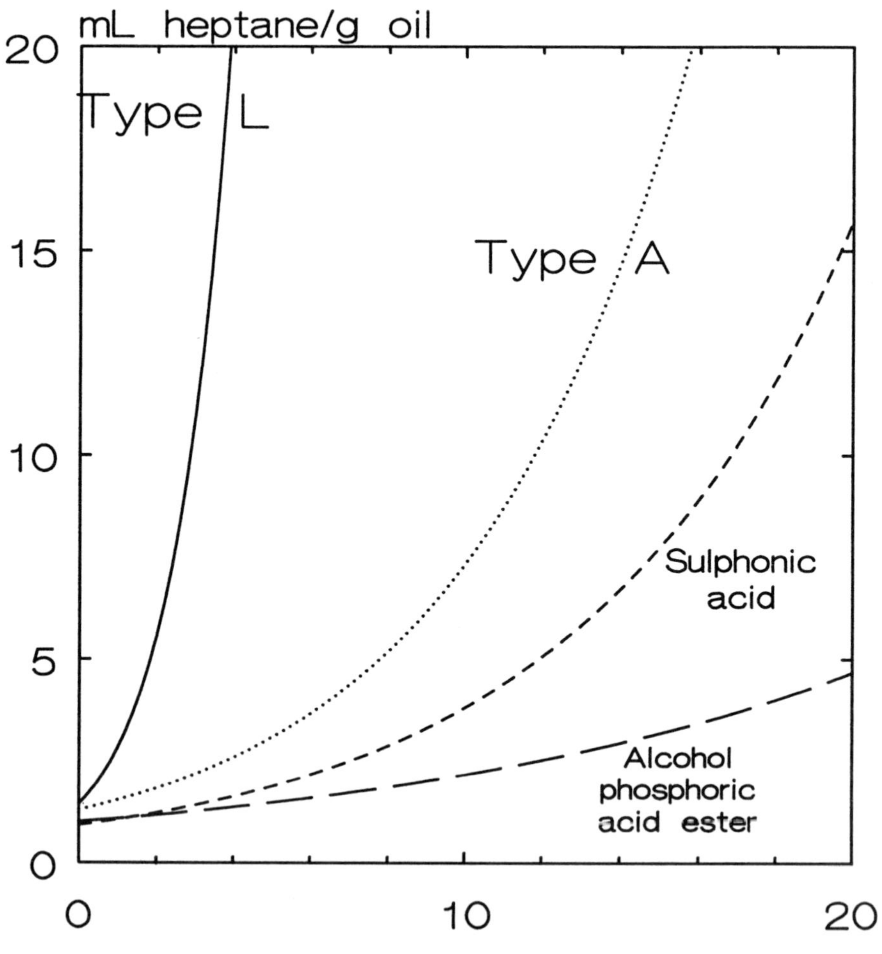
FIG.7:ADDITIVE EFFECT ON ASPHALTENE FLOCCULATION IN HEPTANE TITRATION
mL heptane/g oil
20
15
10
5
0
Type L
Type A
Sulphonic acid
Alcohol phosphoric acid ester
0
10
20
Active matter (ppm x 1000)

operators in the North Sea are rationalising their purchases of rig-cleaning products. These products should fulfil the demands given in Table 5.
These requirements imply that domestic products are <u>not an option</u> and special products had to be developed according to the criteria given in Table 6.

Because of the low temperature of application (seawater, offshore North Sea) nonionics with a low hydrophilic/lipophilic balance (i.e. a low amount of ethylene oxide units per mole) have to be incorporated in the base formulation to ensure a good performance. The rationale behind this is well documented in the literature.[28-30] This performance requirement is incompatible with the requirements of environmental friendliness. It is generally accepted that nonionic surfactants with a large amount of ethylene oxide (EO) in a molecule are less toxic to marine life than their lower molecular-weight counterparts. This sensitivity to only a very small difference in EO content is illustrated in Figure 8. Table 7 summarises the products currently used and tested on a larger scale. The emulsion cleaner, a blend consisting of surfactant and a cleaning solvent, with or without some water, gives upon dilution with sea water a very efficient micro-emulsion to be applied on the surface to be cleaned. Our target is to design such systems that fall in a CATEGORY I classification (product usage up to 100 t/a per installation).

<u>Table 5</u> Cleaning formulations
Rig cleaners - Market requirements

- Preferably be of a high concentration with a low pour point.
- Not contain caustic material or nitrates.
- Meet biodegradation regulations and have a low marine toxicity.
- Not have a flash-point hazard rating.
- Display low foaming, with enhanced detergency, wetting and emulsifying properties.
- Be tolerant to hard or saline water and effective on materials found in oil-based fluids.

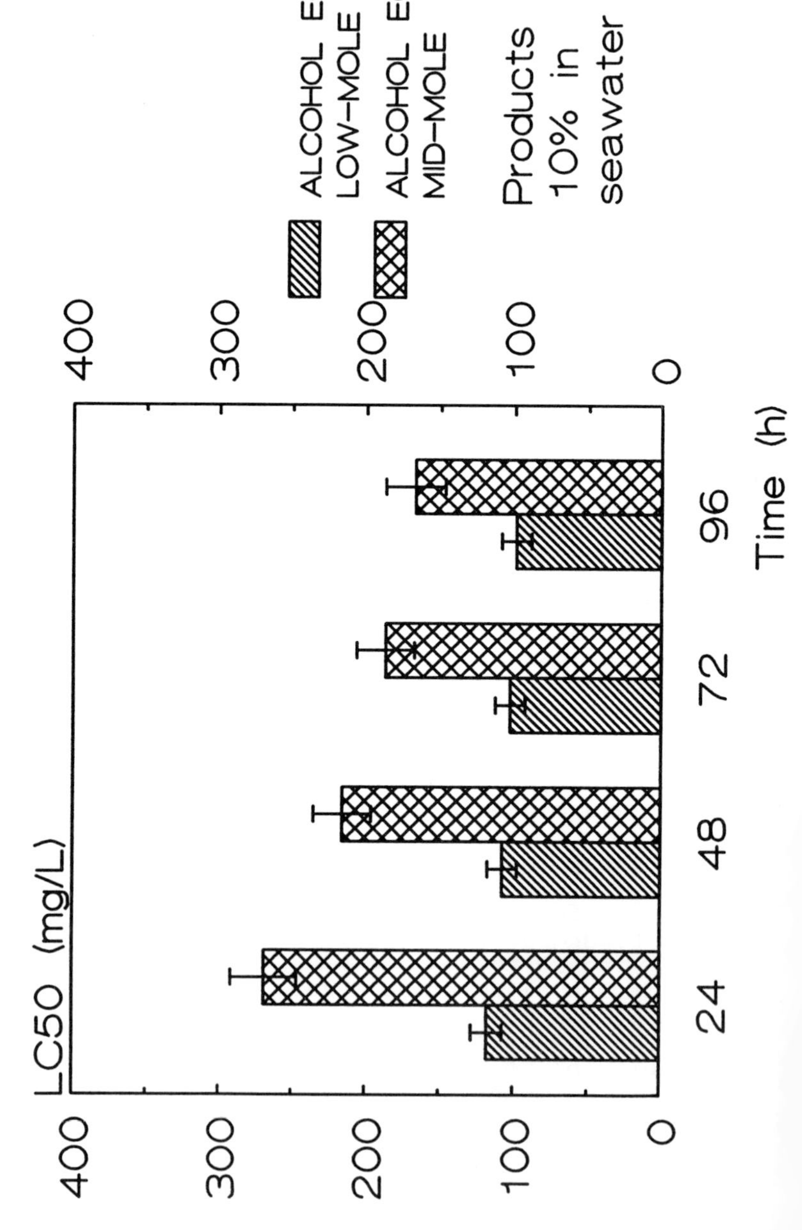
FIG.8:ACUTE TOXICITY of NONIONICS
(Crangon crangon)
LC50 (mg/L)
400
300
200
100
0
24
48
72
96
Time (h)
ALCOHOL EO LOW-MOLE
ALCOHOL EO MID-MOLE
Products 10% in seawater

Table 6 Cleaning formulations
Rig Cleaners - Development targets

- Removal of non-polar soil at low temperatures
- Parameters to be studied:
  Soil type
  Nonionics (number of EO groups)
  Temperature
- Low-temperature handleability

Table 7 Types of cleaners for offshore use

R110 & R120 (DOE/MAFF CATEGORY 1)
Formulations to be used after dilution with cold seawater

R111 (Special application)
Formulation for very heavy soil (a surfactant solvent cleaner); to be applied via e.g. brushing or spraying, with subsequent rinsing

R113 (Specific requirement)
An emulsion cleaner containing a solvent with a high auto-ignition temperature (for use in high-explosion-risk areas on offshore rigs)

4 SURFACTANTS FOR INCREASED OIL PRODUCTION

In oil fields where (infill) drilling can no longer be carried out, enhanced or improved oil recovery is an option. In spite of the considerable effort spent on research and (the sometimes disappointing) field tests during the last two decades, improved oil recovery techniques - maybe apart from polymer flooding - are on the whole still very much in a stage of development. The problems and prospects will be reviewed in detail in a separate paper to be presented by colleagues at this congress.

A special and attractive case is formed by the foam/surfactant enhanced thermal processes (steam soak and steam/hot water drives). The results of our process and product development work, with specially designed surfactants, have already been reported.[31,32] An excellent review of the performance of steam-foam surfactants has been published recently.[33] Results obtained with these surfactants (foamers) in the field have also been reported.[34] The additional oil recovered

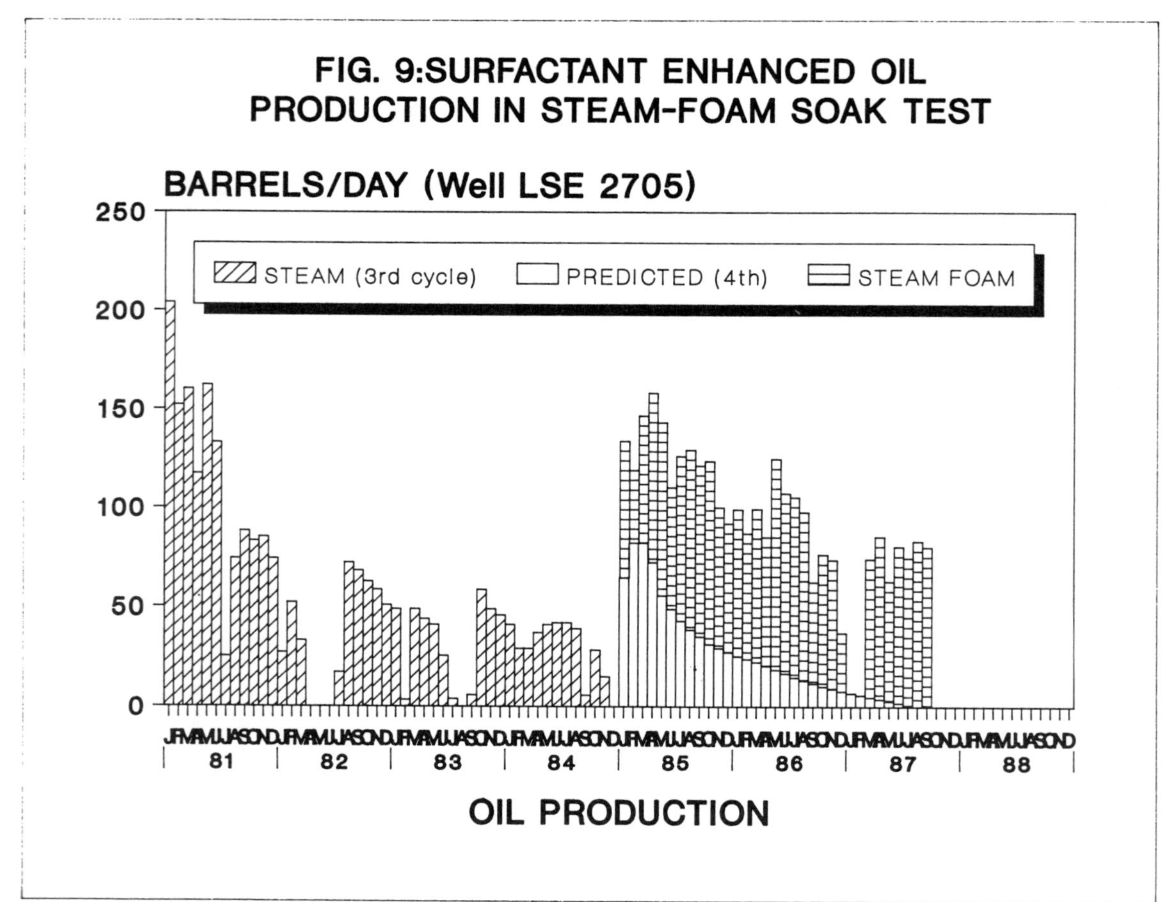
FIG. 9:SURFACTANT ENHANCED OIL
PRODUCTION IN STEAM-FOAM SOAK TEST
BARRELS/DAY (Well LSE 2705)
STEAM (3rd cycle)
PREDICTED (4th)
STEAM FOAM
250
200
150
100
50
0
81
82
83
84
85
86
87
88
OIL PRODUCTION

by such a steam-foam soak test is well illustrated in Figure 9. It has been stated that if the process were used on a commercial scale in the Bolivar coast reservoirs in Venezuela, the cost of the additional oil would amount to 0.6 to 1.6 US$ per barrel.[35]

5 CONCLUSIONS

Surfactants can be used in a wide variety of oil-field operations, varying from well drilling and downhole-engineering to the transport of oil and gas and rig cleaning.

Surfactants can be a useful tool for oil-field operators to improve production and to combat various problems encountered in operation.

Surfactant-enhanced steam injection is an attractive technique to improve the recovery of heavy oils.

REFERENCES

1. C.H. Kline & Co., 'Oil Field Chemicals - North America', 1983.
2. R.C. Parker, "The Market for Chemicals in the Oil Industry", in 'Chemicals in the Oil Industry' (Ed. P.H. Ogden). The Proceedings of a Symposium Organised by the North West Region of the Industrial Division of the Royal Society of Chemistry, University of Manchester, 22nd-23rd March 1983.
3. D. Antheunis, "Changing Chemical Needs of the Oil Producing Industry", in 'Chemicals in the Oil Industry' (Ed. P.H. Ogden). The Proceedings of the 3rd International Symposium, Organised by the North West Region of the Industrial Division of the Royal Society of Chemistry, University of Manchester, 19th-20th April 1988.
4. Anon., Offshore Engineer, October 1990, 25.
5. Anon., Noroil, March 1989, 35.
6. S. Bird and T. Howe, "Environmental Data Requirements and Future Use of Oilfield Chemicals in the North Sea - A Service Company Viewpoint", Paper presented at the Oil Field Chemicals Symposium in Geilo, Norway, March 19-21, 1990.
7. R. Peresich, Offshore, September 1990, 32.
8. G.D.L. Morris, Chemicalweek, August 22, 1990, 28.

9. R.A. Hook, L.W. Cooper and B.R. Payne, World Oil, April 1977, 95 (Part I) and World Oil, May 1977, 83 (Part II).
10. H. Lorenz, World Oil, June 1980, 187-190, 192-193.
11. J. Dupont, Oil & Gas Journal, May 7, 1984, 188, 193-194.
12. J.W. Burman and B.E. Hall, World Oil, November 1987, 31.
13. R. Montman, D.L. Sutton, W.M. Harms and B.G. Mody, Oil & Gas Journal, July 26, 1982, 209.
14. R. Montman, D.L. Sutton, W.M. Harms and B.G. Mody, World Oil, June 1982, 171-174, 176, 178, 181-182, 184, 186.
15. Anon., Oilweek (Calgary, Alberta), March 2, 1987, 38 No. 5, 11.
16. J.K. Borchardt, D.B. Bright, M.K. Dickson and S.L. Wellington, Paper SPE 14394 presented at the 60th Annual Technical Conference and Exposition of the Society of Petroleum Engineers of AIME, Las Vegas, NV, September 22-25, 1985.
17. J. Ali, R.W. Burley and C.W. Nutt, Chem. Eng. Res. Des., 1985, 63 101.
18. S.M. Farouq Ali and R.J. Selby, Oil & Gas Journal, Feb 3, 1986, 57.
19. H.M. Muijs, N.C.M. Beers, R.J Wiersma and P.P.M. Keijzer, "Foaming agents in oil field production", paper presented at the 2nd World Surfactants Congress, Paris, 24-27 May 1988, Proceedings, Vol. IV, 199.
20. J.A. Jensen and F. Friedman, "Physical and chemicals effects of an oil phase on the propagation of foam in porous media", Paper SPE 16375, presented at the SPE California Regional Meeting, Ventura, Ca., April 8-10, 1987.
21. R. Kok, J.T. Bouman and R. Blanco, "Performance Testing of Dishwashing Liquids - Development of a Foam Titration Method", Jorn. Com. Esp. Deterg., 1989, 20 433-444. Paper presented at the XX Meeting of the Spanish Detergents Committee (1989).
22. S.R. Setliff, G.M. Parfitt and T.S. Chilton, "Hydrate mitigation in a deepwater production system", paper SPE 19266 presented at the Offshore Europe 89 Meeting, Aberdeen, 5-8 September 1989.
23. C. Blanc and J. Tournier-Lasserve, World Oil, November 1990, 63.
24. Y.F. Makogon, 'Hydrates of Natural Gas', PennWell Publishing Co., Tulsa, Okla., 1981,

25. Kolloquium 1989 at the Institut für Erdölforschung entitled: "Die Ausfällung erdöleigener Stoffe bei Fördurung und Transport und ihre Bekämpfung", Clausthal-Zellerfeld, 24 November 1989.
26. K.J. Leontaritis and G.A. Mansoori, "Asphaltene flocculation during oil production and processing: A thermodynamic colloidal model", paper SPE 16258 presented at the SPE International Symposium on Oilfield Chemistry, San Antonio, Texas, February 4-6, 1987.
27. K.J. Leontharitis and G.A. Mansoori, J. of Petr. Sc. and Eng., 1988, 1, 229.
28. H.L. Benson, K.R. Cox and J.E. Zweig, Soap/Cosmetics and Chemical Specialities for March, 1985, 35.
29. K.H. Raney, W.J. Benton and C.A. Miller, J. of Colloid and Interface Science, May 1987, 117, No. 1, 282.
30. K.H. Raney and C.A. Miller, J. of Colloid and Interface Science, October 1987, 119, No. 2, 539.
31. P.P.M. Keijzer, H.M. Muijs, R. Janssen-van Rosmalen, D. Teeuw, H. Pino, J. Avila, and L. Rondon: "Application of Steam-Foam in the Tia Juana Field, Venezuela - Laboratory Tests and Field Results," paper SPE 14905 presented at the 1986 SPE/DOE 5th Symposium on EOR, Tulsa. OK, April 20-23.
32. H.M. Muijs, P.P.M. Keijzer and R.J. Wiersma: "Surfactants for Mobility Control in High-Temperature Steam-Foam Applications", paper SPE 17361 presented at the 1988 SPE/DOE 6th Symposium on Enhanced Oil Recovery, Tulsa, OK, April 17-20.
33. D.C. Shallcross, L.M. Castanier and W.E. Brigham, "Characterization of Steam Foam Surfactants Through One-Dimensional Sandpack Experiments", Report DOE/BC/14126-19, May 1990.
34. J.L. Ziritt and O. Rivas, "Evaluation of a cyclic steam-surfactant injection pilot test in the Bolivar Coast, Venezuela", Proceedings of the 4th European Symposium on Enhanced Oil Recovery, Hamburg, October 27-29, 1987, 611.
35. J.L. Ziritt, O. Rivas and G. Bresolin, "Evaluation of a Cyclic Steam-Surfactant Injection Pilot Test in the Bolivar Coast, Venezuela", in Report DOE/BC-89/1/SP - 'Venezuela-MEM/USA-DOE Fossil Energy Report IV-5- Supporting Technology for Enhanced Oil Recovery EOR Thermal Processes', November 1988.

New Developments in the Use of Sulfur Transfer Agents for the Reduction of SO_X Emissions from Fluid Catalytic Cracking Units

J.R. Pearce and P.H. Desai

AKZO CHEMICALS DIVISION, PO BOX 975, 3800 AZ AMERSFOORT, THE NETHERLANDS

ABSTRACT

The basic manner in which sulfur transfer additives work in a Fluid Catalytic Cracking unit remains the same as first described nearly two decades ago. Nonetheless, intensive R&D continues, with the main objective being the development of additives which can function under an expanded range of FCC unit operating conditions. Of special interest are additives which are effective at very high regenerator temperatures where the oxidation of SO_2 is not favorable, and at very low reactor temperature where the slow reduction of metal sulfates can be limiting. In addition, the search for materials with a higher resistance to silica poisoning is an ongoing challenge.

This paper describes some advances which are being made to overcome these limitations and thereby expand the operability "window" of the sulfur transfer additives. In particular, SO_3 absorbents are becoming available which are more effective at very severe regeneration, or under conditions of partial combustion. Furthermore, co-promoters are now commonly used which facilitate the reduction and regeneration of sulfate laden additives, lending them a higher efficiency and longer functional life. Taken together, these improvements allow the use of less additive to achieve the desired flue gas SOx reduction effect, to the obvious cost benefit of the end user.

I INTRODUCTION

The basis of SOx emission reduction from cat-crackers through the use of absorbent additives remains the same as first described in the mid 70's. [1,2] Specifically, about 10% of the sulfur in the incoming reactor feedstock is laid down with the coke deposited on the catalyst. As the coke is subsequently burned off in the catalyst regenerator system, the sulfur is also oxidized and liberated as mixed sulfur oxides (SOx). Without a suitable control method, the SOx would be vented with the regenerator flue gas, an increasingly unacceptable and costly option.

As pointed out in two recent overviews of this field, [3,4] the Sulfur Transfer Additives (STA's for short) consist of absorbents which hitchhike around the system with the circulating catalyst inventory. They generally consist of supported metal oxides which bind SO_3 in the form of metal sulfates which are stable under regenerator conditions. By sequestering the sulfur oxides in this fashion, they can be reduced or nearly eliminated from the regenerator outlet. As the sulfate laden additives are later transported to the reactor side of the FCC unit, the sulfates are reduced and hydrolysed. This liberates the easily recoverable H_2S and rejuvenates the metal oxide absorbent. Since 90% of the total incoming sulfur already leaves the system via this outlet, the added burden is easily accomodated.

Thus, the net effect of sulfur transfer additives is the reduction of SOx in the regenerator flue gas, at the expense of a minor increase in the level of H_2S in the reactor overhead. The overall process is illustrated in Figure 1.

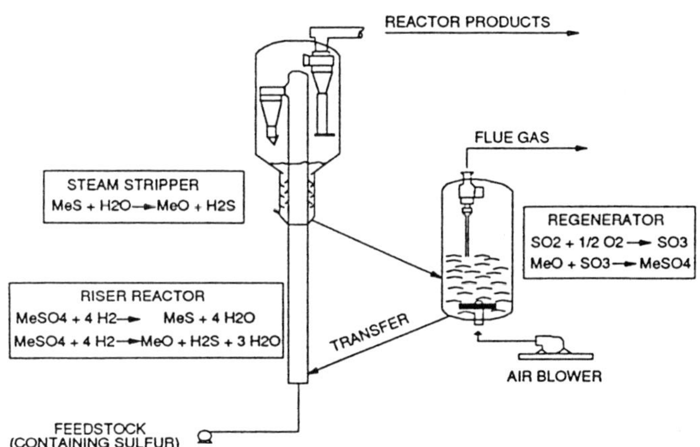

Figure 1: Sulfur transfer reactions in the FCC process

In combination with some practical aspects, the six major requirements of effective STA's are:

1. Capture and hold SOx under regenerator conditions
2. Release SOx under reactor/stripper conditions
3. Acceptable long term stability
4. FCC typical physical properties
5. No adverse effects on overall FCC yields
6. No increase of NOx emissions

There are three other options for reducing or controlling the SOx emissions from the cat-cracker: using lower sulfur crudes, more severe upstream hydrotreating, or flue gas desulfurization. However, it has been pointed out that the use of STA additives is both cheaper and certainly easier.[5] These benefits have led to a high level of interest over the last fifteen years, as evidenced by the density of relevant patents (Figure 2).

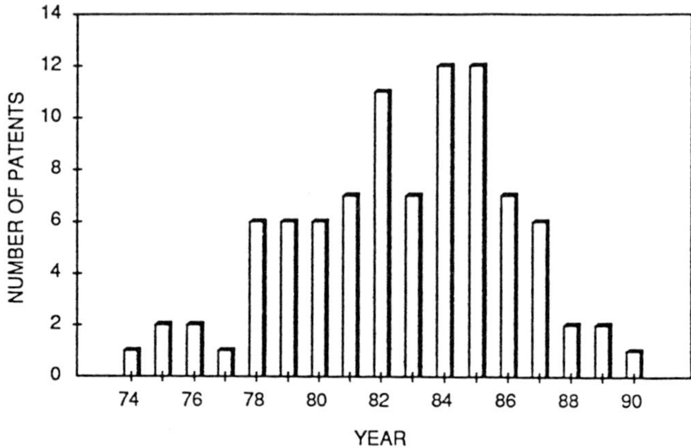

Figure 2: Patent applications for sulfur transfer agents

With so much activity, one might suspect that the technology behind these additives would assume a "mature" status. However, intensive R&D efforts in the field are ongoing at a number of laboratories, and significant performance improvements are frequently announced. The driving force behind the continuing research is the expectation that regulatory constraints on SOx emissions will become more stringent in the coming years. Furthermore, feedstocks must eventually become heavier and carry higher loads of sulfur, so reliance on SOx transfer additives is expected to grow in popularity. Finally, there is also a certain degree of conflict in the literature as to the mechanisms by which the STA's actually operate. This continues to motivate researchers in their quest to reduce this field to some concensus of accepted truth.

It is essential to note that the sulfur transfer process is not truly catalytic, but consists of two sets of stoichiometric reactions which occur independently under the different conditions at the regenerator and reactor sides of the FCC unit. In their simplest form, they are:

REGENERATOR (typically 700-800°C, oxidizing conditions)

$$SO_2 + \frac{1}{2}O_2 \rightarrow SO_3$$

$$MeO + SO_3 \rightarrow MeSO_4 \quad (MeO = METAL\ OXIDE)$$

REACTOR/STRIPPER (typically 500-550°C, reducing conditions)

$$MeSO_4 + H_2 \text{ or } HC \rightarrow MeS + H_2O$$

$$MeS + H_2O \rightarrow MeO + H_2S$$

It is also important to point out that different conditions on the reactor and regenerator sides of the FCC system each impose a unique set of constraints and operating limits on the SOx gettering process. Most notable among these limitations are:

1. Very high temperature regenerator operation (> 750°C) where the SO_3/SO_2 equilibrium becomes unfavorable, and where silica poisoning of the additive is accelerated.[6,7]
2. Partial combustion regeneration where the availability of oxygen for the conversion of SO_2 to SO_3 can be limiting.
3. Low temperature reactor operation where the reduction of accumulated sulfates on the additive is kinetically unfavorable.

Given these differences, it is not surprising that an understanding of the combined SOx reduction process is only possible by studying the effects of each set of individual conditions separately. Above all, it is necessary to fully understand the reason behind each of the constraints so that the STA's can be optimized for application over as wide a range of unit conditions as possible. In response to that challenge, some significant advances have recently been made which can expand the window of operability for sulfur transfer additives. Those advances are the topic of this paper.

II EXPERIMENTAL

The micro-reactor system used for the present studies is fairly typical, and is shown schematically in Figure 3. It consists of a single pass, fluidized bed reactor which contains a total charge of 8 grams of material. This reactor charge was a mixture of a Pt-free equilibrium FCC catalyst to which was added various small amounts (typically 2-5 wt%) of candidate SOx transfer agents. Both materials were pre-steamed for five hours at 788°C to simulate unit deactivation. The FCC and STA were normally steamed separately before blending and testing. In selected cases, however, they were steamed as a physical mixture to look for any mutually harmful interactions between the materials under these severe hydrothermal conditions.

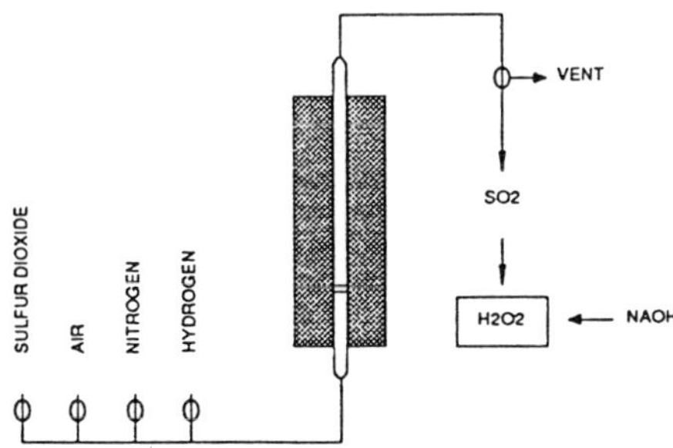

Figure 3: Microreactor for testing STA

The candidate additives were evaluated using a cyclic four-step test consisting of:

1. Contact with 0.2% SO_2 + 2% O_2 (simulated regenerator)
2. Stripping with nitrogen
3. Reduction with dilute hydrogen (simulated reactor)
4. Stripping with nitrogen

During step 1 (simulated regeneration atmosphere), the amount of SO_2 absorbed and retained by the mixed reactor charge was determined by trapping the SO_2 breaking through in dilute H_2O_2, then back titrating with standardized caustic. The delta SO_2 across the reactor was defined as "PERCENT DESOX". Each complete cycle took 20 minutes and could be repeated as desired to examine deactivation phenomena. In general however at least 30 cycles were examined to assure lined-out operation and measurement.

The temperature during the cyclic evaluation could be preselected, but was generally held constant at either 525 or 725°C. This was done to determine the limitations of the chemical reactions at each of the two separate temperature regimes encountered in the FCC system. Thus, the effectiveness of each candidate additive was determined twice: once at 725°C (a typical regenerar temperature) and again at 525° (typical of the reactor side). This test protocol is rather unusual, since the two halves of the reaction are usually studied at their respective temperatures. However, by looking at the system in this unique way, a new and valuable insight was gained into the basic mechanism of the sulfur transfer reaction network.

An illustrative example of such a test is shown in Figure 4. Here we see the important effect of rare earth oxides on the SOx absorption capacity of an additive when tested at 725°. It is clear from this figure that neither the equilibrium FCC itself, or the bare support used for the additive, can absorb much SOx at this temperature. However, a strong increase in the PERCENT DESOX is observed as rare earth is gradually incorporated in the additive. This is a classical response, since the rare earth oxides (particularly ceria) have long been known as effective SOx getters.[5] Indeed, rare earths are presently used in all commercial sulfur transfer agents.

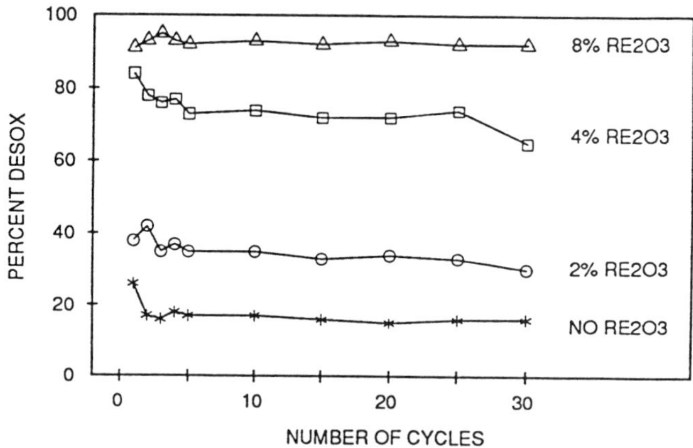

Figure 4: Effect of rare earths on sulfur transfer at typical regenerator temperature (725°C)

The indicated rare earth levels in Figure 4 refer to RE_2O_3 on the additive, which comprises 2.5 wt% of the total reactor charge. The percent DESOX is a measure of the fraction of SO_2 absorbed from the SO_2/O_2 gas stream.

As will be described below, the isothermal DESOX testing protocol was used for the quick evaluation of many, many modifications of the basic sulfur transfer additives. The goal of this research was to find ways to push back the boundaries imposed by the FCC system, and to develop additives which would provide improved activity under an expanded range of unit operating conditions. The method of examining each temperature separately has not yet been described in the literature, and has provided some surprising and valuable insights. Above all, however, the information generated in these studies has allowed us to make attractive improvements in the performance of sulfur transfer additives.

III RESULTS AND DISCUSSION

A. High Temperature Regenerators

One of the fundamental requirements of STA's is their ability to capture and hold SOx under regenerator conditions. The use of rare earth oxides goes a long way in accomplishing this task, as shown above by the dramatic increase in the PERCENT DESOX at 725°C as RE_2O_3 is introduced. This is due primarily to the ability of the rare earths to form stable sulfates at this temperature. Among the rare earth oxides, ceria has been found to be the most effective at the relatively modest 725°C.[3] However, many FCC units operate at significantly higher regenerator temperatures, where it is known that traditional transfer additives do not work particularly well. A good simulation is shown in Figure 5 for an additive containing a relatively high level of RE_2O_3 as the only promoter. The effectiveness of this STA clearly decreases rapidly with temperature until, at 800°C, it is essentially non-functional.

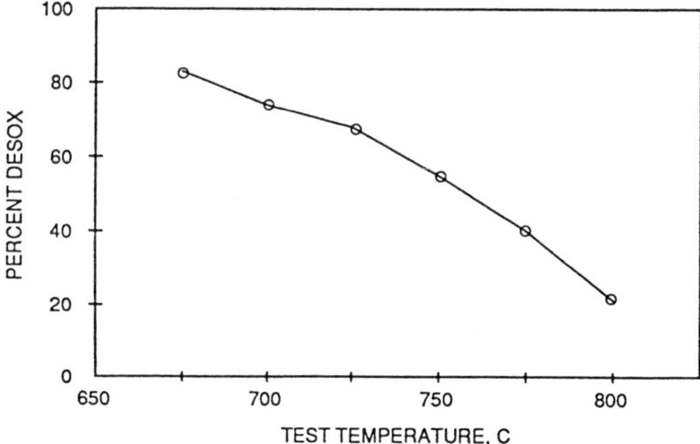

Figure 5: Effect of temperature on conventional STA activity

The additive (Figure 5) contains 8 wt% RE_2O_3 and comprises 2.5 wt% of the total reactor charge. Each data point represents an average of five cycles performed at the indicated temperatures.

It has often been stated that at this high temperature the loss of SOx reduction performance is not so much due to any properties of the additive itself, but to an unfavorable shift of the SO_2/SO_3 equilibrium.

$$SO_2 + \frac{1}{2}O_2 \leftrightarrow SO_3 \qquad \log K_p = (4956/T) - 4.678$$

Recall that one of the essential steps in the transfer mechanism is the exothermic oxidation of SO_2 to SO_3. As shown in Figure 6, this equilibrium becomes unfavorable at high temperature, and would suggest a fundamental limitation to the sulfur gettering process.

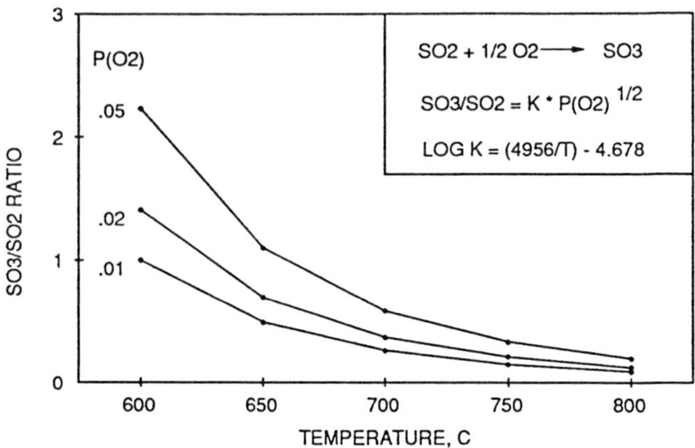

Figure 6: Equilibrium data for SO_2 oxidation

However, the reaction is incomplete as shown, and should really include the SO_3 absorbent metal oxide (MeO) as part of the system.

$$SO_2 + \frac{1}{2}O_2 \rightarrow SO_3 \xrightarrow{+MeO} MeSO_4$$

It follows that if an absorbent is used which is capable of forming a much more stable sulfate adduct, then the steady state concentration of SO_3 can be effectively reduced, thereby mitigating the unfavorable SO_3/SO_2 equilibrium.

Indeed, intensive investigations along this line have identified a package of absorbent metal oxides, which when coupled with an appropriate support, do form a more stable high temperature sulfate than the traditional rare earth oxide mixture. Futhermore, by using this new package, the sulfur transfer activity may thus be maintained at higher than traditional regenerator temperatures. This is shown in Figure 7 where the PERCENT DESOX of an experimental additive containing the new absorbent system is compared to a traditional additive containing a simple supported rare earth oxide. Thus, an expanded window of application may be available for FCC units constrained to operate at high regenerator severity.

The ability of a more powerful SOx absorbent to shift the effective $SO_2 + 1/2\ O_2 \leftrightarrow SO_3$ equilibrium raises another very interesting possibility. Specifically, such a shift should also be reflected in a reduced dependence of the reaction on the partial pressure of oxygen in the system. In principle, this should increase the chances that the transfer additive would provide some SOx reduction under the conditions of partial combustion, where SO_2 must compete with CO for the available oxygen. This possibility is presently under investigation.

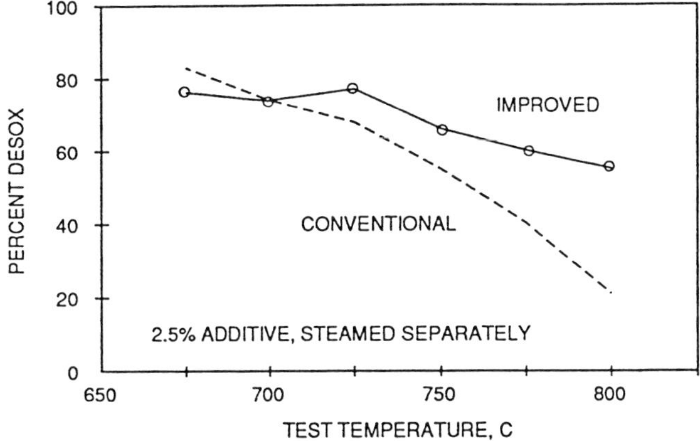

Figure 7: Improved sulfate stability at high temperature

B. Silica Poisoning

In general, the STA's are known to be susceptible to poisoning by silica. However, silica can and does migrate from the cracking catalyst to the STA under the severe hydrothermal conditions in FCC regenerators.[7] The mechanism of the poisoning is thought to be the competitive adsorption of silicon oxides on the same sites that act as SO_3 receptors. This effect is easily simulated in the laboratory by comparing the sulfur transfer activity of an additive which has been steamed either separately, or as a physical mixture with the silica containing FCC catalyst.

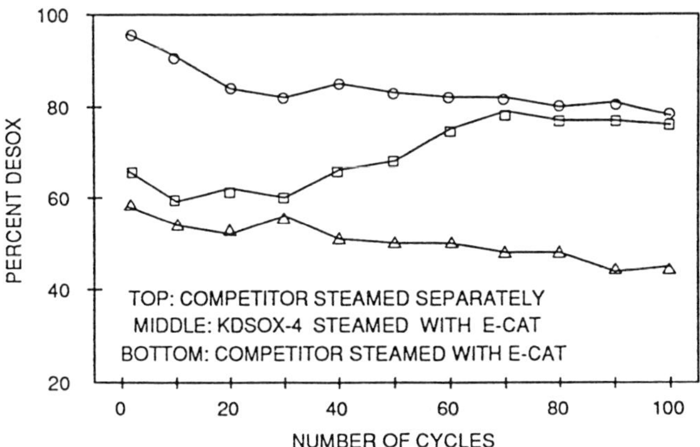

Figure 8: KDSOX-4 recovers from silica poisoning

As an example, Figure 8 shows test results on a competitive STA prepared by rare earth impregnation of a simple alumina support. It can be seen that the sample steamed with the cracking catalyst (bottom curve) is only about half as active as the same material which was steamed separately (top curve), and that the poisoning of this additive is irreversible.

However, support materials are now becoming available which have been developed specifically to work against the silica effect. These supports comprise a unique combination of chemical and pore size characteristics, and are found to act much like silica "sponges". A typical result of using this new generation of STA support is included as the middle curve in Figure 8. While the initial depressed DESOX activity clearly shows the adverse effect of the silica accumulated during the mixed steaming, the catalyst displays an ability to recover its "silica free" activity with time. This behavior is typical of the new porous support material, and can be achieved without an unregulated increase in surface area, which can lead to non-selective cracking as an unwanted side effect. This new support is the basis of Akzo's KDSOX-4 series of sulfur transfer additives.

C. Low Temperature Reactors

Another fundamental requirement of sulfur transfer additives is that the sulfates they form in the regenerator must eventually be reducible on the reactor side of the FCC unit. This can be the limiting factor for units operating at low reactor temperatures and short contact times, where rare earth sulfates can be rather stable. As an example, Figure 9 compares the activity at 525° for a pair of materials with or without rare earths. Although the sulfur transfer activity starts high in both cases, it is unstable and declines rapidly, regardless of the presence of RE_2O_3. This is a characteristic indication that the additive is being self-poisoned by accumulated sulfates.

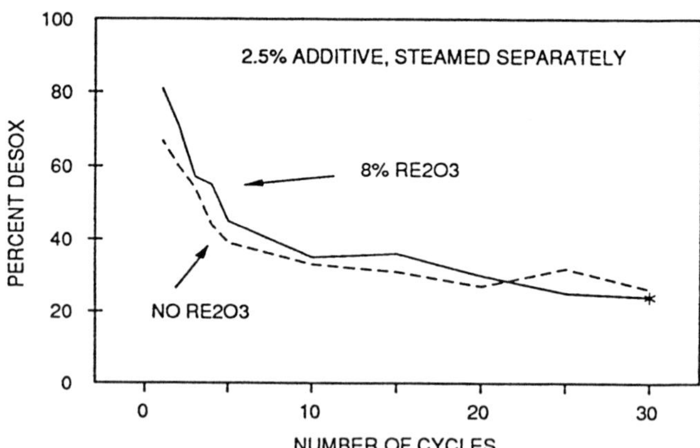

Figure 9: (Lack of) Effect of rare earths on sulfur transfer activity at reactor typical temperature (525°C). Reactor charge contains 2.5 wt% of additive, either with or without rare earth promoter.

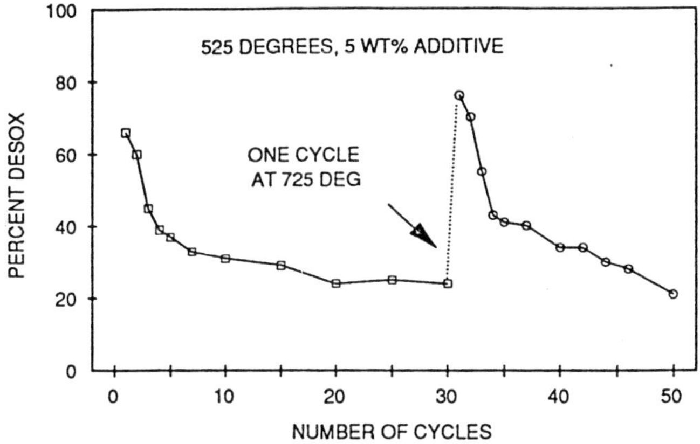

Figure 10: Low temperature instability of STA promoted with rare earth

This poisoning phenomenon is confirmed by the experiment shown in Figure 10. For this test, an additive with 8% RE_2O_3 is operated at 525° until its activity has shown the characteristic decline. At that point, a single four-step cycle is performed at 725°, where the reduction of sulfates is much more facile. Not surprisingly, this high temperature cycle temporarily restores the initial activity of the fresh additive. Upon returning to 525°, however the rapid sulfate deactivation is again observed.
Fortunately, it is possible to overcome this deactivation by the incorporation of a reduction co-promoter as an additional component of the sulfur transfer additives. The characteristic effect on sulfur transfer activity is found in Figure 11, where the PERCENT DESOX at 525° is plotted for a series of samples containing increasing levels of a co-promoter, all other aspects of the additive being identical.

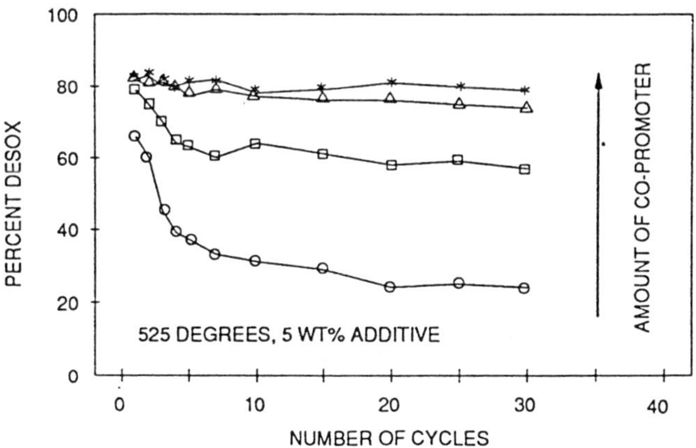

Figure 11: Effect of co-promoter on activity and stability of STA's.

The bottom curve is for an additive without co-promoter, and displays the typical decay of sulfate accumulation. The upper curves show the benefits obtained as the co-promoter is introduced (Figure 11).

It is interesting to note that the low temperature reducibility of the sulfate laden additives could only be isolated and studied independently of other variables by using the special isothermal testing protocol which we employed. This is important since the incorporation of an additional active metal in the FCC system must be done judiciously; only limited amounts can be tolerated without adverse effects on overall unit selectivities.
Careful study of the metal sulfate reduction allows the inclusion of just the minimum amount of reduction co-promoter required to be effective. That amount is a sensitive balance between good sulfur transfer activity and the detrimental effects of an additional active (and sometimes mobile) metal in the FCC catalyst inventory. This balance is a key element in state-of-the-art sulfur transfer technology.

A final demonstration of the manner in which the co-promoter functions can be made using a simple ThermoGravimetric Analysis (TGA) experiment. This is done by first treating additive samples at 725° with the SO_2/O_2 mixture, until they contain a typical level of adsorbed sulfates. The treated additives are next transferred to the TGA apparatus where, in a dilute hydrogen atmosphere, a temperature programmed reduction is performed. The onset of the sulfate decomposition is monitored by a characteristic weight loss.

A typical result of this experiment is depicted in Figure 12 for a pair of samples which differ only in that one contains a reduction co-promoter. The sulfate decomposition for the non-promoted sample only begins in the 525-550° range, which is borderline for some low temperature FCC operations, and might render this particular additive ineffective. However, with the co-promoter present, the sulfate decomposition begins well below 500 degrees. Thus, the modest use of the added co-promoter has the effect of pushing down the low reactor temperature barrier at which the sulfur transfer additive can be profitably employed.

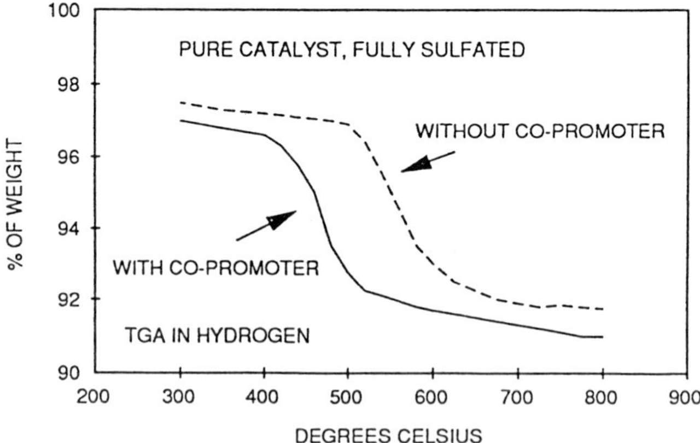

Figure 12: The effect of co-promoter on the sulfate reduction temperature as monitored by Thermo Gravimetric Analysis (TGA).

This data indicate that the sulfate reduction is much easier (lower temperature) when the c promoter is a component of the sulfur transfer additives.

The ultimate proof of the effectiveness of a reduction co-promoter lies in its performance under catcracking conditions. In the next table for example, the SOx reduction activities of several commercial additives are compared in a circulating pilot reactor with a high sulfur (2.9%) VGO feed. The values reported are relative to a base line measurement without any additive present, and were obtained 24 hours after each of these additives were introduced. The measurements were obtained independently at two different reactor temperatures in order to investigate the influence ot the co-promoters on the regenerability and long term stability of the various additives.

Most of the additives have roughly equivalent activity at the higher reactor temperature (545°C) where the reduction of accumulated sulfates is relatively facile. However, it does appear that those materials containing a reduction co-promoter do indeed maintain a higher transfer activity as the reactor temperature is reduced to 515°C, where the sulfate reduction becomes more difficult. Thus, the conclusions made from the earlier micro-reactor experiments are confirmed by the data obtained at the pilot plant scale. More specifically, the use of a co-promoter results in a measurable increase of sulfur transfer activity.

COMPETITIVE SULFUR TRANSFER ADDITIVES IN A CIRCULATING PILOT PLANT COMPARISON

REDUCTION IN TOTAL FLUE GAS SOX (%)

TEMPERATURE	545/520	515/485	
BASE CASE	0	0	
ADDITIVE A	55	25	WITHOUT CO-PROMOTER
ADDITIVE B	60	10	
ADDITIVE C	20	20	
ADDITIVE D	60	50	
ADDITIVE E	65	70	WITH CO-PROMOTER
KDSOX-4	**80**	**60**	

Kuwait VGO, .9033 g/ml @ 50C., 2.93 wt% sulfur
T(regen) = 700C, cat/oil ratio = 6, oxygen = 4%
5.0 wt% additive, 95% FCC, steamed separately

IV CONCLUSIONS

The basic mechanism of the sulfur transfer reaction remains the same as described in the 1970's. However, stretching the limits of conditions under which these additives can be applied is a continuing challenge in the field of sulfur transfer technology. In this respect, significant advances are being made. As described in the foregoing, these include:

1. Introduction of more effective SOx absorbents to improve performance at high temperature and low oxygen concentrations.
2. Development of a carrier for the additives which may counteract the effects of silica poisoning.
3. Incorporation of co-promoters to facilitate the reduction of accumulated sulfates under low temperature reactor conditions.

Taken together, these improvements have two obvious benefits to refiners. First, the sulfur transfer additives may now find application in an increased number of FCC units, operating under a wider range of conditions and constraints. The second benefit will be to those already using transfer additives. Specifically, by making the additives as active, regenerable, and resistant to silica poisoning as possible, the amount required to achieve the desired sulfur oxide emission control can be minimized, to the obvious cost benefit of the end user.

REFERENCES

1. Bertolacini, G. Lehman, and E. Wollaston, United States Patent No. 38331, Sept 10 1974.
2. I. Vasalos, C. Hsieh, E. Strong, and G. D'Souza, Oil & Gas Journal, June 27 1977, p.141.
3. E. Hirschberg and R. Bertolacini in "Fluid Catalytic Cracking", Mario Occelli ed., ACS Symp. Series #375, American Chemical Society, Washington, D.C., Chapter 8, 114 (1988).
4. L. Rheaume and R. Ritter, ibid, Chapter 9. p. 146.
5. (a) D. McArthur, H. Simpson, and K. Baron, Oil & Gas Journal, February 23 1981, p. 55.
   (b) E.T. Habib, Oil & Gas Journal, August 8, 13, p. 111.
6. Evans and Wagman, J. Res. Nat'l Bur. Standards, **49**, 141 (1952).
7. J. Byrne, B. Speronello, and E. Leuenberger, Oil & Gas Journal, October 15 1984, p. 101.

MTBE and TAME: From Speciality Chemicals to Oil Commodities

C-P. Hälsig, R. Gregory, and T. Peacock

BP RESEARCH CENTRE, CHERTSEY ROAD, SUNBURY-ON-THAMES, MIDDLESEX TW16 7LN, UK

1 INTRODUCTION

MTBE and TAME have become well known abbreviations in the petroleum industry and are set to become even more widely known as the production of these two high octane gasoline components increases. Indeed, MTBE is already the fastest growing chemical of the last few years and will be the growth chemical of this decade. MTBE represents methyl tertiary butyl ether, a trivial name for 2-methoxy-2-methylpropane (Figure 1), which is a clear liquid of molecular weight 88 and boiling point 55°C, and contains 18% wt of oxygen. TAME represents tertiary amyl methyl ether, a trivial name for 2-methoxy-2-methylbutane (Figure 1), which is also a clear liquid of molecular weight 102 and boiling point 88°C, and contains 16% wt of oxygen.

The claim of growth chemical of the decade can be substantiated by the present and predicted world production of MTBE and TAME shown in Table 1. The present production of MTBE is of the order of 9M te/a and is expected to increase to 30M te/a by the year 2000. Similarly, TAME is also expected to increase from 600,000 to 3M te/a in the same period, although the estimates for TAME production are less certain. A comparison with some major bulk chemicals, also shown in Table 1, shows the tremendous increase in MTBE production the above estimates represent with the predicted production of MTBE for the year 2000 outstripping the present world propene production and the total amount of gasoline currently sold in the UK.

Today, all commercial MTBE processes react methanol

Table 1 Present and Predicted[a] World Production of MTBE and TAME (Data in million tonnes per annum)

	Year	Production	For Comparison[b] (Data for 1988)	
MTBE	1990	9	Ethylene	56
	1995	21	Propene	27
	2000	30	Benzene	21
			Methanol	20
			Acetic Acid	5
TAME	1990	0.6		
	2000	3	UK Gasoline Pool	24

[a] The average of many estimates in recent publications.
[b] Data mainly from Reference 1.

with isobutene, usually contained in a C_4 stream, to give MTBE (Equation 1) which is separated from the product stream by distillation. Similarly, TAME is produced by the reaction of methanol with 2-methylbut-2-ene and 2-methylbut-1-ene in a C_5 stream, and is also separated by distillation (Equation 2). A third branched pentene isomer, 3-methylbut-1-ene, is also present but is unreactive under processing conditions because it does not contain a fully substituted olefinic carbon atom. After MTBE and TAME production, the unreacted C_4 hydrocarbon streams generally go on to other processing, but the unreacted C_5 hydrocarbon streams are usually added directly to gasoline.

$$(CH_3)_2C{=}CH_2 + CH_3OH = (CH_3)_3COCH_3 \qquad (1)$$

$$\begin{matrix}(CH_3)_2C{=}CHCH_3 \\ \text{or} \\ CH_2{=}C(CH_3)CH_2CH_3\end{matrix} + CH_3OH = (CH_3)_2C(C_2H_5)OCH_3 \qquad (2)$$

2 HISTORICAL DEVELOPMENT OF ETHERS

Early Laboratory Development

Although ethers have been described in chemical literature from ca 1540[2], the first generally applicable synthesis was discovered by Alexander William Williamson in London in 1850[3] who made diethyl ether from potassium ethoxide and ethyl iodide (Equation 3). He was attempting to prepare higher alcohol homologues and was surprised when the reaction produced 'common ether'. The well-known Williamson synthesis has since become a standard laboratory method for ether preparation.

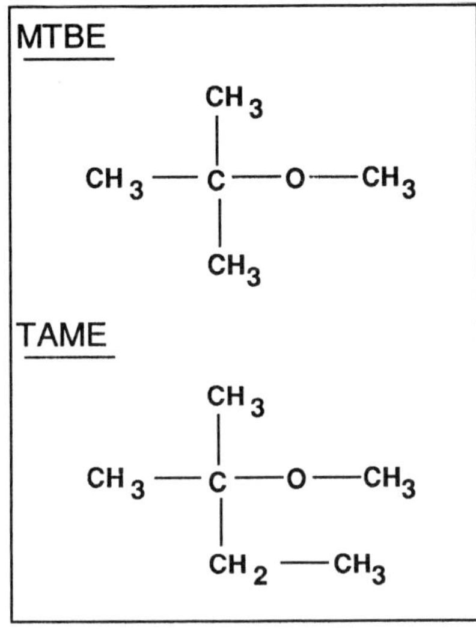

Figure 1 Structure of MTBE and TAME

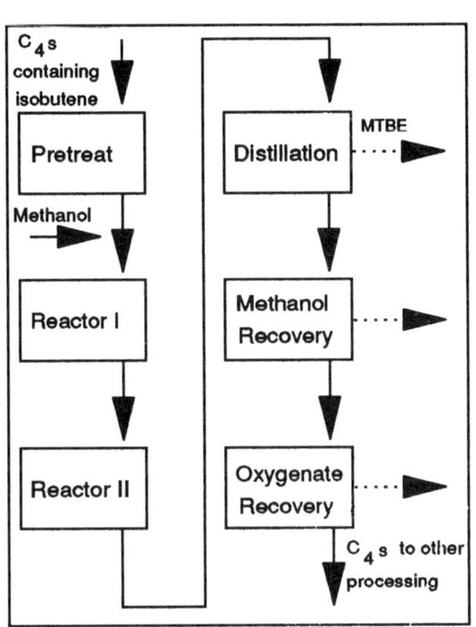

Figure 2 Schematic diagram of the MTBE process

$$C_2H_5OK + C_2H_5I = C_2H_5OC_2H_5 + KI \quad (3)$$

The first synthesis of a tertiary alkyl ether reported in Beilstein was that of t-amyl ethyl ether by Reboul and Truchot in 1867[4] using the Williamson synthesis (Equation 4). MTBE was reported to be first made by Lazinsky and Swadkowsky in 1903[5], again using the Williamson synthesis (Equation 5). Although the Williamson synthesis could be applied to the preparation of tertiary ethers, the yields were poor and the products difficult to purify. In the following year, 1904, the Belgian chemist, Louis Henry[6], increased the yield of MTBE to ca 60% using an improved Williamson synthesis (Equation 6). Henry's method became a standard laboratory route for tertiary ethers. Some improvement in the purity of the MTBE product was achieved in the following years[7,8], but it was not until 1932 that the Americans, J F Norris and G W Rigby, claimed the production of a pure sample of MTBE and other tertiary ethers using the condensation reaction between mixed alcohols[9].

$$(CH_3)_2CHCHBrCH_3 + C_2H_5ONa = (CH_3)_2C(C_2H_5)OC_2H_5 + NaBr \quad (4)$$

$$(CH_3)_3CCl + CH_3OK = (CH_3)_3COCH_3 + KCl \quad (5)$$

$$(CH_3)_3CONa + CH_3I = (CH_3)_3COCH_3 + NaI \quad (6)$$

The condensation reaction between two alcohols using a mineral acid catalyst, usually sulphuric acid, is another well-known laboratory route to ethers. The first tertiary ether made by this route was ethyl tertiary-butyl ether (ETBE) synthesised by the Russian chemist, Mamontow, in 1897[10] from tertiary-butanol and ethanol in the presence of sulphuric acid (Equation 7). The condensation reaction generally gives a mixture of all three possible ethers from mixed alcohols and the Williamson synthesis is normally preferred. For tertiary ethers, however, some workers prefer the condensation route[9] in preference to Henry's method claiming higher yields and a more easily purified product.

$$(CH_3)_3COH + C_2H_5OH = (CH_3)_3COC_2H_5 + H_2O \quad (7)$$

TAME was first synthesised by the Belgian chemist, A Reychler, in 1907[11] by the addition of methanol to 2-methylbut-2-ene (Equation 8). This route is similar to the present commercial one but sulphuric acid was used as catalyst instead of acid resin. Reychler's work was the first reported example of an ether made by the addition of an alcohol to an alkene, but this route was largely ignored in laboratories, presumably because of the relative scarcity of tertiary alkenes. In 1936, T W Evans and K R Edlund[12] working for Shell in the USA reinvestigated Reychler's method and applied it to a series of tertiary alkenes which were becoming increasing available. This was the first time that tertiary ethers had been readily prepared in a pure state and their physical properties were properly determined for the first time. The last major development was in 1949 when Atlantic Refining Company in the USA first used the then recently discovered organic matrix-based ion-exchange resins as solid-acid catalysts instead of sulphuric acid.

$$(CH_3)_2C{=}CHCH_3 + CH_3OH = (CH_3)_2C(C_2H_5)OCH_3 \quad (8)$$

Early Determination of Octane and Other Properties

Before 1936, all ethers were considered to have very poor octane quality. Diethyl ether had been tested by Ricardo in England and shown to have a very poor octane quality[13]. Diisopropyl ether had also been tested[13] and was erroneously found to be poor because no precautions had been taken to prevent the formation of peroxides (violent pro-knock agents) in the ether, which can easily be prevented by the addition of an anti-oxidising agent. In a meeting of the Society of Automotive Engineering (SAE) in W Virginia in 1936, H E Buc (Jersey Chemical Co) and E E Aldrin (US Air Force) stunned colleagues by showing that diisopropyl ether and, by inference, other branched ethers had a high octane quality and were excellent blending agents for aviation gasoline[14]. Following this discovery, further branched ethers were prepared and tested by Jersey in small laboratory engines, but the particular advantages of MTBE were missed. A little later, improved testing at Shell highlighted the superior quality of MTBE[13], and towards the end of the Second World War, further testing of branched ethers for blending in aviation gasoline confirmed this result[15,16].

In the 30 years from 1940 to 1970, little interest was shown in branched ethers. There was a small amount of engine testing as gasoline additives[17,18] and a few publications giving novel preparations and determining physical properties[19,20,21]. The production and octane quality of gasoline were increasing during this period but these were generally met with new processing and lead additives. There was no incentive to make ethers, and the availability of the necessary feedstocks was limited, since these are derived from other processing and are not naturally abundant.

3 MTBE PROCESS DEVELOPMENT

Catalyst Development

The catalyst used by practically all commercial processes are macroporous resins of polystyrene and divinylbenzene matrix with pendant sulphonic acid groups. These resin catalysts, in the shape of 0.3 to 1.5 mm diameter beads, behave as solid acids, and were discovered in 1948 and first used in the MTBE reaction in the following year. Only one plant built in Hungary used another catalyst, sulphuric acid, and this is reputed to have environmental problems. Many other catalysts eg zeolites[22,23] and others such as clays[24] or mixed oxides of wolfram and molybdenum[25] have been proposed and examined but not commercialised for this reaction.

Feedstock Availability

Methanol. Before 1980, methanol was more expensive than gasoline but the advent of large scale methanol plants (700,000 te/a) has elevated methanol into a very cheap, bulk commodity[26,27]. These plants convert natural gas (methane, ethane, and propane) to synthesis gas (a mixture of CO and hydrogen) followed by reaction to methanol, (eg Equation 9). Often CO_2 is added to attain the optimum C:H atomic ratio of 1:4[28]. Methanol's increasing use in the production of MTBE and TAME, about 25% of present output, will soon surpass its present major derivative, formaldehyde, and the predicted increase in MTBE and TAME production is likely to create a tighter methanol supply in the near future.

$$CO + 2H_2 = CH_3OH \qquad (9)$$

Isobutene. There are 4 main sources of isobutene - steam cracker (SC) C_4s, fluidised catalytic cracker (FCC) C_4s, the dehydration of tertiary butanol (TBA), and the isomerisation/dehydrogenation of field butanes.

Ethylene production in steam crackers produces a great deal of other by-products. One such by-product is a C_4 stream which makes up ca 10% wt of the products when a typical light naphtha feedstock is processed. The major component of the C_4 stream, butadiene (ca 40% wt), is a valuable chemical intermediate and is extracted, leaving a stream containing nearly 50%wt of isobutene (Table 2). This stream is the feedstock for the MTBE process; the unreactive C_4s pass through unchanged and can go on to other processing[29].

Refinery FCC units crack heavy gas oils to gasoline and produce a C_4 by-product stream which is of the order of 5 to 10% wt of the products, depending on the severity of the cracking operation. The whole C_4 stream, of which a typical composition is shown in Table 2, contains ca 20% wt of isobutene and is fed directly to the MTBE process. The unreactive C_4s pass on to other processing[29].

Isobutane is a co-reactant with propene in the ARCO propylene oxide (PO) process and is converted to tertiary butyl alcohol (TBA) during the process. The TBA by-product, which is produced with PO in the ratio of 3:1 by weight, is then dehydrated to isobutene, a pure feedstock for the MTBE process (Equation 10). At present, over 50% of MTBE is made from this source of isobutene but this route is not likely to increase significantly in the near future as it is dependent on the PO market.

$$(CH_3)_3COH = (CH_3)_2C{=}CH_2 + H_2O \qquad (10)$$

Field butanes have only recently been considered as a source of isobutene due to the high cost of processing, but this source is the only one capable of meeting expected future demand. This process involves the isomerisation of normal butane to isobutane, which is separated and dehydrogenated to isobutene (Equations 11 and 12).

$$CH_3CH_2CH_2CH_3 = (CH_3)_2CH_2CH_3 \qquad (11)$$

$$(CH_3)_2CH_2CH_3 = (CH_3)_2C{=}CH_2 + H_2 \qquad (12)$$

<u>Isopentenes</u>. There are two sources of isopentenes produced from SC and FCC processing. The SC C_5 stream is typically about 4% wt of the total products. The C_5 stream has a high pentadienes content (ca 40% wt) which, unlike butadiene, has no commercial value and these components are selectively hydrogenated in the C_5 stream to give further amounts of pentenes, and a typical partially hydrogenated stream is shown in Table 2. The active isopentene components comprise only ca 22% wt of the C_5 stream.

A FCC C_5 stream usually comprises ca 10% wt of the total FCC products and can be used directly. A typical composition is shown in Table 2. The active pentenes make up almost 30% of the C_5 feedstock stream. The minor amount of reactive pentadienes present, which can polymerise on the catalyst and shorten life, can be hydrogenated on the initial section of the catalyst bed by incorporating palladium metal and feeding sufficient hydrogen for the purpose[30].

Process Development

In 1970, ANIC (Italy) began to consider a MTBE process, with the aim of using its high blending octane quality to replace lead in gasoline[31,32]. At that time, environmental concerns over the presence of lead components in gasoline were turning into legislation limiting the use of lead. This work resulted in the first commercial MTBE plant in 1973, with a capacity of 100,000 te/a, built in Ravenna, Italy using a SC C_4 stream which, of course, is preferred to a FCC C_4 stream because it contains a greater isobutene content (Table 2).

Table 2 Typical Compositions of Streams Containing Isobutene (% wt)

	SC C_4s[a]	FCC C_4s
Isobutene	45	20
n-Butenes	47	39
Butanes	8	41
Butadiene	tr	tr

	SC C_5s[b]	FCC C_5s
2-Methylbut-2-ene	15.5	18.5
2-Methylbut-1-ene	7	9
3-Methylbut-1-ene	1.5	1.5
n-Pentenes	18	27
Pentanes	26.5	39.5
Cyclopentene	24	2.5
Cyclopentane	7	1
Pentadienes	0.5	1

[a] After butadiene extraction. [b] After selective hydrogenation.

Work at Hüls (Germany) began in 1972 with another reason for making MTBE - as a means of separating butenes - an otherwise difficult process. In this scheme, isobutene is removed from a SC C_4 stream by forming MTBE, making the other butenes easier to separate by distillation, and recovered by MTBE decomposition[33,34,35]. In 1976, Hüls brought a small MTBE plant of 60,000 te/a capacity on stream in Marl, Germany to produce pure butenes but also to add to gasoline to meet the lead reduction which was being introduced into W Germany at that time.

The first MTBE plant in the USA was built in 1979 by ARCO, now the world's leading MTBE producer, in Channelview, Texas making 200,000 te/a using isobutene from TBA[36]. In 1981, the first MTBE plant to be built in a refinery and using a FCC C_4 feedstock was built by Charter-International in Houston, Texas. This 50,000 te/a capacity plant was also the first to use catalytic distillation technology (discussed later). Modern MTBE plants of greater than 500,000 te/a are based on field butane, or on PO production. More field butane plants are being planned worldwide and especially in the USA.

Interest in the production of TAME has only recently developed, and the optimisation of the process has not yet completely evolved. Erdölchemie built the first TAME plant of capacity 1,500 te/a in Germany based on a SC C_5 stream in 1983. This first plant was not built to produce a gasoline additive but to separate pure isopentenes in a similar manner to the first Hüls MTBE process. Another first-time development in this plant was the catalyst which incorporated palladium to hydrogenate the small amount of dienes still present in the feed stream thereby improving catalyst life and product colour[30]. The first plant aimed at producing TAME for the gasoline market was in a mixed MTBE/TAME plant (Etherol Process). This plant was built by BP in Vohburg, Germany in 1986, with a capacity to produce 400,000 te/a of etherified FCC $C_4/C_5/C_6$ product[37]. The first solely TAME plant was built by Lindsey Oil in Killingholme, UK in 1987 producing 200,000 te/a based on a FCC C_5 stream. There are a few other plants producing TAME and several more are in design or under review.

Operating Conditions for MTBE Production

A schematic diagram of a MTBE plant is shown in Figure 2. The C_4 stream is pretreated to remove catalyst poisons, mixed with methanol in a ratio of methanol:isobutene of ca 1.1:1, and fed into the reactors. After reaction the higher-boiling mtbe is removed by distillation, the excess methanol is recovered and recycled, and traces of other oxygenates removed from the C_4 before it can be passed on for other processing.

Practically all MTBE processes have the same operating conditions. Reaction takes place in the liquid phase with a liquid hourly space velocity of 1 to 2. A pressure of greater than 7 bar is employed to ensure that liquid-phase is maintained. The addition reaction of methanol to isobutene is exothermic (δH=-60 KJ/mole), and the fact that heat of reaction has to be removed from the reactors has greatly influenced their design. Two reactors are used in series - the first primary reactor is maintained at ca 70°C where about 80% of the reaction occurs, and the second is generally operated at 50-60°C to maximise isobutene conversion to >98%. The isobutene conversion at equilibrium composition for different temperature and different methanol:isobutene feed ratios is shown in Figure 3. Increased isobutene conversion is attained at lower temperatures where reaction rates are slower, and the balance between conversion and reaction rate is optimised by the two reactor system at these temperatures.

Three different techniques are used to remove heat from the primary reactor - tubular reactors with

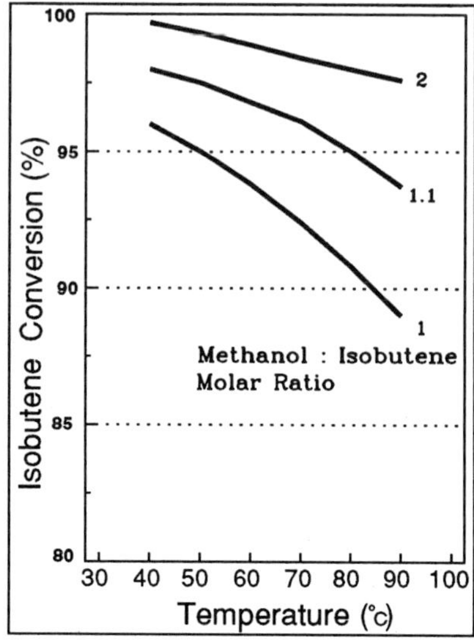

Figure 3 Variation of the equilibrium conversion of isobutene with temperature

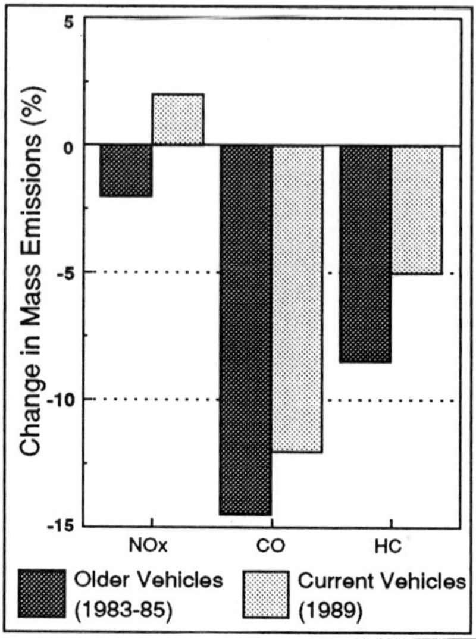

Figure 4 Effects on automotive emissions by the addition of MTBE (15% vol) to gasoline

circulating cooling water, dilution with the recycling of a cooled product stream, and the boiling-point reactor which ingeniously controls heat by vaporising and condensing the reactants. The finishing reactor can be a large cool fixed-bed or a catalytic distillation. Catalytic distillation has become increasingly employed in MTBE plants and consists of bales of catalyst centrally placed in a distillation column[38]. Reaction of methanol and isobutene occurs on the catalyst forming MTBE, which with its higher boiling point, descends down the column and disturbs the equilibrium composition in the reaction zone allowing further isobutene conversion. Practically complete isobutene conversion can be attained using this technique.

4 DRIVING FORCE FOR MTBE AND TAME PRODUCTION

Automotive exhaust emissions, and to a lesser extent gasoline evaporative losses, have long been known to cause atmospheric pollution. In the early seventies, concern about this pollution resulted in legislation to reduce the lead content in gasolines, and is still progressing today. The reduction of lead left

Table 3 Some Gasoline Related Properties of The Most Likely Oxygenated Gasoline Components

	MTBE	TAME	MeOH	EtOH
Oxygen Content (% wt)	18	16	50	35
Amount in Gasoline (% vol) corr. to 2.7% wt Oxygen (gasoline density = 0.75)	15	16.9	5.1	7.4
Approx. Blending Octane Quality in Unleaded Gasoline				
RON	118	111	133	130
MON	100	98	99	96
Blending Reid Vapour Pressure (psi)[39]	9	<3	12	10

refiners with an octane deficiency. In Europe, blending MTBE or other high octane oxygenates was one solution to this problem and also provided flexibility. At the same time in the USA, oxygenates were seen as a means to extend crude oil supplies. As other oxygenated contenders were shown to have major problems, eg methanol is extracted from gasoline with water and both methanol and ethanol have high blending vapour pressures in gasoline, interest in MTBE increased because of its excellent compatibility with gasoline. Some pertinent properties of MTBE and TAME are compared with other possible oxygenates for gasoline blending in Table 3. MTBE and TAME have high octane qualities, and low blending vapour pressures. It gradually became clear that gasolines containing oxygenates reduced exhaust emissions from internal combustion engines. For example, MTBE has been shown to reduce the amounts of carbon monoxide and hydrocarbons in exhaust emissions[40] from engines fitted with catalytic converters, shown in Figure 4.

In the late seventies, a few companies in the USA with a potential MTBE supply saw a lucrative market as a gasoline octane enhancer. The non-oil based additives that can be used in US gasolines are controlled by the US Environmental Protection Agency (EPA) established by the 1970 Clean Air Act. In 1979, the EPA granted a waiver to allow 7% vol of MTBE to be added to gasoline and only recently, in 1988, was this increased to 15% vol. Some areas in the USA with severe air pollution have imposed local restrictions on the composition of gasoline, and some enterprising companies introduced reformulated gasolines in these areas.

More recently, the 1990 Amendments to the Clean Air Act in the USA will cause a dramatic increase in future MTBE and TAME production. The Clean Air Act sets legal standards for the quality of air. Certain areas in the USA have air polluted with ozone and carbon monoxide above these legislated levels. Carbon monoxide is produced directly from automotive exhausts, and ozone is formed by the action of sunlight on hydrocarbon and nitrogen oxides emissions. For areas which have ozone and carbon monoxide above legal levels (so-called ozone non-attainment areas and CO non-attainment areas), new laws will require reformulation of the gasoline sold there in an attempt to reduce pollution. The main points are summarised in Table 4.

The regulations in the CO non-attainment areas will have the greatest effect on demand for oxygenates with gasolines set to have an oxygen content of 2.7% wt. It is expected that the oxygen requirement will be supplied by 2 major components, MTBE and ethanol, although there is likely to be some contribution from ETBE, methanol/TBA mixtures and TAME. If the required oxygen were supplied only by MTBE then the requirement for MTBE would be some 1.5M te/a.

MTBE is the favoured oxygenate component for reformulated gasolines because of its superior gasoline compatibility. The dramatic developments in gasoline compositions in the USA as a result of the Clean Air Act

Table 4 USA Clean Air Act Requirements for Gasolines Sold in Certain Areas of the USA

CO Non-attainment Areas	Ozone Non-attainment Areas
Areas where CO in air above legal limit. Includes 41 cities, 29% of US gasoline demand.	Nine areas where ozone in air above legal limit. 19% of US gasoline demand.
Must contain 2.7% wt Oxygen	Must contain 2.0% wt Oxygen
Beginning before Nov.1st 1992 for minimum 4 month period	Beginning Jan.1st 1995 Benzene 1% vol, Aromatics 25% vol Volatiles 15% reduction by 1995 25% reduction by 2000 No lead

are likely to spread to the rest of the world. Thus, the prediction of the enormous increase in MTBE production is firmly based.

REFERENCES

1. Basic Petrochemicals Production Statistics 1984-1989, APPE, ECIF, Brussels, Belgium, 1990.

2. Beil. 1,314.

3. A. W. Williamson, Philosophical Magazine, 1850, 37(iii), 350.

4. Beil. 1,389.

5. Beil. 1,381.

6. L. Henry, Rec. Trav. Chim., 1904, 23, 326.

7. G. M. Bennett and W. G. Philip, J. Chem. Soc., 1928, 1930.

8. P. A. K. Clusius, J. Chem. Soc., 1930, 2607.

9. J. F. Norris and G. W. Rigby, J. Amer. Chem. Soc., 1932, 54, 2088.

10. W. Mamontow, J. Russ. Phys. Chem. Soc., 1897, 29, 230.

11. A. Reychler, Bull. Soc. Chim. Belg., 1907, 21, 71.

12. T. W. Evans and K. R. Edlund, Ind. Eng. Chem., 1936, 28, 1186.

13. R.Schlaifer and S. D. Heron, 'Development of Aircraft Engines and Fuels', Andover Press Ltd, Andover, Massachusetts, 1950, Page 609.

14. H. E. Buc and E. E. Aldrin, S.A.E.Trans, 1936, 234.

15. H. C. Barnett and J. W. Slough, Natl. Advisory Comm. Aeronautics Wartime Rept., 1944, ACR No E4H10, 6 pp.

16. H. C. Barnett, C. L. Meyer and A. W. Jones, Natl. Advisory Comm. Aeronautics Wartime Rept., 1944, ACR No E4HO3, 10 pp.

17. I. L. Drell and J. R. Branstetter, Natl. Advisory Comm. Aeronautics, 1950, Tech. Note No 2070, 58 pp.

18. D. G. Roddick and L. B. Scott, 1962, Ger. 1,127,141.

19. W. T. Olson, H. F. Hipsher, C. M. Buess, I. A. Goodman, I. Hart, J. H. Lamneck Jr., and L. C. Gibbons, J. Amer. Chem. Soc., 1947, 69, 2451.

20. F. W. McLafferty, Anal. Chem., 1957, 29, 1782.

21. H. A. Ory, Anal. Chem., 1960, 32, 509.

22. P. Chu and G. H.Kühl, Ind. Eng. Chem. Res., 1987, 26, 365.

23. S. I. Pien and W. J. Hatcher, "Synthesis of Methyl Tertiary Butyl Ether on HZSM-5 Zeolite," Annual Meeting of A.I.Ch.E., 1988.

24. British Petroleum Patent, 1980, EP 0 031 687.

25. A. K. K. Lee and A. Al-Jarallah, Chem. Econ. Eng. Rev., 1986, 18(9), 25.

26. A. B. Stiles, A.I.Ch.E., 1977, 23, 362.

27. J. R. Crocco, "Will There Be Enough Methanol," 1991 NPRA Annual Meeting, March 17-19, 1991, San Antonio, Texas.

28. A.B. Stiles, Al. Ch. E., 1977, 23, 362.

29. G. Emmrich and K. Lackner, Hydrocarbon Processing, Jan. 1989, 71.

30. Prichard, "Novel Catalyst Widens Octane Opportunities", NPRA Annual Meeting, 1987, San Antonio, Texas, March 29-31.

31. G. Pecci and T. Floris, Hydrocarbon Processing, Dec., 1977, 98.

32. F. Ancillotti, M. M. Maura, and E. Pescarollo, J. Catal., 1977, 46, 49.

33. F. Obenaus and W. Droste, Erdöl und Kohle Erdgas, 1980, 33(6), 271.

34. A. Convers, B. Juguin, and B. Torck, Hydrocarbon Processing, Mar. 1981, 95.

35. Chemische Werke Hüls, "Methyl tertiary Butyl Ether as Gasoline Component, Der Lichtbogen, 1976, No 181, 4.

36. R. Landau, G. A. Sullivan, and D. Brown, Chem. Tech., Oct., 1979, 602.

37. C-P. Hälsig, B. Schleppinghoff, and D. J. Westlake, "Erdolchemie and BP Ether Technology More Scope for Octane Boost", NPRA Annual Meeting, 1986, Los Angeles, California.

38. W. P. Stadig, Chemical Processing, 1987, Feb.

39. American Petroleum Institute, "Alcohols and Ethers," API Publication 4261, 1988, Second Edition, July.

40. K. L. Rock, R. O. Dunn, and D. J. Makovec, "Automotive Fuels for an Improved Environment. How Does MTBE Contribute?", NPRA Annual Meeting, 1991, San Antonio, Texas.

Zeolites and Their Use in Petroleum Refining

J. Weitkamp and S. Ernst

INSTITUTE OF CHEMICAL TECHNOLOGY I, UNIVERSITY OF STUTTGART, PFAFFENWALDRING 55, D-W-7000 STUTTGART 80, GERMANY

1 INTRODUCTION

The first large scale application of a zeolite catalyst in petroleum refining dates back about thirty years. At this time, rare earth exchanged X-type zeolites started to replace the formerly used amorphous silica-alumina catalysts in fluid catalytic cracking (FCC) of vacuum gas oil. [1-3] This resulted in considerably improved yields of the desired gasoline and, hence, in a much more efficient utilization of the crude oil.

Today, FCC is by far the single most important process which makes use of zeolite catalysts. [4,5] Besides, however, zeolitic catalysts are employed in a variety of other refinery processes, [6] such as hydrocracking of vacuum gas oil, [7] the isomerization of light gasoline [8] and dewaxing of distillate fuels or basestocks for lubricating oils. [9] In all these applications, the zeolite catalyst is used in an acidic form. In some instances, this acidic form is modified by a hydrogenation/dehydrogenation component, whereby a bifunctional catalyst results. Among the advantages of zeolites over conventional solid catalysts are their strictly regular pore size and geometry, their stability up to high temperatures and their acid properties which can be tailored to a large extent. Moreover, the pore width of zeolites is in the same order as molecular dimensions which enables shape or size selective catalysis. So far, the number of crystallographically different zeolite catalysts used in petroleum refining has been remarkably small, just faujasite, mordenite, ZSM-5 and, perhaps, erionite are employed on an industrial scale. This situation could change, however, in the future.

Another important application of zeolites in petroleum refining is their use as selective adsorbents in purification and separation processes. Practically all these adsorbents are based on zeolite A, and of particular interest is the separation of n-alkanes from iso-alkanes on zeolite 5A which acts as a molecular sieve.

The present paper intends to give an overview on these applications of zeolites in modern petroleum refining. After an introduction into the building principles and crystallographic structures of zeolites, their large scale applications in petroleum refining are discussed. In a final paragraph, the potential of zeolites for future refinery processes will be outlined.

2 ZEOLITES AND ZEOLITE-LIKE MATERIALS

The history of zeolites (Table 1) began as early as 1756 when the Swedish baron *A. Cronstedt* discovered the mineral stilbite. Upon heating, the material released the water adsorbed in its pores and the solid particles were ebullated by the water vapour, hence Cronstedt coined the term zeolite (boiling stone). Modern zeolite science was initiated in 1938 by *R.M. Barrer* [10,11] who, in his pioneering work, began to synthesize zeolites and placed zeolite science on a physico-chemical basis. Considerable progress was made in the hydrothermal synthesis of zeolites around 1950 by *R.M. Milton, D.W. Breck et al.* [12] at Union Carbide Corporation. They succeeded in synthesizing zeolites A and X. In 1962, zeolite X exchanged with trivalent rare earth cations, was introduced commercially as a catalyst in FCC. Soon afterwards, rare earth Y zeolites were recognized to be more stable and superior. The economic benefit brought about by these new cracking catalysts was so tremendous that the introduction of zeolites into FCC has often been referred to as a true technical revolution. The next outstanding event was the discovery of zeolite ZSM-5 by researchers at Mobil Oil Corporation. Today, ZSM-5 is considered as the prototype of shape selective catalysts and employed in an impressive number of large scale processes, both

Table 1 Landmarks in zeolite science and technology.

- 1756:	A. Cronstedt discovers stilbite. Upon heating, the mineral desorbs water vapour and is ebullated (zeolite = "boiling stone")
- ca. 1940:	R.M. Barrer starts with hydrothermal synthesis of zeolites
- ca. 1950:	Union Carbide Corp. (R.M. Milton, D.W. Breck and others) succeeds in synthesizing zeolites A and X
- 1962:	Mobil Oil Corp. introduces rare earth X (REX) containing cracking catalysts. Shortly afterwards, REX is replaced by REY
- ca. 1970:	Mobil Oil Corp. discovers zeolite ZSM-5
- ca. 1980:	Union Carbide Corp. (E.M. Flanigen, S.T. Wilson and others) discovers the family of microporous alumophosphates ($AlPO_4$ s, SAPOs etc.)
- 1988:	M.E. Davis recognizes that VPI-5 is a super-large pore molecular sieve material

in petroleum refining and in the manufacture of petrochemical commodities, such as ethylbenzene or para-xylene. In the early 1980's *E.M.Flanigen, S.T. Wilson et al.* [13-15] at Union Carbide Corporation discovered a whole family of microporous alumophosphates ($AlPO_4$s) and chemically related materials (e. g. silicoalumophosphates, SAPOs) which may contain a variety of other elements from the periodic table on tetrahedral positions in the lattice. While, up to now, no commercial applications of these new materials are known, their potential as catalysts in petroleum refining is considered to be high. [16] For a long time, all known zeolites could be classified into small pore materials (pores formed by eight tetrahedra), medium pore materials (pores formed by ten tetrahedra) and large pore materials (pores formed by twelve tetrahedra). A few years ago, however, *M.E. Davis and coworkers* recognized, [17,18] that the alumophosphate VPI-5 possesses 18-membered ring pores. Since then, the generic term *"super-large pore materials"* has been in use for all microporous solids whose pores are formed by more than twelve tetrahedra. One potential application for such materials is catalytic upgrading of the bulkier molecules occuring in the residua of crude oil distillation.

The scientific definition of the term "zeolite" continues to be a matter of debate, [19-23] especially among mineralogists and crystallographers. For practical purposes, however, the following broadest definition seems to be appropriate: *A zeolite is a microporous crystalline material with a framework forming regularly shaped channels and/or cages of molecular dimensions (ca. 0.3 to 1.5 nm cf. Figure 1). These channels and cages contain water which, upon heating, can be*

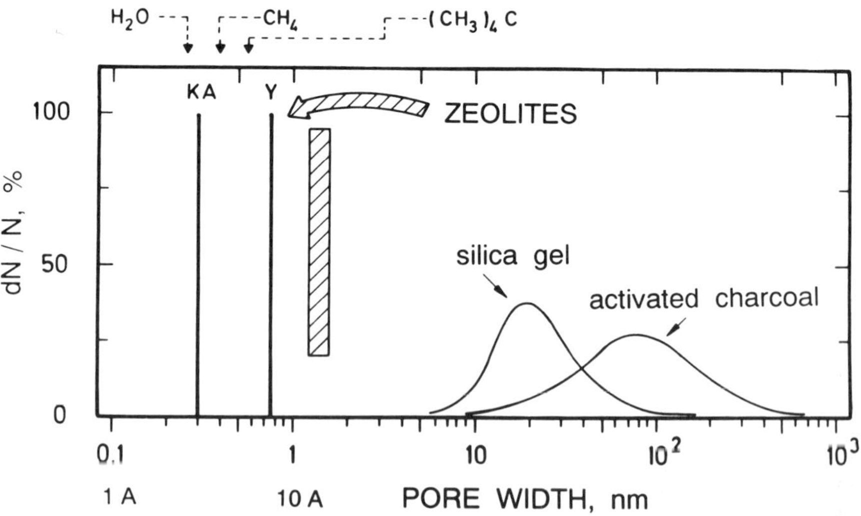

Figure 1 Schematic representation of the pore size distribution in zeolites and conventional porous solids and dimensions of some simple molecules after *D.W. Breck.* [12]

desorbed without collapse of the structure. Moreover, the channels may contain small and exchangeable cations which compensate the negative framework charge. This definition encompasses all chemical compositions, in particular, it is not confined to alumosilicates. The science and application of all materials covered by the above definition have been the subject of the International Zeolite Conferences organized regularly by the International Zeolite Association (IZA).

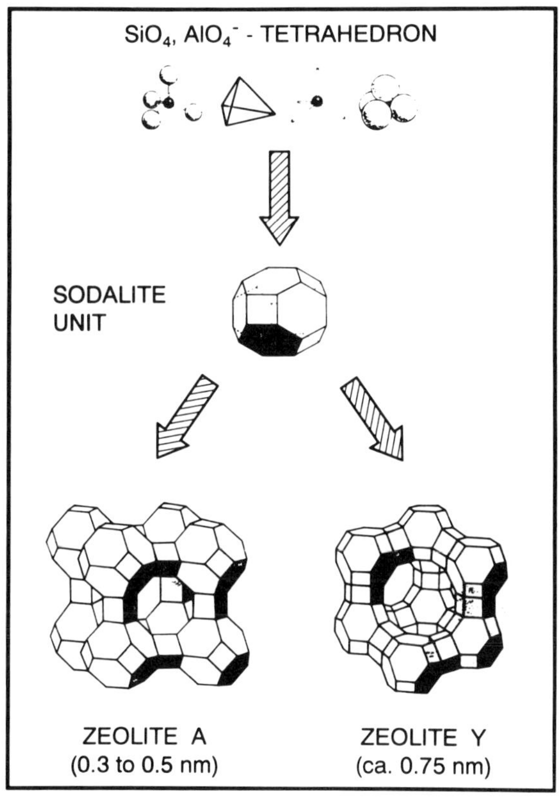

Figure 2 Building principle of the framework and structures of zeolites A and Y (faujasite).

The general building principle of zeolites and two important structures are depicted in Figure 2. The smallest building units are tetrahedra in which a central metal ion like Si^{4+} or Al^{3+} (often referred to as the T-atom) is tetrahedrally surrounded by four O^{2-} ions. A given oxygen ion is shared by two adjacent tetrahedra. This way, subunits are formed which consist of a limited number of linked tetrahedra. One example is the sodalite unit shown in Figure 2, which is composed of 24 tetrahedra. In this and similar drawings, the vertices (or corners) represent positions of the T-atoms while two adjacent tetrahedra are linked together along an edge via a common oxygen ion. The six quadratic and eight hexagonal faces of the sodalite unit circumscribe a near-spherical cage, the so-called sodalite or β-cage with a free diameter of 0.66 nm.

If sodalite units are linked together at their six quadratic planes via quadratic prisms, the structure of zeolite A results. It contains large cages (the so-called α-cages), again with a near-spherical shape and a free diameter of 1.14 nm. Each of these cages is connected with six neighbouring α-cages via 8-membered ring (8-MR) windows with a crystallographic diameter of 0.41 nm. Actually, the effective diameter and, hence, the critical pore width of zeolite A can be manipulated by proper choice of the nature of the cations which are required to compensate the negative lattice charge brought about by each $AlO_{4/2}^-$-tetrahedron. The effective pore widths of zeolites KA (or, synonymously, 3 Å), NaA (or 4 A) and CaA (or 5 A) are ca. 0.3, 0.4 and 0.5 nm, respectively.

If, on the other hand, sodalite units are linked together at four of their eight hexagonal planes via hexagonal prisms (cf. Figure 2), the structure of zeolite Y results. Zeolite Y is isostructural with the mineral faujasite and zeolite X, an aluminium-rich variant of zeolite Y. It contains again large, near-spherical cages (the so-called supercages) with a free diameter of 1.3 nm. Each supercage is connected tetrahedrally with four neighbouring supercages via 12-membered ring (12-MR) windows with a crystallographic diameter of 0.74 nm. For most molecules, except very bulky ones, zeolite Y offers a spacious cage and pore system through which they can diffuse without hindrance.

Figure 3 shows the building principle and channel system of zeolite ZSM-5, the most important microporous material with 10-membered ring (10-MR)

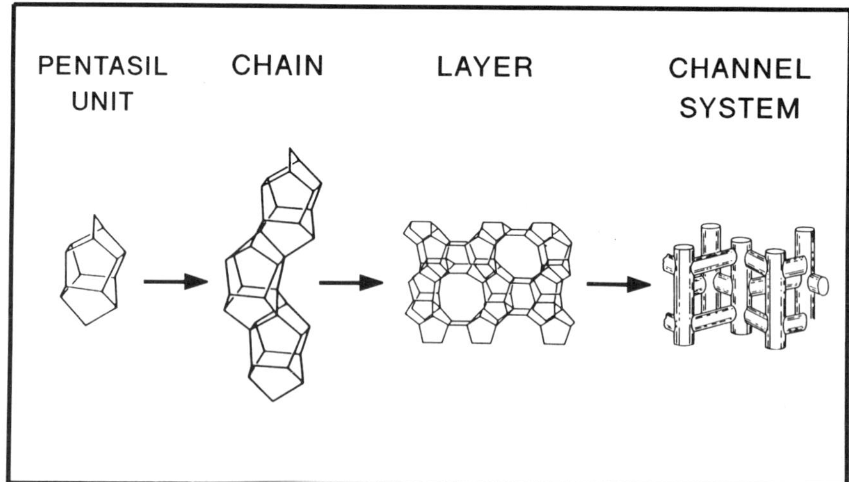

Figure 3 Building principle and pore system of zeolite ZSM-5.

pores. This time, 14 tetrahedra form a subunit (referred to as the pentasil unit) with eight faces which are all pentagonal. Linear linkage of pentasil units results in chains. By mirror imaging of these chains into the second dimension, layers are generated which already contain the 10-MR pore opening. Stacking these layers together into the third dimension via inversion finally gives the framework structure of ZSM-5. It contains two sets of 10-MR-channels, running perpendicularly to each other through the lattice. One set of channels is straight with a slightly elliptical cross section (0.51 x 0.55 nm), whereas the second set of sinusoidal (or zigzag) channels has an almost circular cross section (0.53 x 0.56 nm).

Beside silicon and aluminium, a variety of other elements can occur on tetrahedral lattice positions. Of particular importance is the family of alumophosphates (cf. Figure 4). [13-15] Some members of this large family possess structures which were already known from the alumosilicate zeolites. An example is SAPO-42 which is isostructural with zeolite A. [24] There are other members, however, for which no structural analogues are known so far among the alumosilicates. An impressive example is VPI-5, the first super-large pore zeolitic material. [17,18] The cross section of its channels (cf. Figure 5) which are formed by an 18-membered ring exhibit a diameter of ca. 1.2 nm. VPI-5 tends to undergo a solid state transformation into $AlPO_4$-8, even at relatively low temperatures around 100 °C. [25,26] $AlPO_4$-8 is itself an interesting super-large pore (14-MR) material. The cross section of its channels is elliptical with dimensions of 0.79 x 0.87 nm. [27]

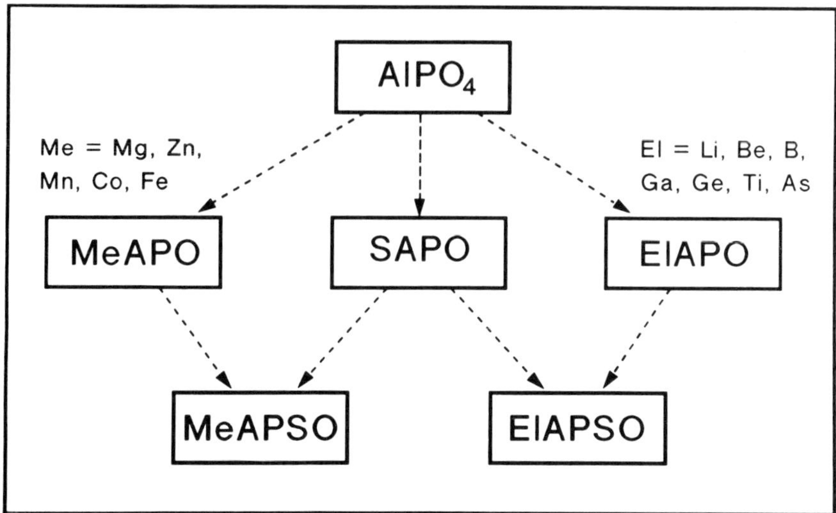

Figure 4 The family of alumophosphates ($AlPO_4$s), silicoalumophosphates (SAPOs) and related materials after *E.M. Flanigen et al.* [15]

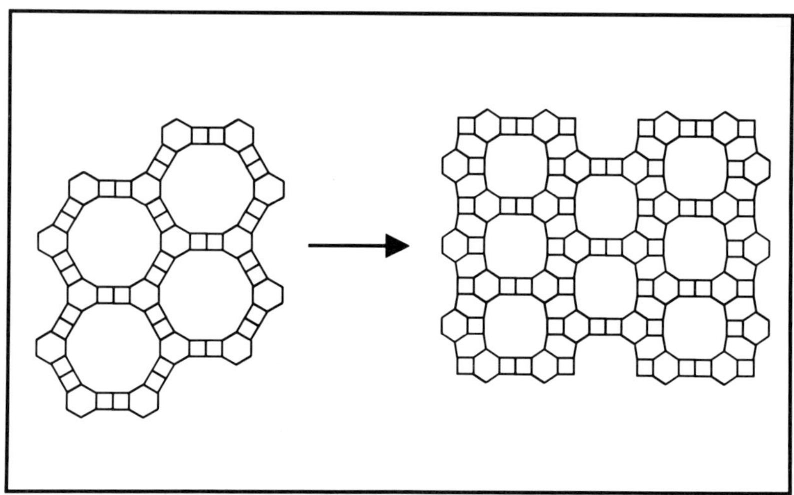

Figure 5 Solid state transformation of VPI-5 into $AlPO_4$-8.

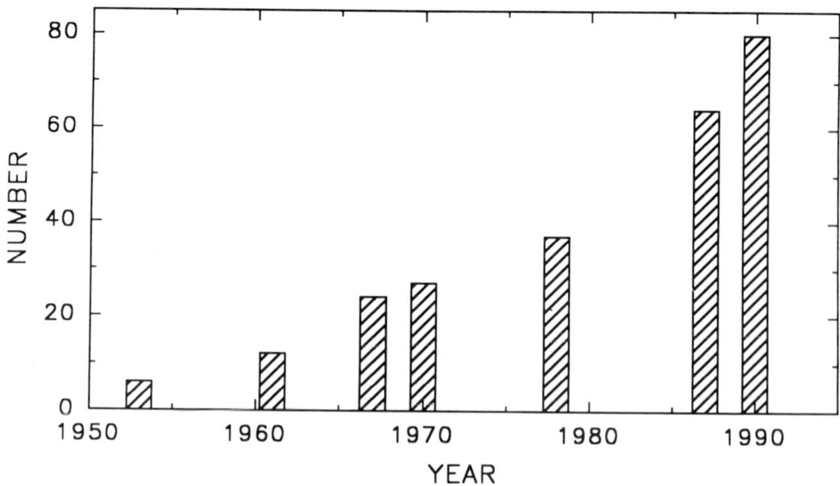

Figure 6 Number of known zeolite structures after *W.M. Meier*.[28]

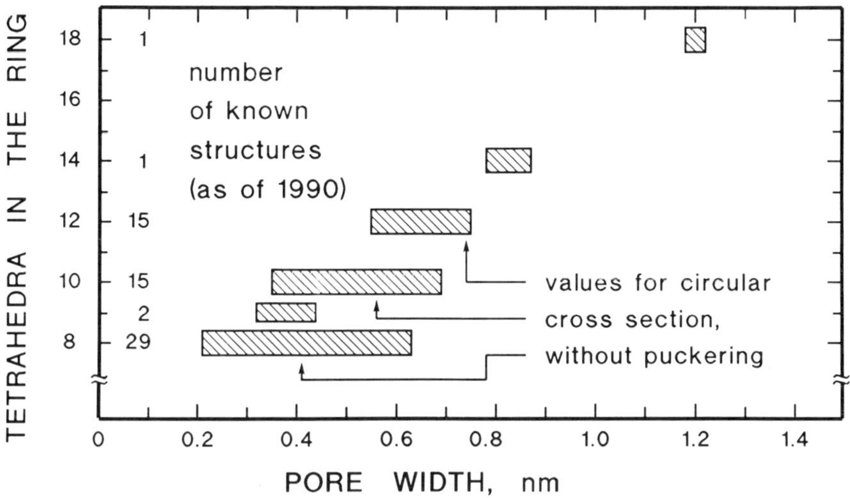

Figure 7 Range of crystallographic pore diameters covered by zeolites after *W.M. Meier.*[28]

While the number of zeolites used in commercial refinery processes is relatively small, an impressive number of new zeolites has been discovered in recent years. Today, ca. 80 crystallographically different zeolites are known (Figure 6) and have received approval by the Structure Commission of the International Zeolite Association.[29] The search for novel applications of these fascinating materials in catalysis and adsorption has just begun. In toto, they cover a broad range of crystallographic pore widths (Figure 7) and it remains to be seen how many of these new zeolites will find commercial applications in the future.

3 PETROLEUM REFINING PROCESSES USING ZEOLITES

Molecular Sieve Separations

n-Alkanes are readily adsorbed on zeolite 5 A (zeolite A exchanged with Ca^{2+} ions) whereas the bulkier iso-alkanes, cycloalkanes etc. do not have access to the cavities of this narrow pore zeolite. The molecular sieve effect is demonstrated with n-hexane and 3-methylpentane as model compounds in Figure 8. Under the conditions applied in this study, n-hexane is quantitatively removed from the gas stream for more than 75 min, while 3-methylpentane breaks through immediately.

The molecular sieve separation is applied in two areas of petroleum refining.

n-Alkanes with their poor octane numbers are removed from light gasoline, which is essentially the C_5/C_6-hydrocarbon stream of a refinery. The Union Carbide ISOSIV process which is carried out in the gas phase, [30-32] usually brings about an increase in the research octane number of ca. 5 to 10. [6,8] Another application is UOP's Molex process [33] which has been used in refineries for a long time for the removal of C_{10}- to C_{20}-n-alkanes from kerosene and gas oils. The n-alkanes in this carbon number range are raw materials for the production of linear alkylbenzenesulfonates (LABS) which

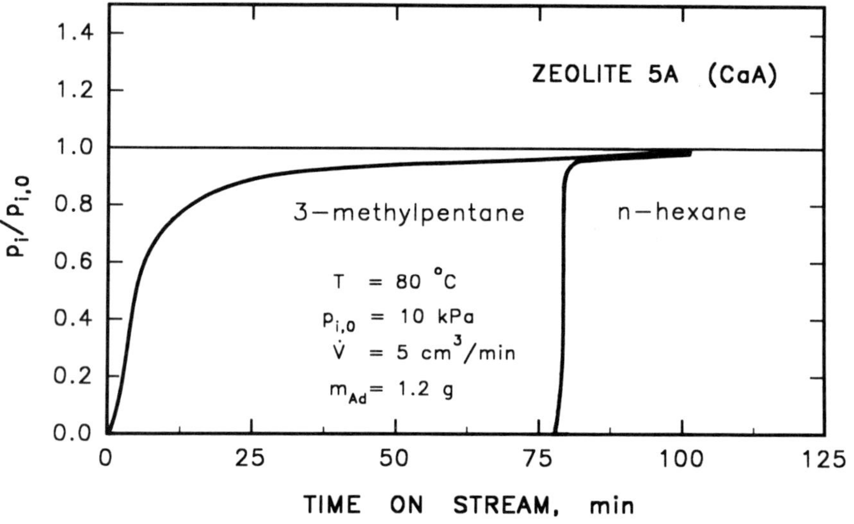

Figure 8 Laboratory scale demonstration of the molecular sieve separation of n-hexane and 3-methylpentane on zeolite 5A (breakthrough curves measured in a fixed-bed adsorber with N_2 as carrier gas; $p_{i,0}$ is the partial pressure at the adsorber inlet, p_i the partial pressure at the adsorber outlet).

are biodegradeable components in detergents. At the same time, removal of n-alkanes from kerosene or gas oil improves their cold flow properties. In the Molex process, the hydrocarbons are in the liquid phase, and use is made of UOP's Sorbex technology. [34]

Isomerization of Light Gasoline

Instead of removing the low octane n-alkanes from light gasoline, they can be catalytically isomerized into the more valuable iso-alkanes. In recent years this skeletal isomerization of the light gasoline components has experienced a remarkable boom in connection with the more stringent lead legislation. The catalytic isomerization of alkanes is carried out in the presence of hydrogen, on bifunctional catalysts comprising both an acidic and a hydrogenation/dehydrogenation component.[35] An example for such catalytic systems is zeolite mordenite in its Bronsted acid form loaded with small amounts (ca. 0.3 to 0.5 wt.-%) of platinum. Pt/H-mordenite is used as a commercial isomerization catalyst in the Shell Hysomer process.[36,37] Mordenite is a large pore (12-MR) zeolite with unidimensional, non-intersecting channels. The cross section of its pores is slightly elliptical with dimensions of 0.65 x 0.70 nm. The isomerization of alkanes is a typical equilibrium reaction, and hence, a complete conversion of the n-alkanes cannot be achieved. Therefore, in many large-scale units, the Shell Hysomer process is combined with an ISOSIV separation, and the unconverted n-alkanes are recycled. This process configuration is usually referred to as *Total Isomerization Process (TIP)*. Upgrading light gasolines by TIP usually results in an increase of the research octane number of 15 to 20.[8,37-40]

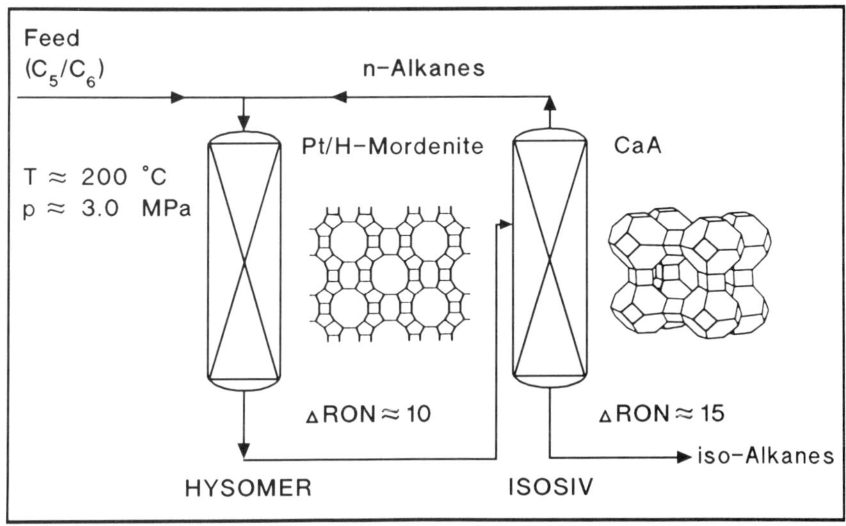

Figure 9 Isomerization of light gasoline by the combined Hysomer and ISOSIV processes (Total Isomerization Process, TIP).

Fluid Catalytic Cracking (FCC)

Fluid catalytic cracking is the single most important application of zeolite catalysis. The world-wide process capacity is in the order of $500 \cdot 10^6$ t/a. Typically, FCC converts vacuum gas oil into LPG hydrocarbons, gasoline, heavier fractions (referred to as light and heavy cycle oil) and coke, but selected residua have also been converted by FCC, either as such or in admixture with vacuum gas oil. A simplified process scheme of a modern FCC unit is shown in Figure 10.[3-5, 41]

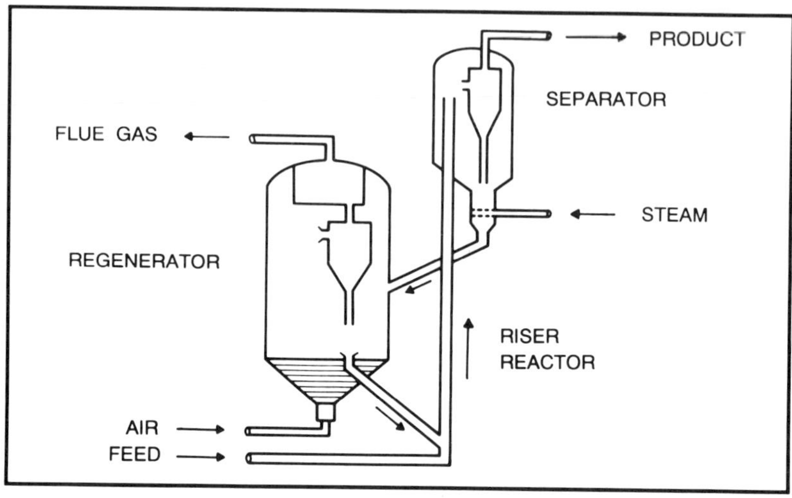

Figure 10 Fluid catalytic cracking unit with a riser reactor.

The powdered catalyst (particle size around 60 μm) which contains ca. 20 to 40 wt.-% of an acid Y type zeolite in a non-zeolitic matrix, is circulated continuously through the unit. At the bottom of the riser reactor, the heated (T$\approx$ 700 °C) catalyst is brought into contact with the preheated feed oil, whereby the oil is further heated to the reaction temperature of ca. 500 °C and vapourized. The catalyst/hydrocarbon mixture flows upwards inside the riser reactor where the endothermic cracking reactions occur. The residence time is in the order of a few seconds. It is a salient feature of catalytic cracking that, beside the volatile hydrocarbon products, coke is formed and deposits inside the catalyst pores which brings about a deactivation. In the separator, the gaseous products are removed and sent to the distillation train whereas the coked catalyst is stripped with steam to desorb higher hydrocarbons. The catalyst is then transferred to the regenerator, where the coke is burnt off in a fluidized bed at elevated temperatures around 700 °C. This way, the coke content of the catalyst is strongly reduced to ca. 0.1 wt.-% or even lower, depending on the operating conditions in the regenerator. Moreover, from an engineering point of view, the combustion of coke, i. e. the product with the lowest value, furnishes large amounts of heat which can be efficiently used inside the unit for heating and vapourizing the vacuum gas oil and for supplying the enthalpy for the endothermic cracking reactions in the riser. The reactivated and heated catalyst powder is withdrawn continuously from the regenerator and flows downwards to the bottom of the riser reactor, whereby the cycle is closed.

Typical distributions of the cracked products obtained from vacuum gas oils on catalysts of different generations are shown in the left-hand part of Figure 11. It is evident that, compared to the amorphous SiO_2-Al_2O_3 type catalysts used thirty years ago, the zeolites give considerably larger yields of gasoline which is the product of the highest value. Today, the typical gasoline yield is 50 to 55 wt.-%. A look at the typical compositions of the gasolines produced on amorphous SiO_2-Al_2O_3 and zeolites (Figure 11, right-hand part) immediately reveals the following salient features: The gasoline obtained with zeolitic cracking catalysts contains much more aromatics and paraffins, at the expense of naphthenes and olefins. These compositional differences have significant repercussions on the octane number of the gasoline. With the amorphous catalysts the research octane number (RON) and motor octane number (MON) typically amounted to 95 and 82, respectively; upon introducing REY catalysts, RON and MON dropped to ca. 88 - 90 and 78 - 81, respectively. [3,8,43] With the advent of lead phase-out, these relatively poor octane numbers of FCC gasoline became a problem. Later, it was found that the use of dealuminated Y type zeolites, in which the density of acid sites is reduced, results in enhanced research octane numbers, probably due to increased olefin contents, and somewhat lower coke yields. For all these reasons, the rare earth content in

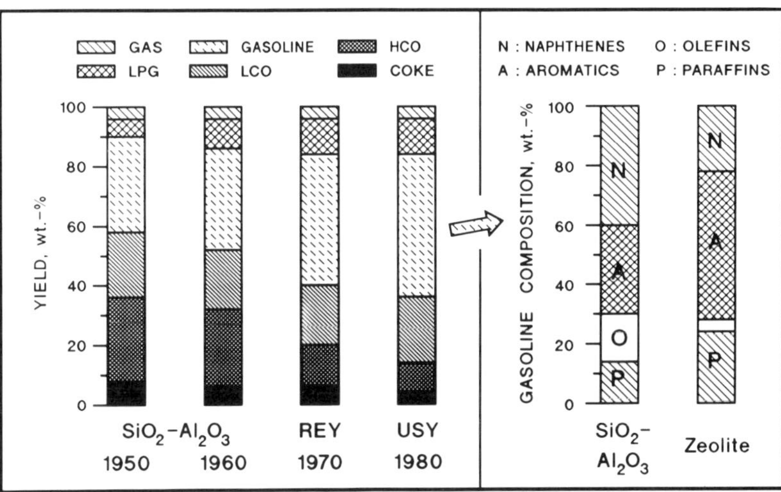

Figure 11 Typical product yields in catalytic cracking of vacuum gas oil (LPG: liquefied petroleum gas; LCO: light cycle oil; HCO: heavy cycle oil; REY: zeolite Y containing rare earth cations; USY: ultrastable zeolite Y); adapted from refs. 3, 42 and 43.

FCC catalysts has been declining in recent years. All these developments led to a significant increase in the average RON of FCC gasoline; the MON values, however, continue to be a problem, and large efforts are being undertaken by catalyst manufacturers and the petroleum industry to solve or mitigate this MON problem. [44-47]

A completely different approach for improving the octane number of FCC gasoline is the addition of small amounts of a zeolitic additive, viz. HZSM-5, either to the FCC catalyst or separately to the FCC unit. The medium pore zeolite ZSM-5 (cf. Figure 3) in its acid form acts as a *shape selective co-catalyst.*

One reaction catalyzed by HZSM-5 in the FCC unit is the selective cracking of straight-chain components with their very poor octane numbers. While, as a whole, the effect of HZSM-5 co-catalysts on the chemistry of FCC is complex, their seems to be agreement as to its octane enhancing potential: RON and MON can be increased by ca. 1 to 3 points, [48-50] but taking into account that the principle effect of the co-catalyst is to remove undesired components, it is clear that the gain in octane rating is inevitably accompanied by a loss in the gasoline yield.

The yield and octane number of the gasoline are just two process characteristics which can be influenced by the properties of the zeolite catalyst. Over the years, an impressive number of additional catalyst related improvements could be achieved. One example is the diminution of the concentration of pollutants in the flue gas from the regenerator. For example, the emission of carbon monoxide can be almost completely avoided by introducing small amounts (in the order of 1 wt.-ppm) of a noble metal like platinum into the zeolite cavities in an appropriate manner. On such catalysts, CO is completely oxidized to CO_2 inside the regenerator, but the noble metal does not exert any detrimental effect on the cracking reactions in the riser reactor. [51-53]

Another example is the so-called SO_x-transfer. Typically, some 5 to 10 wt.-% of the sulfur introduced into the FCC unit with the vacuum gas oil end up in the coke deposits on the catalyst while ca. 50 wt.-% are converted into H_2S in the riser reactor. H_2S is conveniently removed from the gaseous hydrocarbon products without excessive costs. The sulfur in the coke, however, is converted into SO_x in the regenerator, upon combustion of the coke. Removal of SO_x from the large flue gas streams would be very expensive. An alternative for cleaning the flue gas is SO_x-transfer which works on special metal oxides added to the catalyst. In the oxidative environment, these additives bind SO_x, presumably under the formation of the metal sulfate. Eventually, the sulfur is thus transported with the circulating catalyst into the riser reactor with its reducing atmosphere. Either in the riser or upon stripping in the separator, the sulfur is released as H_2S, leaves the separator together with the hydrocarbon products, and this additional H_2S can usually be handled without difficulties in the existing gas treating section. [54-57] SO_x-transfer is an impressive example for cost efficient environmental protection through innovative catalysis.

From an economic point of view, it is particularly attractive to convert distillation residues by catalytic cracking, and this has, in fact, been practiced in many units, mostly by adding a certain portion of residua to conventional vacuum gas oil feeds. At least three process variants, however, were especially designed for catalytic cracking of residual oils, viz. the *Heavy Oil Cracking* process, [58,59] the *Reduced Crude Conversion* (*RCC*) process [60,61] and *Total's* process. [62,63] Distillation residues contain asphaltenes and metal compounds, especially of vanadium and nickel. Both the process configuration and the cracking catalyst have to cope with these ingredients. With a residue feed, the coke yield is usually higher than with vacuum gas oils. In some residue crackers, the coke is burnt off in two consecutive stages. In the first stage, which is operated at a lower temperature, combustion of those coke components takes place which are relatively rich in hydrogen; in the second stage, a higher temperature of combustion is applied to burn the remaining graphite-like coke. This way, the simultaneous action of water vapour and excessive temperatures on the catalyst are avoided which helps to prevent a structural collapse of the zeolite. Special zeolite catalysts with a low coke selectivity have also been designed.[64,65] Vanadium and nickel, introduced into the FCC unit with the feed oil, deposit on the catalyst and bring about a number of harmful effects.[66] In the first place, vanadium destroys the zeolite whereas nickel shows undesired catalytic actions, such as the formation of molecular hydrogen and additional coke. Several approaches have been undertaken to combat these and other detrimental effects of the metals. Among these strategies are the use of metal passivators, e.g. antimony compounds for the passivation of nickel[67] or tin compounds for the passivation of vanadium,[68,69] and the design of metal traps in the catalyst formulation which prevent the migration of vanadium to the zeolite particles.[70]

Hydrocracking

Hydrocracking is the most serious competitor for FCC. It produces gasoline, jet fuel and diesel fuel from vacuum gas oil. Hydrocracking is environmentally very clean and, in the proper process configuration, enables a high degree of product flexibility, i.e., in an existing plant, the yields of the above-mentioned products can be adjusted to market demands within broad limits. On the other hand, both the investment and operating costs of hydrocracking are high, mainly due to the high pressure, the consumption of hydrogen and the expensive catalyst, especially if the latter contains a noble metal. Hydrocracking makes use of bifunctional catalysts, e.g. palladium (in small amounts of 0.1 to 1 wt.-%) on ultrastable zeolite Y or sulfided NiO-MoO_3 on an acid form of zeolite Y (CoO can be used instead of NiO and WO_3 instead of MoO_3). It is estimated that the world-wide process capacity of hydrocrackers operated with zeolite catalysts is in the order of $50 \cdot 10^6$ t/a, i.e. ca. one tenth of the FCC capacity. Besides, non-zeolitic hydrocracking catalysts, viz. amorphous SiO_2-Al_2O_3 loaded with a hydrogenation/dehydrogenation component continue to be in use.[7,37,71-75]

There are various process configurations[7,74] and as an example, the so-called single stage process is shown in Figure 12. The vacuum gas oil feed and an excess of hydrogen are heated to the process temperature (300 to 400 °C) under high pressure (100 to 150 bar). This mixture is then passed through two fixed-bed reactors. In the first reactor which contains a non-zeolitic hydrotreating catalyst of the CoO-MoO_3/Al_2O_3 type, the organic sulfur and nitrogen compounds are converted into H_2S and NH_3, respectively. Without any intermediate separation step, the resulting mixture is sent to the second reactor where the bifunctional hydrocracking catalyst is placed. After removal of the gases, including H_2S and NH_3, with a recycle of excess hydrogen, the hydrocracked products are fractionated.

The achievements in the field of zeolitic hydrocracking catalysts, though very impressive for the expert and beneficial to the process economics, were perhaps less spectacular than those in catalytic cracking. It might be, however, that the relative importance of hydrocracking increases in the future, especially

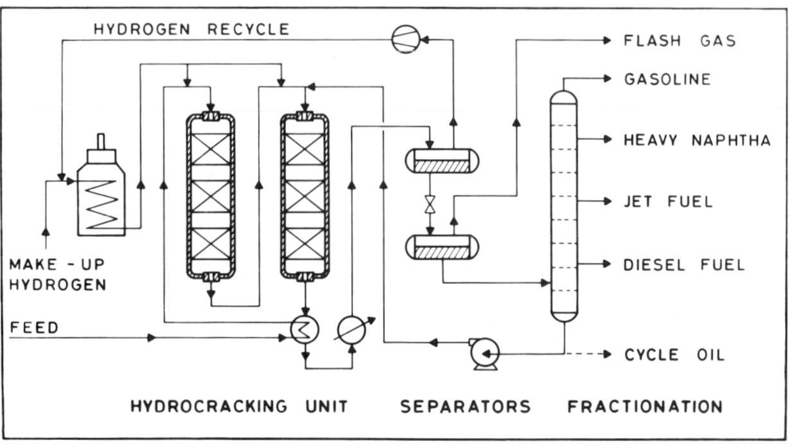

Figure 12 Single stage hydrocracking unit.

in highly industrialized areas, mainly on account of its pollution-free operation. Such a development would, no doubt, foster the search for new hydrocracking catalysts based on molecular sieve zeolites.

Shape Selective Hydrocracking

Since zeolite catalysts possess pore widths in the order of molecular dimensions, they enable shape selective catalysis. [9,76-79] One mode of shape selective catalysis, generally referred to as reactant shape selectivity, is of particular importance in petroleum refining. Selective hydrocracking of the molecules with the smallest dimensions, out of the complex mixture of substances present in an oil fraction, is done on an industrial scale for two different purposes:

Catalytic reforming of heavy naphtha is one of the most important refinery processes. The objective is to increase the octane number, mainly by making aromatics from naphthenes and alkanes. Inevitably, however, there are some n-alkanes and slightly branched iso-alkanes left in the product of reforming, and these components have very bad octane numbers. In the earlier Selectoforming process (Figure 13), the n-alkanes were selectively hydrocracked on Ni/H-erionite, a small pore (8-MR) zeolite. [76,80-82] In the more recent M-Forming process, a medium pore zeolite based on HZSM-5 is employed, which selectively converts both the n-alkanes and the moderately branched iso-alkanes. [9,83,84] M-Forming, therefore, results in a higher octane gain than Selectoforming, of course at the price of a somewhat lower gasoline yield. The products of selective hydrocracking are predominantly propane and butanes which can be easily separated from the heavy naphtha. The shape selective hydrocracking catalyst can be placed at the exit of the last reforming reactor, as shown in Figure 13, or in a separate reactor downstream of the reforming unit.

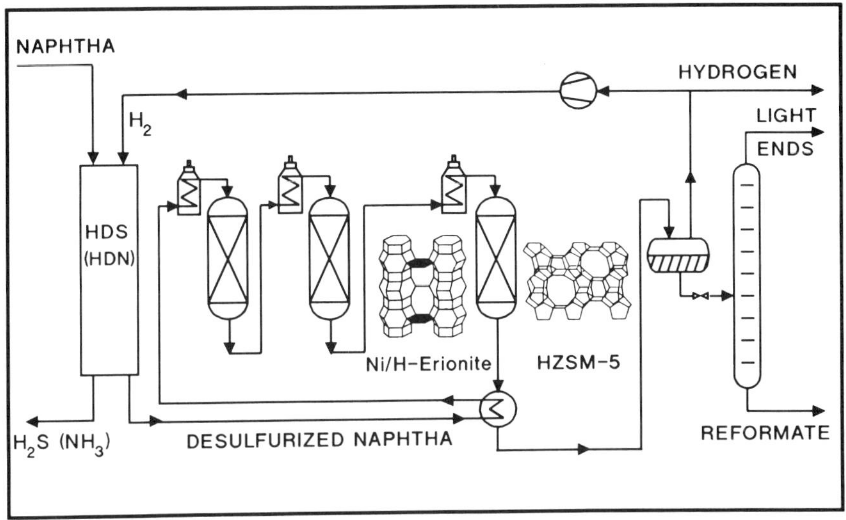

Figure 13 Incorporation of Selectoforming or M-Forming into catalytic reforming of heavy naphtha.

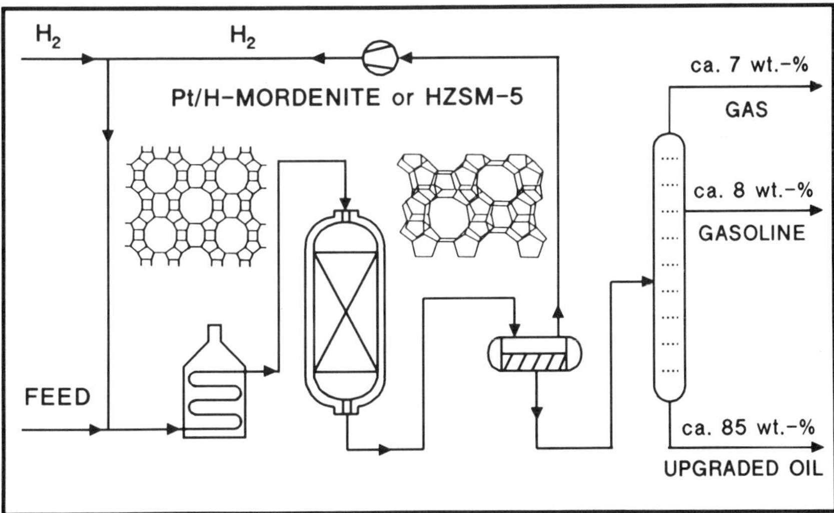

Figure 14 Hydrodewaxing of heavy gas oils or lubricating oils.

In heavier petroleum fractions, especially lubricating oils and diesel fuel, the n-alkanes and slightly branched alkanes are responsible for unfavourable cold flow properties, expressed for instance by the pour point or the cloud point. The principle of shape selective hydrocracking is applied to such fractions as well (Figure 14). Pt/H-mordenite has been employed as catalyst in a process developed by BP.[85,86] Mobil Oil Corporation prefers catalysts based on zeolite HZSM-5, and the corresponding processes were named Mobil Distillate Dewaxing (MDDW) and Mobil Lube Dewaxing (MLDW).[9,87,88]

4 A LOOK TO THE FUTURE

A number of additional refinery processes which rely on zeolite catalysis are ready for commercial application. Among them is the conversion of LPG hydrocarbons (C_3/C_4) into aromatics. The cyclar process, jointly developed by BP and UOP, and Mobil's M2-Forming process are available for this application.[89,90] Modified ZSM-5 type zeolites seem to give the best results.

The MOGD (Mobil Olefin to Gasoline or Distillate) process was designed for the manufacture of gasoline or middle distillates from light olefins. In the distillate mode, the process is operated under elevated pressures (above 25 bar) and at relatively low temperatures (in the order of 200 °C), and long-chain iso-olefins are the main products,[9] whereas in the gasoline mode, higher reaction temperatures are applied which are favourable for the formation of aromatics.[91,92]

Nonacidic forms of zeolite L modified by platinum, e.g., Pt/KL or Pt/BaL, have been found to possess excellent catalytic properties in the aromatization of C_6- or C_7-alkanes or naphtha.[93] A process named Aromax has been developed by Chevron Research Co.,[94] and it will be interesting to watch whether zeolites can conquer this broad area of catalysis and petroleum refining.

Removing waxy components from lubricating oils or diesel fuel by shape selective hydrocracking (cf. Figure 14) necessarily brings about a yield loss since the undesired hydrocarbons are converted into gasoline and LPG. Another approach is to transform the waxy paraffins into branched isomers. Indeed, long-chain alkanes can be isomerized on bifunctional zeolites with a strong hydrogenation component, such as Pt/CaY or Pd/LaY in the presence of hydrogen.[95-97] On such catalysts, however, hydrocracking reactions strongly interfere at elevated conversions, hence the selectivity for isomerization is unsatisfactory. Bifunctional forms of zeolite Beta have been found to give better isomerization selectivities, at least with model alkanes.[98,99] Systematic research along this line is very desirable since dewaxing of heavy petroleum fractions by isomerization rather than by shape selective hydrocracking is economically attractive.

Another refinery process which urgently calls for zeolite catalysts is the alkylation of isobutane with light alkenes. Today, alkylation gasoline is manufactured with liquid catalysts, viz. concentrated sulfuric acid or anhydrous hydrogen fluoride. There is a strong incentive for replacing these existing processes by a new technology which is environmentally more acceptable and safer.[97] Acid faujasites, in a fresh state, were indeed found to catalyze isobutane/olefin alkylation.[100,101] Their time-on-stream behaviour is, however, unsatisfactory.[97] There is a huge incentive for the discovery of a good alkylation catalyst among the new zeolitic materials (Figure 6) synthesized in recent years.

ACKNOWLEDGEMENT

The authors gratefully acknowledge financial support of their research in the fields of zeolite materials science and petroleum refining reactions by the following funding institutions: The German Science Foundation (Deutsche Forschungsgemeinschaft), Max Buchner-Forschungsstiftung and Fonds der Chemischen Industrie.

REFERENCES

1. C.J. Plank, Chemtech, 1984, **14**, 243.
2. C.J. Plank, in: 'Heterogeneous Catalysis – Selected American Histories' (B.H. Davis and W.P. Hettinger, Jr., eds.), ACS Symposium Series, Vol. 222, p. 253, American Chemical Society, Washington, D.C., 1983.
3. P.B. Venuto and E.T. Habib, Jr., Catal. Rev.-Sci. Eng., 1978, **18**, 1.
4. P.B. Venuto and E.T. Habib, Jr., 'Fluid Catalytic Cracking with Zeolite Catalysts', Marcel Dekker, New York, Basel, 1979.
5. J. Biswas and I.E. Maxwell, Appl. Catal., 1990, **63**, 197.
6. A.P. Bolton, in: 'Zeolite Chemistry and Catalysis' (J.A. Rabo, ed.), ACS Monograph 171, p. 714, American Chemical Society, Washington, D.C., 1976.
7. J.W. Ward, in: 'Preparation of Catalysts III' (G. Poncelet, P. Grange and P.A. Jacobs, eds.), Studies in Surface Science and Catalysis, Vol. 16, p. 587, Elsevier, Amsterdam, Oxford, New York, 1983.
8. J.E. Naber, W.H.J. Stork, P.M.M. Blauwhoff and K.J.W. Groeneveld, Erdöl, Erdgas, Kohle, 1991, **107**, 124.
9. N.Y. Chen, W.E. Garwood and F.G. Dwyer, 'Shape Selective Catalysis in Industrial Applications', Marcel Dekker, New York, Basel, 1989.
10. R.M. Barrer, 'Zeolites and Clay Minerals as Sorbents and Molecular Sieves', Academic Press, London, New York, San Francisco, 1978.
11. R.M. Barrer, 'Hydrothermal Chemistry of Zeolites', Academic Press, London, New York, 1982.
12. D.W. Breck, 'Zeolite Molecular Sieves – Structure, Chemistry and Use', Wiley, New York, 1973.
13. S.T. Wilson, B.M. Lok, C.A. Messina and E.M. Flanigen, Proc. 6th Intern. Zeolite Conf. (D. Olson and A. Bisio, eds.), p. 97, Butterworths, Guildford, 1984.

14. E.M. Flanigen, B.M. Lok, R.L. Patton and S.T. Wilson, in: 'New Developments in Zeolite Science and Technology' (Y. Murakami, A. Iijima and J.W. Ward, eds.), Studies in Surface Science and Catalysis, Vol. 28, p. 103, Kodansha, Tokyo, and Elsevier, Amsterdam, Oxford, New York, Tokyo, 1986.
15. E.M. Flanigen, R.L. Patton and S.T. Wilson, in: 'Innovation in Zeolite Materials Science' (P.J. Grobet, W.J. Mortier, E.F. Vansant and G. Schulz-Ekloff, eds.), Studies in Surface Science and Catalysis, Vol. 37, p. 13, Elsevier, Amsterdam, Oxford, New York, Tokyo, 1988.
16. J.A. Rabo, R.J. Pellet, P.K. Coughlin and E.S. Shamshoum, in: 'Zeolites as Catalysts, Sorbents and Detergent Builders – Applications and Innovations' (H.G. Karge and J. Weitkamp, eds.), Studies in Surface Science and Catalysis, Vol. 46, p. 1, Elsevier, Amsterdam, Oxford, New York, Tokyo, 1989.
17. M.E. Davis, C. Saldarriaga, C. Montes, J. Garces and C. Crowder, Nature, 1988, **331**, 698.
18. M.E. Davis, C. Saldarriaga, C. Montes, J. Garces and C. Crowder, Zeolites, 1988, **8**, 362.
19. L.V.C. Rees, Nature, 1982, **296**, 491.
20. F. Liebau, Zeolites, 1983, **3**, 191.
21. J.V. Smith, Zeolites, 1984, **4**, 309.
22. F. Liebau, H. Gies, R.P. Gunawardane and B. Marler, Zeolites, 1986, **6**, 373.
23. A. Dyer, 'An Introduction to Zeolite Molecular Sieves', p. 1, Wiley, Chichester, New York, Brisbane, Toronto, Singapore, 1988.
24. B.M. Lok, C.A. Messina, R.L. Patton, R.T. Gajek, T.F. Cannan and E.M. Flanigen, J. Am. Chem. Soc., 1984, **106**, 6092.
25. K.Sørby, R. Szostak, J.G. Ulan and R. Gronsky, Catal. Letters, 1990, **6**, 209.
26. R. Löw, S. Ernst, A. Kiss, P. Kleinschmit and J. Weitkamp, Chem.-Ing.-Tech., 1991, **63**, in press.
27. R.M. Dessau, J.L. Schlenker and J.B. Higgins, Zeolites, 1990, **10**, 522.
28. W.M. Meier, Proc. of the Intern. Symposium on Catalysis and Adsorption by Zeolites, Leipzig, Germany, August 20-23, 1990, Elsevier, Amsterdam, Oxford, New York, Tokyo, in press.
29. W.M. Meier and D.H. Olson, 'Atlas of Zeolite Structure Types', 2nd edition, Butterworths, London, 1987.
30. D.M. Ruthven, Chem. Eng. Progr., 1988, **84** (No. 2), 42.
31. R.V. Jasra and S.G.T. Bhat, Separation Science and Technology, 1988, **23**, 945.
32. D.B. Broughton, in: 'Kirk-Othmer: Encyclopedia of Chemical Technology' (H.F. Kirk, D.H. Othmer et al., eds.), Wiley, New York, Chichester, Brisbane, Toronto, 3rd edition, Vol. 1, 1978, p. 563.
33. D.M. Ruthven, 'Principles of Adsorption and Adsorption Processes', Wiley, New York, 1984.
34. J.A. Johnson and A.R. Oroskar, in 'Zeolites as Catalysts, Sorbents and Detergent Builders – Applications and Innovations' (H.G. Karge and J. Weitkamp, eds.), Studies in Surface Science and Catalysis, Vol. 46, p. 451, Elsevier, Amsterdam, Oxford, New York, Tokyo, 1989.

35. J. Weitkamp, in 'Hydrocracking and Hydrotreating' (J.W. Ward and S.A. Qader, eds.), ACS Symposium Series, Vol. 20, p. 1, American Chemical Society, Washington, D.C., 1975.
36. W.C. van Zijl Langhout, in: 'Proc. 9th World Petroleum Congress', Applied Science Publishers, London, Vol. 5, 1975, p. 197.
37. I.E. Maxwell, Catalysis Today, 1987, **1**, 385.
38. W.C.J. Quick and J. Collins, Erdöl, Kohle-Erdgas-Petrochem., 1972, **25**, 706.
39. M.F. Symoniak and T.C. Holcombe, Hydrocarbon Process., 1983, **62** (No. 5), 62.
40. T.C. Holcombe and M.C. Pederson, Oil Gas J., 1983, **81** (No. 47), 76.
41. D. Decroocq, 'Catalytic Cracking of Heavy Petroleum Fractions', Editions Technip, Paris, 1984.
42. P.B. Weisz, Chemtech, 1973, **3**, 498.
43. H. Knab, C. Mielicke and E. Rosum, Chem.-Ing.-Tech., 1982, **54**, 79.
44. J. Scherzer, 'Octane-Enhancing Zeolitic FCC Catalysts', Marcel Dekker, New York and Basel, 1990.
45. G.C. Edwards, K. Rajagopalan, A.W. Peters, G.W. Young and J.E. Creighton, in 'Fluid Catalytic Cracking, Role in Modern Refining' (M.L. Occelli, ed.), ACS Symposium Series, Vol. 375, p. 101, American Chemical Society, Washington, D.C., 1988.
46. J. Biswas and I.E. Maxwell, Appl. Catal. 1990, **58**,1.
47. J. Biswas and I.E. Maxwell, Appl. Catal. 1990, **58**, 19.
48. D.A. Pappal and P.H. Schipper, in 'The Hydrocarbon Chemistry of FCC Naphtha Formation' (H.J. Lovink and L.A. Pine, eds.), p. 121, Editions Technip, Paris, 1990.
49. P.H. Schipper, F.G. Dwyer, P.T. Sparrell, S. Mizrahi and J.A. Herbst, in 'Fluid Catalytic Cracking, Role in Modern Refining' (M.L. Occelli, ed.), ACS Symposium Series, Vol. 375, p. 64, American Chemical Society, Washington, D.C., 1988.
50. J. Biswas, H. Kaak, J. van der Griend and I. Maxwell, in 'Akzo Catalysts Symposium 88' (H.J. Lovink, ed.), p. F-12, Akzo Chemicals, Amersfoort, 1988.
51. A.W. Chester, A.B. Schwartz, W.A. Stover and J.P. McWilliams, Preprints, Div. Petrol. Chem., Am. Chem. Soc., 1979, **24**, 624.
52. L.L. Upson, Preprints, Div. Petrol. Chem., Am. Chem. Soc., 1979, **24**, 632.
53. A.W. Chester, A.B. Schwartz, W.A. Stover and J.P. McWilliams, Chemtech, 1981, **11**,50.
54. E.T. Habib, Oil Gas J., 1983, **81** (No. 32),111.
55. J.W. Byrne, B.K. Speronello and E.L. Leuenberger, Oil Gas J., 1984, **82** (No. 42), 101.
56. E.H. Hirschberg and R.J. Bertolacini, in 'Fluid Catalytic Cracking, Role in Modern Refining' (M.L. Occelli, ed.), ACS Symposium Series, Vol. 375, p. 114, American Chemical Society, Washington, D.C., 1988.
57. L. Rheaume and R.E. Ritter, in 'Fluid Catalytic Cracking, Role in Modern Refining' (M.L. Occelli, ed.), ACS Symposium Series, Vol. 375, p. 146, American Chemical Society, Washington, D.C., 1988.
58. J.B. Rush, Chem. Eng. Progr., 1981, **77** (No. 12), 29.
59. J.B. Rush, Hydrocarbon Process., 1981, **60** (No. 9), 113.

60. O.J. Zandona, L.E. Busch, W.P. Hettinger, Jr. and R.P. Krock, Oil Gas J., 1982, **80** (Nr. 12), 82.
61. W.P. Hettinger, Jr., in: 'Fluid Catalytic Cracking, Role in Modern Refining' (M.L. Occelli, ed.), ACS Symposium Series, Vol. 375, p. 308, American Chemical Society, Washington, D.C., 1988.
62. R.R. Dean, J.-L. Mauleon and W.S. Letzsch, Oil Gas J., 1982, **80** (No. 40), 75.
63 R.R. Dean, J.-L. Mauleon and W.S. Letzsch, Oil Gas J., 1982, **80** (No. 41), 168.
64. I.A. Vasalos, E.R. Strong, C.K.R. Hsieh and G.J. D'Souza, Oil Gas J., 1977, **75** (No.26), 141.
65. R.J. Campagna, A.S. Krishna and S.J. Yanik, Oil Gas J., 1983, **81** (No. 44), 128.
66. D.F. Tatterson and R.L. Mieville, Ind. Eng. Chem. Res., 1988, **27**, 1595.
67. D.L. McKay and B.J. Bertus, Preprints, Div. Petrol. Chem., Am. Chem. Soc., 1979, **24**, 645.
68. A.R. English and D.C. Kowalczyk, Oil Gas J., 1984, **82** (No. 29), 127.
69. M.W. Anderson, M.L. Occelli and S.L. Suib, J. Catal., 1990, **122**, 374.
70. C.J. Groenenboom, in: 'Zeolites as Catalysts, Sorbents and Detergent Builders - Applications and Innovations' (H.G. Karge and J. Weitkamp, eds.), Studies in Surface Science and Catalysis, Vol. 46, p. 99, Elsevier, Amsterdam, Oxford, New York, Tokyo, 1989.
71. T.Y. Yan, Ind. Eng. Chem., Proc. Des. Dev., 1983, **22**, 154.
72. J.W. Ward, in: 'Applied Industrial Catalysis' (B.E. Leach, ed.), Vol. 3, p. 271, Academic Press, Orlando, San Diego, New York, London, Toronto, Montreal, Sydney, Tokyo, 1984.
73. P.J. Nat, Erdöl, Kohle - Erdgas - Petrochem., 1989, **42**, 447.
74. J.W. Ward, in: 'Catalysis in Petroleum Refining 1989' (D.L. Trimm, S. Akashah, M. Absi-Halabi and A. Bishara, eds.), Studies in Surface Science and Catalysis, Vol. 53, p. 417, Elsevier, Amsterdam, Oxford, New York, Tokyo, 1990.
75. R.F. Sullivan and J.W. Scott, in: 'Heterogeneous Catalysis - Selected American Histories'(B.H. Davis and W.P. Hettinger, Jr., eds.), ACS Symposium Series, Vol. 222, p. 293, American Chemical Society, Washington, D.C., 1983.
76. S.M. Csicsery, in: 'Zeolite Chemistry and Catalysis' (J.A. Rabo, ed.), ACS Monograph 171, p. 680, American Chemical Society, Washington, D.C., 1976.
77. P.B. Weisz, Pure Appl. Chem., 1980, **52**, 2091.
78. S.M. Csicsery, Zeolites, 1984, **4**, 202.
79. J. Weitkamp, S. Ernst, H. Dauns and E. Gallei, Chem.-Ing.-Tech., 1986, **58**, 623.
80. N.Y. Chen, J. Maziuk, A.B. Schwartz and P.B. Weisz, Oil Gas J., 1968, **66** (No. 47), 154.
81 S. D. Burd, Jr. and J. Maziuk, Oil Gas J., 1972, **70** (No. 27), 52.
82. K. Becker, H. John, K.-H. Steinberg, M. Weber and K.-H. Nestler, in: 'Catalysis on Zeolites' (D. Kalló and Kh.M. Minachev, eds.), p. 515, Akadémiai Kiadó, Budapest, 1988.
83. N.Y. Chen, W.E. Garwood and R.H. Heck, Ind. Eng. Chem. Res., 1987, **26**, 706.

84. H. Heinemann, in: 'Catalysis - Science and Technology' (J.R. Anderson and M. Boudart, eds.), Vol 1, p.1 , Springer-Verlag, Berlin, Heidelberg, New York, 1981.
85. R.N. Bennett, G.J. Elkes and G.J. Wanless, Oil Gas J., 1975, **73** (No. 1), 69.
86. J.D. Hargrove, G.J. Elkes and A.H. Richardson, Oil Gas J., 1979, **77** (No. 3), 103.
87. S.P. Donnelly and J.R. Green, Oil Gas J., 1980, **78** (No. 43), 77.
88. K.W. Smith, W.C. Starr and N.Y. Chen, Oil Gas J., 1980, **78**, (No. 21), 75.
89. J.R. Mowry, R.F. Anderson and J.A. Johnson, Oil Gas J., 1985, **83** (No. 48), 128.
90. N.Y. Chen and T.Y. Yan, Ind. Eng. Chem., Proc. Des. Dev., 1986, **25**, 151.
91. S.A. Tabak and F.J. Krambeck, Hydrocarbon Process, 1985, **64** (No.9), 72.
92. R.J. Quann, L.A. Green, S.A. Tabak and F.J. Krambeck, Ind. Eng. Chem. Res., 1988, **27**, 565.
93. T.R. Hughes, W.C. Buss, P.W. Tamm and R.L. Jacobson, in: 'New Developments in Zeolite Science and Technology'(Y. Murakami, A. Iijima and J.W. Ward, eds.), Studies in Surface Science and Catalysis, Vol. 28, p. 725, Kodansha, Tokyo, and Elsevier, Amsterdam, Oxford, New York, Tokyo, 1986.
94. D.V. Law, P.W. Tamm and C.M. Detz, Energy Progress, 1987, **7**, 215.
95. J. Weitkamp, Ind. Eng. Chem., Prod. Res. Dev., 1982, **21**, 550.
96. J. Weitkamp, W. Gerhardt and P.A. Jacobs, Proc. of the Intern. Symposium on Zeolite Catalysis, p. 261. Siófok, Hungary, May 13-16, 1985.
97. J. Weitkamp, Proc. of the Intern. Symposium on Catalysis and Adsorption by Zeolites, Leipzig, Germany, August 20-23, 1990, Elsevier, Amsterdam, Oxford, New York, Tokyo, in press.
98. J. Weitkamp, R. Kumar and S. Ernst, Chem.-Ing.-Tech., 1989, **61**, 731.
99. S. Ernst, R. Kumar and J. Weitkamp, in 'Catalysis - Concepts and Applications' (B. Viswanathan and A. Meenakshisundaram, eds.), p. OR 11-1, Proc. of the 9th National Symposium on Catalysis, Madras, India, December 15-17, 1988.
100. J. Weitkamp, Proc. 7th Intern. Zeolite Conf. (L.V.C. Rees, ed.), p. 858, Heyden, London, Philadelphia, Rheine, 1980.
101. J. Weitkamp, in: 'Catalysis by Zeolites' (B. Imelik, C. Naccache, Y. Ben Taarit, J.C. Vedrine, G. Coudurier and H. Praliaud, eds.), Studies in Surface Science and Catalysis, Vol. 5, p. 65, Elsevier, Amsterdam, Oxford, New York, 1980.

Subject Index